ALLE · ZEIT · WACH
1842

Springer Series in Synergetics

Editor: Hermann Haken

Synergetics, an interdisciplinary field of research, is concerned with the cooperation of individual parts of a system that produces macroscopic spatial, temporal or functional structures. It deals with deterministic as well as stochastic processes.

1 **Synergetics** An Introduction 3rd Edition
By H. Haken
2 **Synergetics** A Workshop
Editor: H. Haken
3 **Synergetics** Far from Equilibrium
Editors: A. Pacault, C. Vidal
4 **Structural Stability in Physics**
Editors: W. Güttinger, H. Eikemeier
5 **Pattern Formation by Dynamic Systems and Pattern Recognition**
Editor: H. Haken
6 **Dynamics of Synergetic Systems**
Editor: H. Haken
7 **Problems of Biological Physics**
By L. A. Blumenfeld
8 **Stochastic Nonlinear Systems**
in Physics, Chemistry, and Biology
Editors: L. Arnold, R. Lefever
9 **Numerical Methods in the Study of Critical Phenomena**
Editors: J. Della Dora, J. Demongeot, B. Lacolle
10 **The Kinetic Theory of Electromagnetic Processes** By Yu. L. Klimontovich
11 **Chaos and Order in Nature**
Editor: H. Haken
12 **Nonlinear Phenomena in Chemical Dynamics**
Editors: C. Vidal, A. Pacault
13 **Handbook of Stochastic Methods**
for Physics, Chemistry, and the Natural Sciences
2nd Edition By C. W. Gardiner
14 **Concepts and Models of a Quantitative Sociology** The Dynamics of Interacting Populations By W. Weidlich, G. Haag
15 **Noise-Induced Transitions** Theory and Applications in Physics, Chemistry, and Biology
By W. Horsthemke, R. Lefever
16 **Physics of Bioenergetic Processes**
By L. A. Blumenfeld
17 **Evolution of Order and Chaos** in Physics, Chemistry, and Biology Editor H. Haken
18 **The Fokker-Planck Equation**
By H. Risken
19 **Chemical Oscillations, Waves, and Turbulence**
By Y. Kuramoto
20 **Advanced Synergetics** 2nd Edition
By H. Haken
21 **Stochastic Phenomena and Chaotic Behaviour in Complex Systems** Editor: P. Schuster
22 **Synergetics – From Microscopic to Macroscopic Order** Editor: E. Frehland
23 **Synergetics of the Brain**
Editors: E. Başar, H. Flohr, H. Haken, A. J. Mandell
24 **Chaos and Statistical Methods**
Editor: Y. Kuramoto
25 **Dynamics of Hierarchical Systems**
An Evolutionary Approach By J. S. Nicolis
26 **Self-Organization and Management of Social Systems**
Editors: H. Ulrich, G. J. B. Probst
27 **Non-Equilibrium Dynamics in Chemical Systems** Editors: C. Vidal, A. Pacault
28 **Self-Organization** Autowaves and Structures Far from Equilibrium
Editor: V. I. Krinsky
29 **Temporal Order**
Editors: L. Rensing, N. I. Jaeger
30 **Dynamical Problems in Soliton Systems**
Editor: S. Takeno
31 **Complex Systems – Operational Approaches** in Neurobiology, Physics, and Computers
Editor: H. Haken
32 **Dimensions and Entropies in Chaotic Systems**
Quantification of Complex Behavior
Editor: G. Mayer-Kress
33 **Selforganization by Nonlinear Irreversible Processes**
Editors: W. Ebeling, H. Ulbricht
34 **Instabilities and Chaos in Quantum Optics**
Editors: F. T. Arecchi, R. G. Harrison
35 **Nonequilibrium Phase Transitions in Semiconductors** Self-Organization Induced by Generation and Recombination Processes
By E. Schöll
36 **Temporal Disorder in Human Oscillatory Systems**
Editors: L. Rensing, U. an der Heiden, M. C. Mackey
37 **The Physics of Structure Formation**
Theory and Simulation
Editors: W. Güttinger and G. Dangelmayr
38 **Computational System – Natural and Artificial**
Editor: H. Haken

Hermann Haken

Advanced Synergetics

Instability Hierarchies of Self-Organizing Systems and Devices

With 105 Figures

Springer-Verlag Berlin Heidelberg New York
London Paris Tokyo

Professor Dr. Dr. h. c. Hermann Haken

Institut für Theoretische Physik der Universität Stuttgart, Pfaffenwaldring 57/IV,
D-7000 Stuttgart 80, Fed. Rep. of Germany and
Center for Complex Systems, Florida Atlantic University,
Boca Raton, FL 33431, USA

1st Edition 1983
Corrected 2nd printing 1987

ISBN 3-540-12162-5 Springer-Verlag Berlin Heidelberg New York
ISBN 0-387-12126-5 Springer-Verlag New York Heidelberg Berlin

Library of Congress Cataloging in Publication Data. Haken, H. Advanced synergetics. (Springer series in synergetics ; v. 20) 1. System theory. 2. Self-organizing systems. I. Title. II. Series. Q295.H34 1983 003 83-340

Typesetting: K + V Fotosatz, Beerfelden
Printing: Druckhaus Beltz, 6944 Hemsbach/Bergstr.
Binding: J. Schäffer GmbH & Co. KG, 6718 Grünstadt
2153/3150-543210

to Edith,
Maria, Karin, and Karl-Ludwig

Preface to the Corrected Second Printing

When I wrote this book, I included results intended to appeal to a wide readership of people wishing to learn about the basic concepts and mathematical tools of synergetics and its applications in a variety of fields ranging from physics, chemistry and biology to the engineering sciences. To this end I chose a style, in which rigorous mathematical results are presented in a way that is also accessible to non-mathematicians. I am pleased to notice that this choice of style has been widely approved. Indeed, the first edition of this book sold out in a relatively short time, and it has also been translated into Chinese, Japanese, and Russian. These are countries in which enormous technological effort is presently being made. I have refrained from enlarging the book but preferred only to correct a few misprints.

Stuttgart, July 1987 *Hermann Haken*

Preface to the First Edition

This text on the interdisciplinary field of synergetics will be of interest to students and scientists in physics, chemistry, mathematics, biology, electrical, civil and mechanical engineering, and other fields. It continues the outline of basic concepts and methods presented in my book *Synergetics. An Introduction*, which has by now appeared in English, Russian, Japanese, Chinese, and German. I have written the present book in such a way that most of it can be read independently of my previous book, though occasionally some knowledge of that book might be useful.

But why do these books address such a wide audience? Why are instabilities such a common feature, and what do devices and self-organizing systems have in common? Self-organizing systems acquire their structures or functions without specific interference from outside. The differentiation of cells in biology, and the process of evolution are both examples of self-organization. Devices such as the electronic oscillators used in radio transmitters, on the other hand, are man-made. But we often forget that in many cases devices function by means of processes which are also based on self-organization. In an electronic oscillator the motion of electrons becomes coherent without any coherent driving force from the outside; the device is constructed in such a way as to permit specific collective motions of the electrons. Quite evidently the dividing line between self-organizing systems and man-made devices is not at all rigid. While in devices man sets certain boundary conditions which make self-organization of the components possible, in biological systems a series of self-imposed conditions permits and directs self-organization. For these reasons it is useful to study devices and self-organizing systems, particularly those of biology, from a common point of view.

In order to elucidate fully the subtitle of the present book we must explain what instability means. Perhaps this is best done by giving an example. A liquid in a quiescent state which starts a macroscopic oscillation is leaving an old state and entering a new one, and thus loses its stability. When we change certain conditions, e. g. power input, a system may run through a series of instabilities leading to quite different patterns of behavior.

The central question in synergetics is whether there are general principles which govern the self-organized formation of structures and/or functions in both the animate and the inanimate world. When I answered this question in the affirmative for large classes of systems more than a decade ago and suggested that these problems be treated within the interdisciplinary research field of "synergetics", this might have seemed absurd to many scientists. Why should systems consisting of components as different as electrons, atoms, molecules, photons, cells, animals, or even humans be governed by the same principles when they organize themselves to form electrical oscillations, patterns in fluids, chemical waves, laser beams, organs, animal societies, or social groups? But the past decade has brought an abundance of evidence indicating that this is, indeed, the case, and many individual examples long known in the literature could be subsumed under the unifying concepts of synergetics. These examples range from biological morphogenesis and certain aspects of brain function to the flutter of airplane wings; from molecular physics to gigantic transformations of stars; from electronic devices to the formation of public opinion; and from muscle contraction to the buckling of solid structures. In addition, there appears to be a remarkable convergence of the basic concepts of various disciplines with regard to the formation of spatial, temporal, and functional structures.

In view of the numerous links between synergetics and other fields, one might assume that synergetics employs a great number of quite different concepts. This is not the case, however, and the situation can best be elucidated by an analogy. The explanation of how a gasoline engine functions is quite simple, at least in principle. But to construct the engine of a simple car, a racing car, or a modern aircraft requires more and more technical know-how. Similarly, the basic concepts of synergetics can be explained rather simply, but the application of these concepts to real systems calls for considerable technical (i. e. mathematical) know-how. This book is meant to serve two purposes: (1) It offers a simple presentation of the basic principles of instability, order parameters, and slaving. These concepts represent the "hard core" of synergetics in its present form and enable us to cope with large classes of complex systems ranging from those of the "hard" to those of the "soft" sciences. (2) It presents the necessary mathematical know-how by introducing the fundamental approaches step by step, from simple to more complicated cases. This should enable the reader to apply these methods to concrete problems in his own field, or to search for further analogies between different fields.

Modern science is quite often buried under heaps of nomenclature. I have tried to reduce it to a minimum and to explain new words whenever necessary. Theorems and methods are presented so that they can be applied to concrete problems; i. e. constructions rather than existence proofs are given. To use a concept of general systems theory, I have tried to present an *operational* approach. While this proved feasible in most parts of the book, difficulties arose in connection with *quasi-periodic* processes. I have included these intricate problems (e. g., the bifurcation of tori) and my approach to solving them not only because they are at the frontier of modern mathematical research, but also because we are confronted with them again and again in natural and man-made systems. The chapters which treat these problems, as well as some other difficult chapters, have been marked with an asterisk and can be skipped in a first reading.

Because this book is appearing in the Springer Series in Synergetics, a few words should be said about how it relates to other books in this series. Volume 1, *Synergetics. An Introduction*, dealt mainly with the first instability leading to spatial patterns or oscillations, whereas the present book is concerned with all sorts of instabilities and their sequences. While the former book also contained an introduction to stochastic processes, these are treated in more detail in C. W. Gardiner's *Handbook of Stochastic Methods*, which also provides a general background for the forthcoming volumes *Noise-Induced Transitions* by W. Horsthemke and R. Lefever and *The Fokker-Planck Equation. Methods of Solution and Applications* by H. Risken. These three books cover important aspects of synergetics with regard to fluctuations. The problem of multiplicative noise and its most interesting consequences are treated only very briefly in the present book, and the interested reader is referred to the volume by Horsthemke and Lefever. The Fokker-Planck equation, a most valuable tool in treating systems at transition points where other methods may fail, is discussed in both my previous book and this one, but readers who want a comprehensive account of present-day knowledge on that subject are referred to the volume by Risken. Finally, in order to keep the present book within a reasonable size, I refrained from going into too far-reaching applications of the methods presented. The application of the concepts and methods of synergetics to sociology and economy are given thorough treatment in W. Weidlich and G. Haag's book, *Concepts and Models of a Quantitative Sociology*. Finally, Klimontovich's book, *The Kinetic Theory of Electromagnetic Processes,* gives an excellent account of the interaction of charged particles with electromagnetic fields, as well as the various collective phenomena arising from this interaction.

While the present book and the others just mentioned provide us with the theoretical and mathematical basis of synergetics, experiments on self-organizing systems are at least equally important. Thus far, these experiments have only been treated in conference proceedings in the Springer Series in Synergetics; it is hoped that in future they will also be covered in monographs within this series. In conclusion, it should be pointed out that synergetics is a field which still offers great scope for experimental and theoretical research.

I am grateful to Dipl. Phys. Karl Zeile and Dr. Arne Wunderlin for their valuable suggestions and their detailed and careful checking of the manuscript and the calculations. My thanks also go to Dr. Herbert Ohno, who did the drawings, and to my secretary, Mrs. U. Funke, who meticulously typed several versions of the manuscript, including the formulas, and without whose tireless efforts this book would not have been possible. I gratefully acknowledge the excellent cooperation of the members of Springer-Verlag in producing this book. I wish to thank the Volkswagenwerk Foundation, Hannover, for its very efficient support of the synergetics project, within the framework of which a number of the results presented in this book were obtained over the past four years.

Stuttgart, January 1983 *Hermann Haken*

Contents

1. Introduction

1.1 What is Synergetics About?

Synergetics deals with systems composed of many subsystems, which may be of quite different natures, such as electrons, atoms, molecules, cells, neurons, mechanical elements, photons, organs, animals or even humans. In this book we wish to study how the cooperation of these subsystems brings about spatial, temporal or functional structures on macroscopic scales. In particular, attention will be focused on those situations in which these structures arise in a self-organized fashion, and we shall search for principles which govern these processes of self-organization irrespective of the nature of the subsystems. In the introduction, we present typical examples of disorder – order or order – order transitions in various fields, ranging from physics to sociology, and give an outline of the basic concepts and the mathematical approach.

1.2 Physics

In principle, all phase transitions of physical systems in thermal equilibrium such as the liquid – gas transition, the ferromagnetic transition or the onset of superconductivity fall under the general definition of processes dealt with by synergetics. On the other hand, these phenomena are treated intensively by the theory of phase transitions, in particular nowadays by renormalization group techniques on which a number of books and review articles are available (we recommend the reader to follow up the list of references and further reading at the end of this book while he reads the individual sections). Therefore, these phenomena will not be considered and instead attention will be focused on those modern developments of synergetics which are not covered by phase transition theory.

1.2.1 Fluids: Formation of Dynamic Patterns

Fluid dynamics provides us with beautiful examples of pattern formations of increasing complexity. Because the formation of a pattern means that the former state of fluid cannot persist any longer, i.e., that it becomes *unstable*, the phenomena of pattern formation are often called *instabilities*. Consider as a first

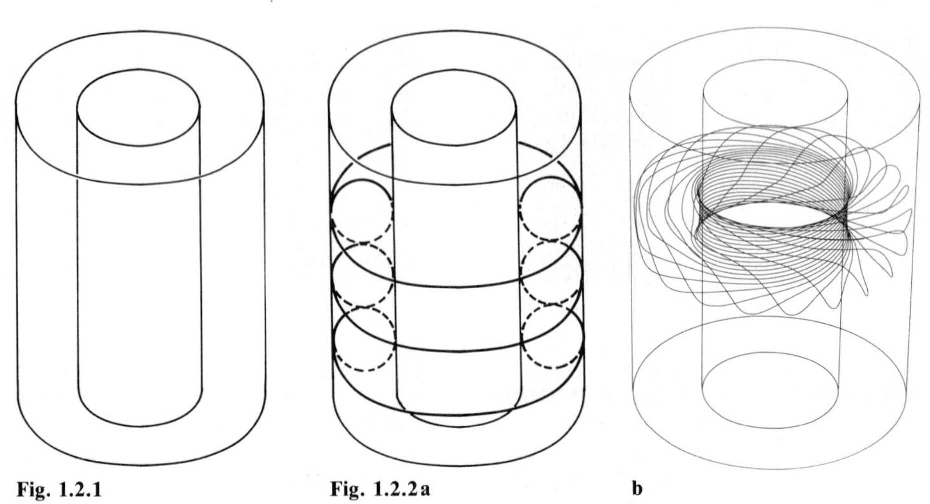

Fig. 1.2.1 Fig. 1.2.2 a b

Fig. 1.2.1. Scheme of experimental setup for the study of the Taylor instability. A liquid is filled in-between two coaxial cylinders of which the outer one is transparent. The inner cylinder can be rotated at a fixed speed

Fig. 1.2.2. **(a)** Schematic diagram of the formation of rolls. **(b)** A computer calculation of the trajectory of a particle of the fluid within a roll. In the case shown two rolls are formed but only the motion of a particle within the upper roll is presented. [After K. Marx: Diplom thesis, University of Stuttgart (1982)]

example the Taylor instability and its subsequent instabilities. In these experiments (Fig. 1.2.1) the motion of a liquid between coaxial cylinders is studied. Usually one lets the inner cylinder rotate while the outer cylinder is kept fixed, but experiments have also been performed where both cylinders rotate. Here we shall describe the phenomena observed with fixed outer cylinders but at various rotation speeds of the inner cylinder. At slow rotation speeds the fluid forms coaxial streamlines. This is quite understandable because the inner cylinder tries to carry the fluid along with it by means of the friction between the fluid and the cylinder. When the rotation speed is increased (which is usually measured in the dimensionless Taylor number), a new kind of motion occurs. The motion of the fluid becomes organized in the form of rolls in which the fluid periodically moves outwards and inwards in horizontal layers (Fig. 1.2.2a, b).

With further increase of the Taylor number, at a second critical value the rolls start oscillations with one basic frequency, and at still more elevated Taylor numbers with two fundamental frequencies. Sometimes still more complicated frequency patterns are observed. Eventually, at still higher Taylor numbers chaotic motion sets in. As is shown in Fig. 1.2.3, the evolving patterns can be seen directly. Furthermore, by means of laser light scattering, the velocity distribution and its Fourier spectrum have been measured (Fig. 1.2.4a–c). In particular cases with increasing Taylor number, a sequence of newly developing frequencies which are just 1/2, 1/4, 1/8, 1/16 of the fundamental frequency is observed. Since half a frequency means a double period of motion, this phenomenon is called period doubling. There are several features which are quite typical for self-organizing systems. When we change an external parameter, in

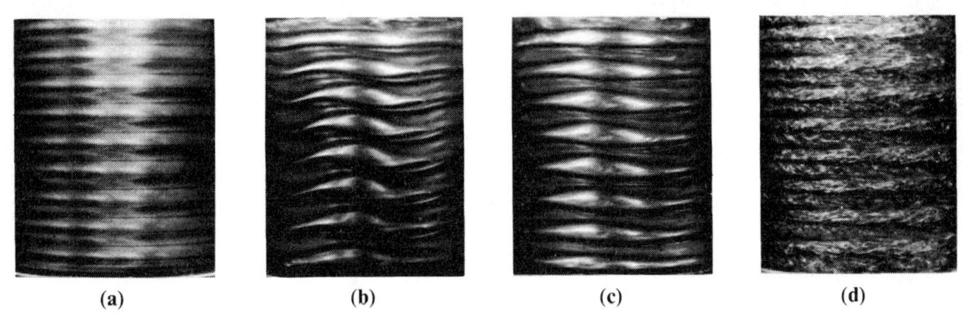

(a) (b) (c) (d)

Fig. 1.2.3. Instability hierarchy of the Taylor instability. **(a)** Formation of rolls. **(b)**The rolls start an oscillation. **(c)** A more complicated oscillatory motion of the rolls. **(d)** A chaotic motion. [After H. L. Swinney, P. R. Fenstermacher, J. P. Gollub: In *Synergetics, A Workshop*, ed. by H. Haken, Springer Ser. Synergetics, Vol. 2 (Springer, Berlin, Heidelberg, New York 1977) p. 60]

the present case the rotation speed, the system can form a hierarchy of patterns, though these patterns are not imposed on the system by some means from the outside. The patterns can become more and more complicated in their spatial and temporal structure.

Another standard type of experiment leads to the convection instability (or Bénard instability) and a wealth of further instabilities. Here, a fluid layer in a vessel with a certain geometry such as cylindrical or rectangular is used. The fluid is subject to gravity. When the fluid layer is heated from below and kept at a certain constant temperature at its upper surface, a temperature difference is established. Due to this temperature difference (temperature gradient), a vertical flux of heat is created. If the temperature gradient is small, this heat transport occurs microscopically and no macroscopic motion of the fluid is visible. When the temperature gradient is increased further, suddenly at a critical temperature gradient a macroscopic motion of the fluid forming pronounced patterns sets in. For instance, the heated liquid rises along stripes, cools down at the upper surface and falls down again along other stripes so that a motion in the form of rolls occurs. The dimension of the rolls is of the order of the thickness of the fluid layer, which in laboratory systems may range from millimeters to several centimeters.

The same phenomenon is observed in meteorology where cloud streets with dimensions of several hundred meters occur. When in a rectangular geometry the temperature gradient, which in dimensionless units is measured by the Rayleigh number, is increased further, the rolls can start an oscillation, and at a still higher Rayleigh number an oscillation of several frequencies can set in. Finally, an entirely irregular motion, called turbulence or chaos, occurs.

There are still other routes from a quiescent layer to turbulence, one such route occurring by period-doubling. When the Rayleigh number is increased, a rather complex motion occurs in which at well-defined Rayleigh numbers the period of motion is repeatedly doubled (Fig. 1.2.5a – c). In cylindrical containers concentric rolls can be observed or, if the symmetry with respect to the horizontal middle plane is violated, hexagonal patterns occur. Also, transitions between rolls and hexagons and even the coexistence of rolls and hexagons can be observed (Fig. 1.2.6). Other observed patterns are rolls which are oriented at rec-

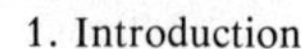

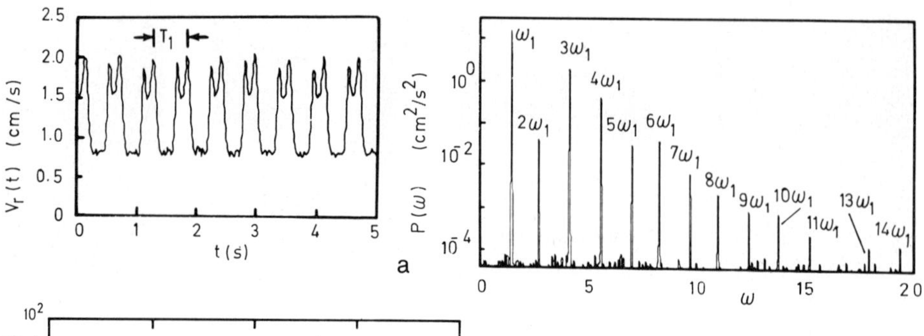

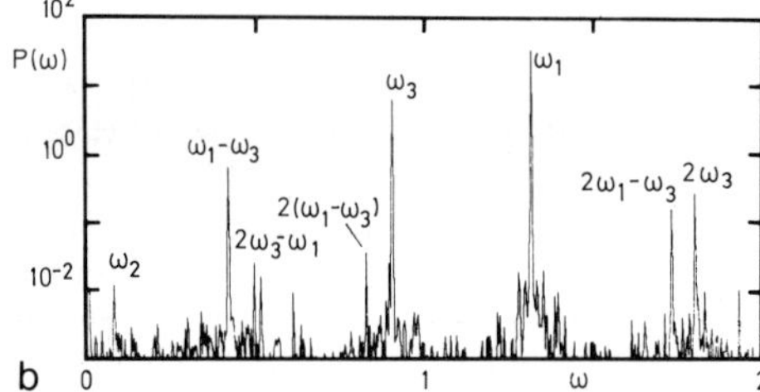

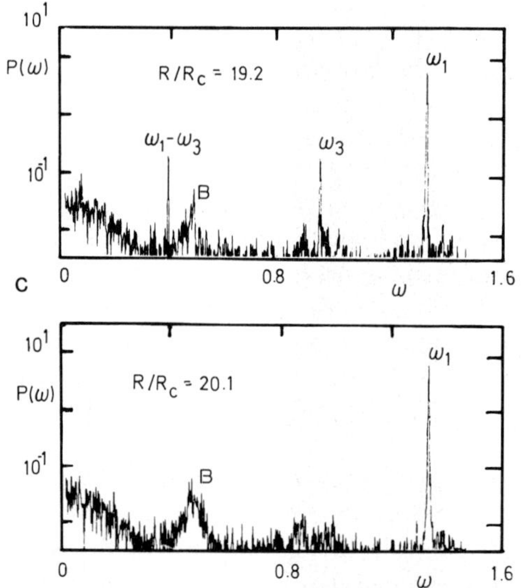

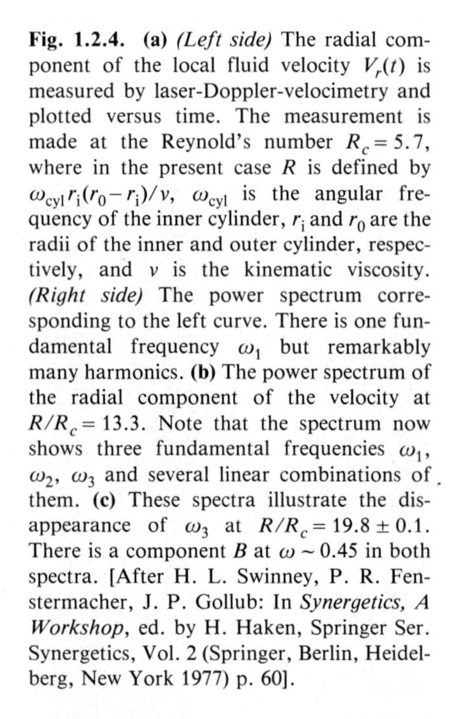
Fig. 1.2.4. **(a)** *(Left side)* The radial component of the local fluid velocity $V_r(t)$ is measured by laser-Doppler-velocimetry and plotted versus time. The measurement is made at the Reynold's number $R_c = 5.7$, where in the present case R is defined by $\omega_{\rm cyl} r_i (r_0 - r_i)/\nu$, $\omega_{\rm cyl}$ is the angular frequency of the inner cylinder, r_i and r_0 are the radii of the inner and outer cylinder, respectively, and ν is the kinematic viscosity. *(Right side)* The power spectrum corresponding to the left curve. There is one fundamental frequency ω_1 but remarkably many harmonics. **(b)** The power spectrum of the radial component of the velocity at $R/R_c = 13.3$. Note that the spectrum now shows three fundamental frequencies ω_1, ω_2, ω_3 and several linear combinations of them. **(c)** These spectra illustrate the disappearance of ω_3 at $R/R_c = 19.8 \pm 0.1$. There is a component B at $\omega \sim 0.45$ in both spectra. [After H. L. Swinney, P. R. Fenstermacher, J. P. Gollub: In *Synergetics, A Workshop*, ed. by H. Haken, Springer Ser. Synergetics, Vol. 2 (Springer, Berlin, Heidelberg, New York 1977) p. 60].

tangles and which thus form a rectangular lattice (Fig. 1.2.7). At elevated Rayleigh numbers a carpet-like structure is found. Many patterns can also exhibit a number of imperfections (Fig. 1.2.8). In order to obtain most clearcut results, in modern experiments the aspect ratio, i. e., the horizontal dimension of the vessel divided by the vertical dimension, is taken small, i. e., of the order of unity. At higher aspect ratios the individual transitions can very quickly follow after one another or can even coexist. Another class of experiments, where the fluid is heated from above, leads to the "Marengo instability". Further, in the atmospheres of the earth and other planets where gravitation, rotation and heating act jointly, a wealth of patterns is formed.

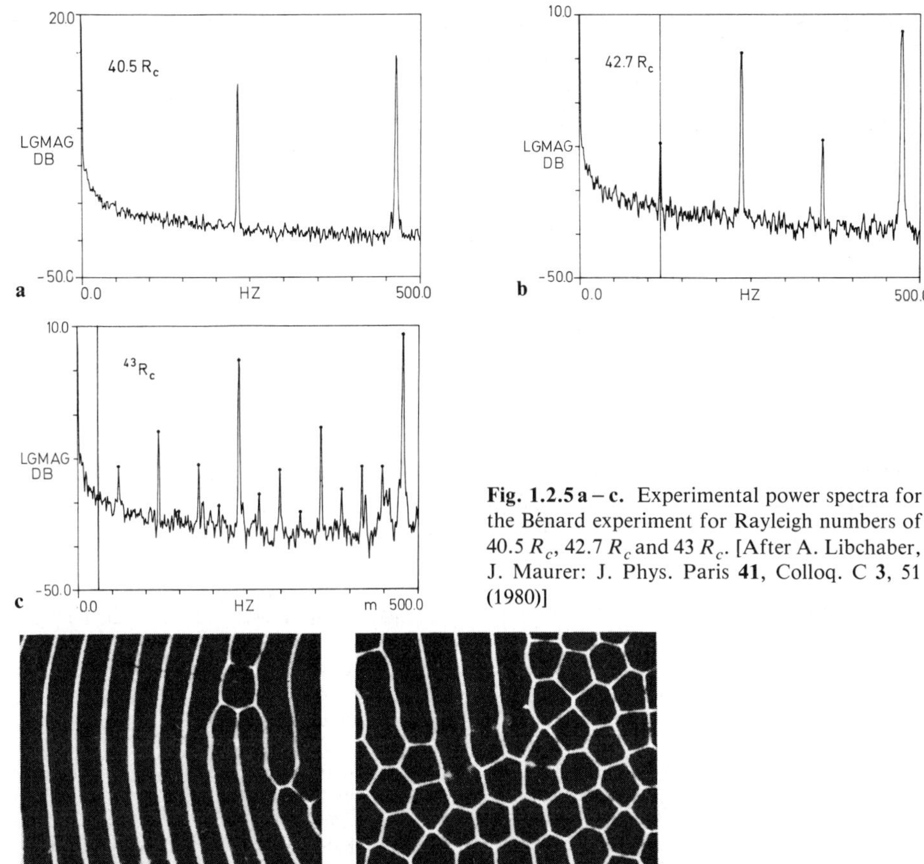

Fig. 1.2.5 a – c. Experimental power spectra for the Bénard experiment for Rayleigh numbers of 40.5 R_c, 42.7 R_c and 43 R_c. [After A. Libchaber, J. Maurer: J. Phys. Paris **41**, Colloq. C **3**, 51 (1980)]

Fig. 1.2.6

Fig. 1.2.7

Fig. 1.2.6. These photographs show the surface of a liquid heated from below. The coexistence of hexagons and rolls is clearly visible. [After J. Whitehead: private communication]

Fig. 1.2.7. Photograph of the surface of a liquid heated from below. The rectangular lattice formed by two rectangular roll systems is clearly visible. [After J. Whitehead: In *Fluctuations, Instabilities and Phase Transitions,* ed. by T. Riste (Plenum, New York 1975)]

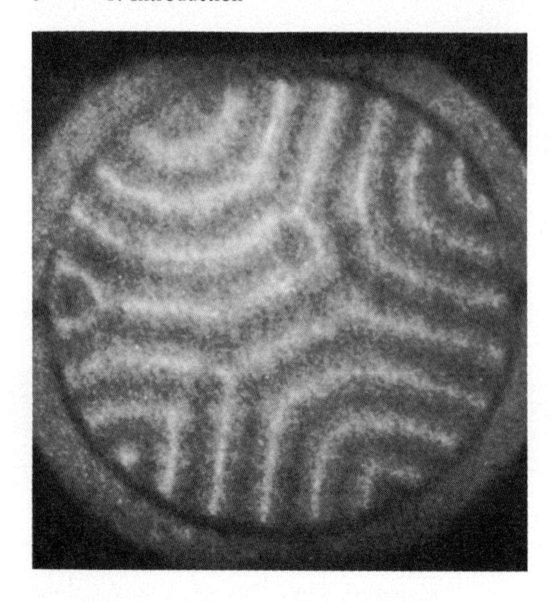

Fig. 1.2.8. Photograph of the surface of a liquid heated from below in a circular vessel. The roll systems are developed only partly and form imperfections at their intersection points. [After P. Bergé: In *Chaos and Order in Nature,* ed. by H. Haken, Springer Ser. Synergetics, Vol. 11 (Springer, Berlin, Heidelberg, New York 1981) p. 14]

A further class of patterns of important practical interest concerns the motion of fluids or gases around moving objects, such as cars, airplanes and ships. Again specific patterns give rise to various effects. Let us now turn to a quite different field of physics, namely lasers.

1.2.2 Lasers: Coherent Oscillations

Lasers are certain types of lamps which are capable of emitting coherent light. Since I have treated the laser as a prototype of a synergetic system in extenso in my book *Synergetics, An Introduction* [1], I shall describe here only a few features particularly relevant to the present book.

A typical laser consists of a crystal rod or a glass tube filled with gas, but the individual experimental arrangement is not so important in the present context. What is most important for us are the following features: when the atoms the laser material consists of are excited or "pumped" from the outside, they emit light waves. At low pump power, the waves are entirely uncorrelated as in a usual lamp. Could we hear light, it would sound like noise to us.

When we increase the pump rate to a critical value, the noise disappears and is replaced by a pure tone. This means that the atoms emit a pure sinusoidal light wave which in turn means that the individual atoms act in a perfectly correlated way – they become self-organized. When the pump rate is increased beyond a second critical value, the laser may periodically emit very intense and short pulses. In this way the following instability sequence occurs (Fig. 1.2.9a – c):

$$\text{noise} \rightarrow \left\{\begin{array}{l}\text{coherent oscillation}\\ \text{at frequency } \omega_1\end{array}\right\} \rightarrow \left\{\begin{array}{l}\text{periodic pulses at frequency } \omega_2\\ \text{which modulate oscillation}\\ \text{at frequency } \omega_1\end{array}\right\}$$

i.e.,

$$\text{no oscillation} \rightarrow 1 \text{ frequency} \rightarrow 2 \text{ frequencies.}$$

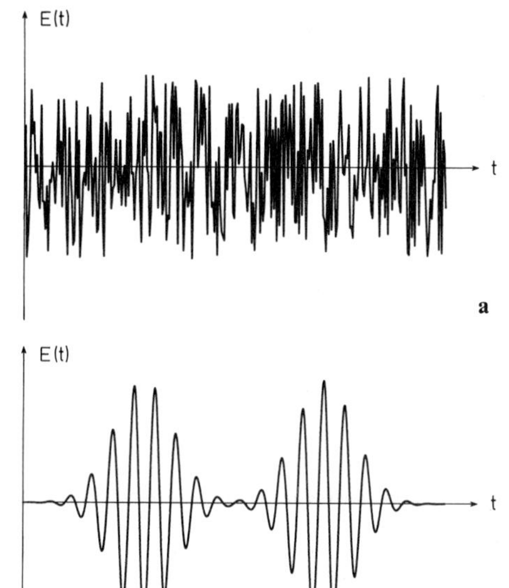

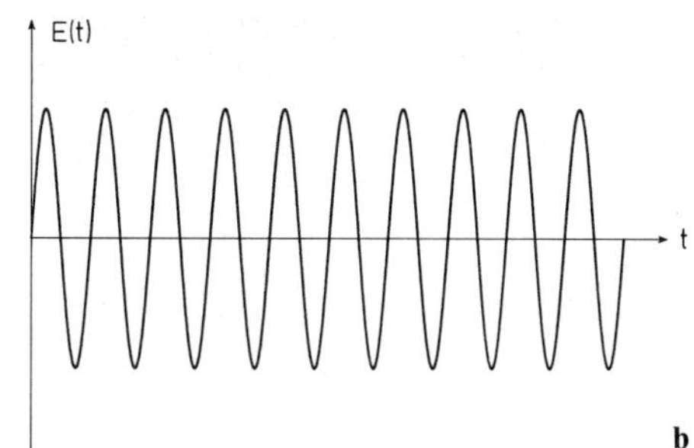

Fig. 1.2.9. **(a)** Schematic plot of the field strength $E(t)$ as a function of time in the case of emission from a lamp. **(b)** The field strength $E(t)$ as a function of time of the light field emitted from a laser. **(c)** Schematic plot of the field strength $\mathrm{E}(t)$ as a function of time in the case of ultrashort laser pulses

Under different conditions the light emission may become "chaotic" or "turbulent", i.e., quite irregular. The frequency spectrum becomes broadband.

If a laser is not only pumped but also irradiated with light from another laser, a number of interesting effects can occur. In particular, the laser can acquire one of two different internal states, namely one with a high transmission of the incident light, or another with a low transmission. Since these two states are stable, one calls this system bistable. Bistable elements can be used as memories and as logical devices in computers.

When a laser is coupled to a so-called saturable absorber, i.e., a material where light transmissivity becomes very high at a sufficiently high light intensity, a number of different instabilities may occur: bistability, pulses, chaos.

The laser played a crucial role in the development of synergetics for various reasons. In particular, it allowed detailed theoretical and experimental study of the phenomena occurring within the transition region, lamp ↔ laser, where a surprising and far-reaching analogy with phase transitions of systems in thermal equilibrium was discovered. This analogy includes a symmetry-breaking instability, critical slowing down and critical fluctuations. The results show that close to the transition point of a synergetic system fluctuations play a crucial role. (These concepts are explained in [1], and are dealt with here in great detail.)

1.2.3 Plasmas: A Wealth of Instabilities

A plasma consists of a gas of atoms which are partly or fully deprived of their electronic cloud, i.e., ionized. Thus a plasma can be characterized as a gas or fluid composed of electrically charged particles. Since fluids may show quite different instabilities it is not surprising that instabilities occur in plasmas also.

Fig. 1.2.10. Calculated lines of constant vertical velocity of one-component plasma heated from below and subjected to a vertical homogeneous magnetic field. [After H. Klenk, H. Haken: Acta Phys. Austriaca **52**, 187 (1980)]

Because of the charges and their interaction with electromagnetic fields, new types of instabilities may occur, and indeed in plasmas a huge variety of instabilities are observed. For example, Fig. 1.2.10 presents a theoretical result on the formation of a velocity pattern of a plasma which is heated from below and subjected to a constant vertical magnetic field. It is beyond the scope of this book to list these instabilities here. Besides acquiring ordered states, a plasma can also migrate from one instability to another or it may suffer a violent instability by which the plasma state collapses. A study of these phenomena is of utmost importance for the construction of fusion reactors and for other fields, e.g., astrophysics.

1.2.4 Solid-State Physics: Multistability, Pulses, Chaos

A few examples of phenomena are given, where under a change of external conditions qualitatively new phenomena occur.

a) The Gunn Oscillator. When a relatively small voltage is applied to a sample of gallium arsenide (GaAs), a constant current obeying Ohm's law is generated. When the voltage is increased, at a critical voltage this constant current is replaced by regular pulses. With a still higher voltage irregular pulses (chaos) can also be observed.

b) Tunnel Diodes. These are made of semiconductors doped with well-defined impurity atoms, so that energy bands (in particular the valence and the conduction band) are deformed. When an electric field is applied, a tunnel current may flow and different operating states of the tunnel diode may be reached, as indicated in Fig. 1.2.11.

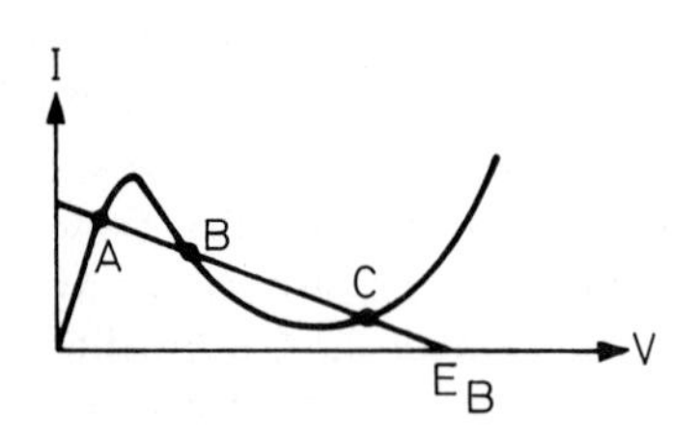

Fig. 1.2.11. Different operating states A, B, C of a tunnel diode where the current I is plotted versus applied voltage V. The states A and C are stable, whereas B is unstable. This picture provides a typical example of a bistable device with stability points A and C

c) Thermoelastic Instabilities. Qualitative changes of behavior on macroscopic scales can also be observed in *mechanics*, for instance in thermoelastic instabilities. When strain is applied to a solid, beyond the elastic region at critical parameter values of the strain qualitatively new phenomena may occur, for instance, acoustic emission.

d) Crystal Growth. Crystals may exhibit structure at two different length scales. At the microscopic level the atoms (or molecules) form a regular lattice with fixed spacings between the lattice sites. The order at the microscopic level can be made visible by X-ray or electron diffraction, which reveals the regular arrangement of atoms or molecules within a lattice. The lattice constant provides one length scale. The other length scale is connected with the macroscopic form of a crystal, e.g., the form of the snow crystal. While it is generally adopted that the lattice structure can be explained by the assumption that the atoms or molecules acquire a state of minimal free energy, which at least in principle can be calculated by means of quantum mechanics, the explanation of the macroscopic shapes requires a different approach. Here we have to study the processes of crystal growth. It is this latter kind of problem which concerns synergetics.

1.3 Engineering

1.3.1 Civil, Mechanical, and Aero-Space Engineering: Post-Buckling Patterns, Flutter, etc.

Dramatic macroscopic changes of systems caused by change of external parameters are well known in engineering. Examples are: the bending of a bar under load as treated by Euler in the eighteenth century, the breakdown of bridges beyond a critical load, deformations of thin shells under homogeneous loads (Fig. 1.3.1), where for instance hexagons and other post-buckling patterns can be observed. Mechanical instabilities can also occur in a dynamical fashion, for instance the flutter of wings of airplanes.

1.3.2 Electrical Engineering and Electronics: Nonlinear Oscillations

Radio waves are generated by electromagnetic oscillators, i.e., circuits containing radio tubes or transistors. The coherent electromagnetic oscillations of these devices can be viewed as an act of self-organization. In the non-oscillating state, the electrons move randomly since their motion is connected with thermal noise, whereas under suitable external conditions self-sustained oscillations may be achieved in which macroscopic electric currents oscillate at a well-defined frequency in the electronic circuit. When oscillators are coupled to each other, a number of phenomena common also to fluid dynamics occur, namely frequency locking, period doubling or tripling (Fig. 1.3.2), and chaos, i.e., irregular emission. An important role is played by the question whether a system of coupled oscillators can still act as a set of oscillators with individual frequencies, or whether entirely new types of motion, e.g., chaotic motion, occur.

Fig. 1.3.1. A thin metallic shell is put under an internal subpressure. This photograph shows the postbuckling pattern where the individual hexagons are clearly visible. [After R. L. Carlson, R. L. Sendelbeck, N. J. Hoff: Experimental Mechanics **7**, 281 (1967)]

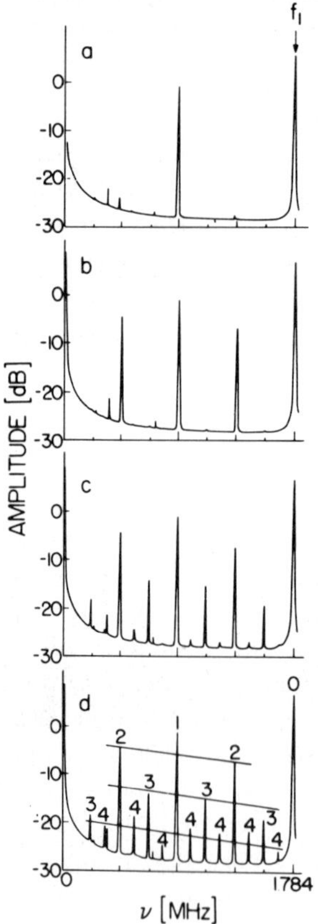

Fig. 1.3.2 a – d. These power spectra of an electronic device with a certain nonlinear capacitance clearly show a series of period doublings (from top to bottom) when the control parameter is increased. [After P. S. Linsay: Phys. Rev. Lett. **47**, 1349 (1981)]

1.4 Chemistry: Macroscopic Patterns

In this field synergetics focuses its attention on those phenomena where macroscopic patterns occur. Usually, when reactants are brought together and well stirred, a homogeneous end product arises. However, a number of reactions may show temporal or spatial or spatio-temporal patterns. The best known example is the Belousov-Zhabotinsky reaction. Here $Ce_2(SO_4)_3$, $KBrO_3$, $CH_2(COOH)_2$, H_2SO_4 as well as a few drops of ferroine (redox indicator) are mixed and stirred.

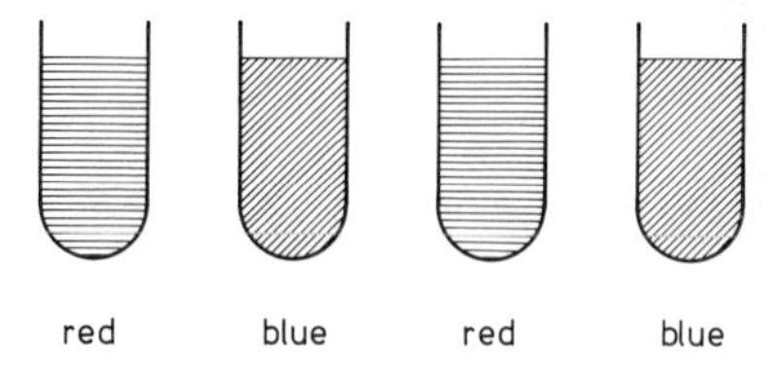

Fig. 1.4.1. Example of a chemical oscillation (schematic)

In a continuously stirred tank reactor, oscillations of the composition may occur, as can be seen directly from the periodic change of color from red to blue (Fig. 1.4.1). In a closed system, i.e., without input of new materials, these oscillations eventually die out. On the other hand, when new reactants are added continuously and the final products are continuously removed, the oscillations can be sustained indefinitely. Nowadays more than a dozen systems are known showing chemical oscillations. If the influx concentration of one reactant is changed, a sequence of different kinds of behavior can be reached, e.g. an alternation between oscillations and chaotic behavior (Figs. 1.4.2, 3). If the chemicals are not stirred, spatial patterns can develop. Examples are provided by the Belousov-Zhabotinsky reaction in the form of concentric waves (Fig. 1.4.4) or spirals. Some of the oscillating reactions can be influenced by light (photochemistry) where either frequency entrainment or chaotic states can be reached. Quite another class of macroscopic patterns formed by a continuous influx of

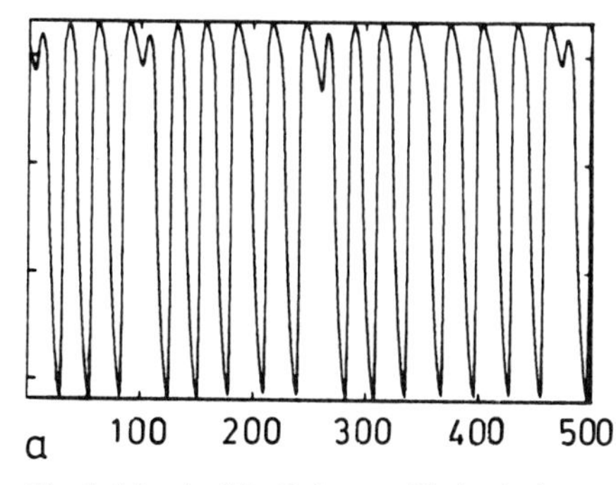

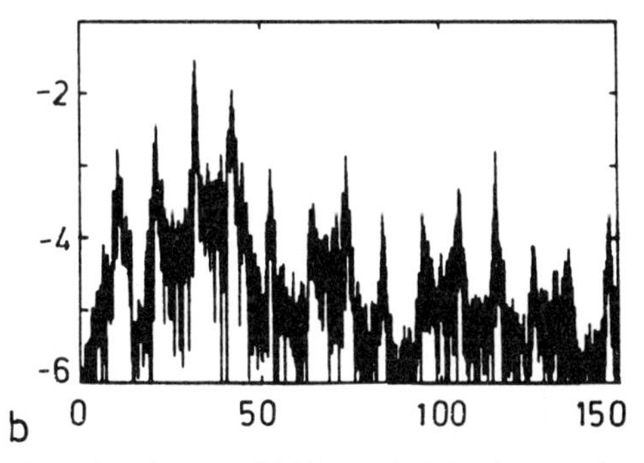

Fig. 1.4.2 a, b. The Belousov-Zhabotinsky reaction in a chaotic state. **(a)** The optical density record (arbitrary units) versus time [s]. **(b)** The corresponding power spectral density versus frequency in semi-logarithmic plot. [After C. Vidal: In *Chaos and Order in Nature*, ed. by H. Haken, Springer Ser. Synergetics, Vol. 11 (Springer, Berlin, Heidelberg, New York 1981) p. 69]

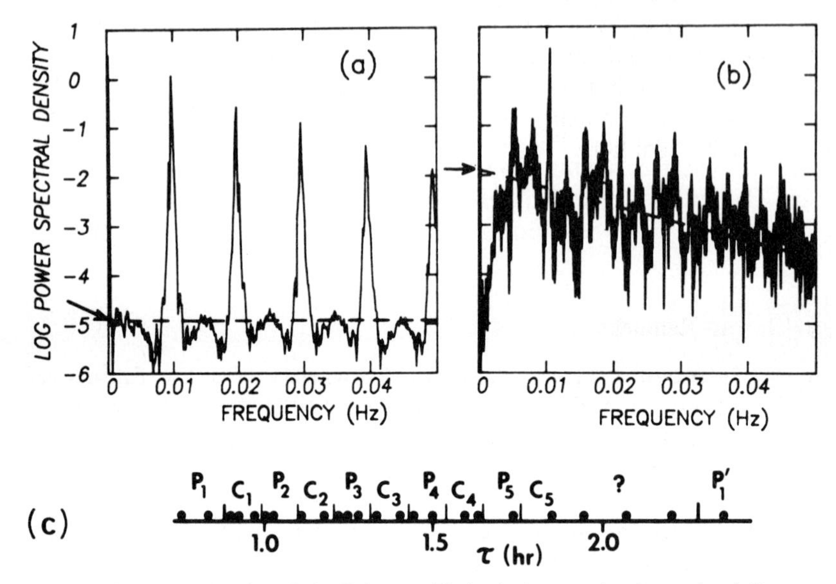

Fig. 1.4.3a–c. Dynamics of the Belousov-Zhabotinsky reaction in a stirred flow reactor. Power spectra of the bromide ion potential are shown **(a)** for a periodic regime, and **(b)** a chaotic regime for different residence times τ where $\tau =$ reactor volume/flow rate. All other variables were held fixed. The plot **(c)** shows the alternating sequence between periodic and chaotic regimes when the residence rate is increased. [After J. S. Turner, J. C. Roux, W. D. McCormick, H. L. Swinney: Phys. Lett. **85A**, 9 (1981) and in *Nonlinear Problems: Presence and Future*, ed. by A. R. Bishop (North-Holland, Amsterdam 1981)]

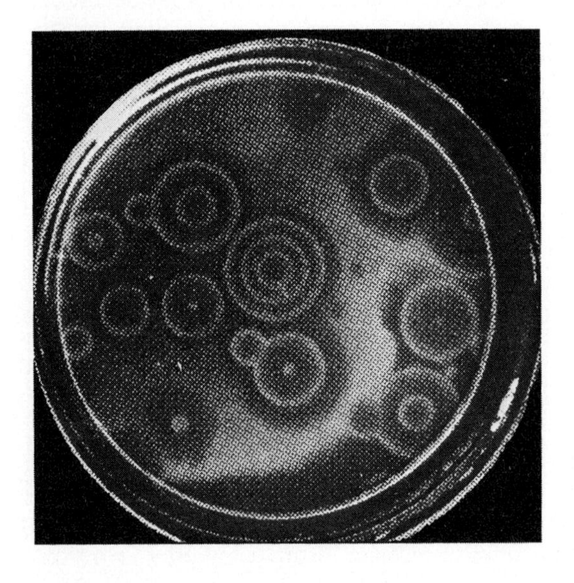

Fig. 1.4.4. Concentric waves of the Belousov-Zhabotinsky reaction. [After M. L. Smoes: In *Dynamics of Synergetic Systems*, ed. by H. Haken, Springer Ser. Synergetics, Vol. 6 (Springer, Berlin, Heidelberg, New York 1980) p. 80]

matter consists of flames. Hopefully the concepts of synergetics will allow us to get new insights into these phenomena known since men observed fire. Finally, it is expected that methods and concepts of synergetics can be applied to chemistry at the molecular level also, in particular to the behavior of biological molecules. But these areas are still so much in the developmental phase that it would be premature to include them here.

1.5 Biology

1.5.1 Some General Remarks

The animate world provides us with an enormous variety of well-ordered and well-functioning structures, and nearly nothing seems to happen without a high degree of cooperation of the individual parts of a biological system. By means of synergetic processes, biological systems are capable of "upconverting" energy transformed at the molecular level to macroscopic energy forms. Synergetic processes manifest themselves in muscle contraction leading to all sorts of movement, electrical oscillations in the brain (Fig. 1.5.1 a, b), build up of electric voltages in electric fish, pattern recognition, speech, etc. For these reasons, biology is a most important field of research for synergetics. At the same time we must be aware of the fact that biological systems are extremely complex and it will be wise to concentrate our attention on selected problems, of which a few examples in the following are mentioned.

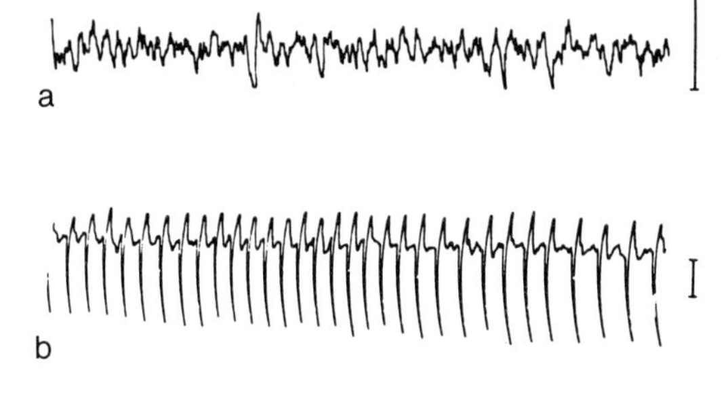

Fig. 1.5.1 a, b. Two examples of an electroencephalogram. **(a)** Normal behavior. **(b)** Behavior in a epileptic seizure. [After A. Kaczmarek, W. R. Adey, reproduced by A. Babloyantz: In *Dynamics of Synergetic Systems,* ed. by H. Haken, Springer Ser. Synergetics, Vol. 6 (Springer, Berlin, Heidelberg, New York 1980) p. 180]

1.5.2 Morphogenesis

The central problem of morphogenesis can be characterized as follows. How do the originally undifferentiated cells know where and in which way to differentiate? Experiments indicate that this information is not originally given to the individual cells but that a cell within a tissue receives information on its position from its surroundings, whereupon it differentiates. In experiments with embryos, transplantation of a cell from a central region of the body into the head region causes this cell to develop into an eye. These experiments demonstrate that the cells do not receive their information how to develop from the beginning (e.g., through their DNA), but that they receive that information from the position within the cell tissue. It is assumed that the positional information is provided by a chemical "pre-pattern" which evolves similarly to the patterns described in the section on chemistry above. These pattern formations are based on the cooperation of reactions and diffusion of molecules.

It is further assumed that at sufficiently high local concentration of these molecules, called morphogenes, genes are switched on, leading to a differentiation of their cells. A number of chemicals have been found in hydra which are good candidates for being activating or inhibiting molecules for the formation of heads or feet. While a detailed theory for the development of organs, e.g., eyes, along these lines of thought is still to be developed, simpler patterns, e.g., stripes on furs, rings on butterfly wings, are nowadays explained by this kind of approach.

1.5.3 Population Dynamics

The phenomena to be explained are, among others, the abundance and distribution of species. If different species are supported by only one kind of food, etc., competition starts and Darwin's rules of the *survival of the fittest* apply. (Actually, a strong analogy to the competition of laser modes exists.) If different food resources are available, *coexistence* of species becomes possible.

Species may show *temporal oscillations.* At the beginning of the twentieth century, fishermen of the Adriatic Sea observed a periodic change of numbers of fish populations. These oscillations are caused by the "interaction" between predator and prey fish. If the predators eat too many prey fish, the number of the prey fish and thus eventually also the number of the predators decrease. This in turn allows an increase of the number of prey fish, which then leads to an increase of the number of predators, so that a cyclic change of the population occurs. Other populations, such as certain insect populations, may show *chaotic variations* in their numbers.

1.5.4 Evolution

Evolution may be viewed as the formation of new macroscopic patterns (namely new kinds of species) over and over again. Models on the evolution of biomolecules are based on a mathematical formulation of Darwin's principle of the

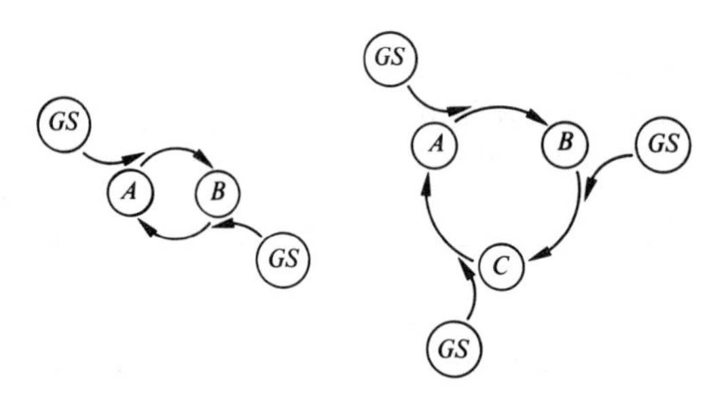

Fig. 1.5.2. Examples of the Eigen-Schuster hypercycles. Two kinds of biomolecules A and B are multiplied by autocatalysis but in addition the multiplication of B is assisted by that of A and vice versa. Here GS indicates certain ground substances from which the molecules are formed. The right-hand side of this picture presents a cycle at which three kinds of biomolecules A, B, C participate

survival of the fittest. It is assumed that biomolecules multiply by autocatalysis (or in a still more complicated way by cyclic catalysis within "hypercycles") (Fig. 1.5.2). This mechanism can be shown to cause selections which in combination with mutations may lead to an evolutionary process.

1.5.5 Immune System

Further examples for the behavior of complex systems in biology are provided by enzyme kinetics and by antibody-antigen kinetics. For instance, in the latter case, new types of antibodies can be generated successively, where some antibodies act as antigens, thus leading to very complex dynamics of the total system.

1.6 Computer Sciences

1.6.1 Self-Organization of Computers, in Particular Parallel Computing

The problem here is to construct a computer net in which computation is distributed among the individual parts of the computer network in a self-organized fashion (Fig. 1.6.1) rather than by a master computer (Fig. 1.6.2). The distribution of tasks corresponds to certain patterns of the information flux. While continuous changes of the total task may be treated by synergetics' methods, considerable research has still to be put into the problem of how the computer net can cope with qualitatively new tasks.

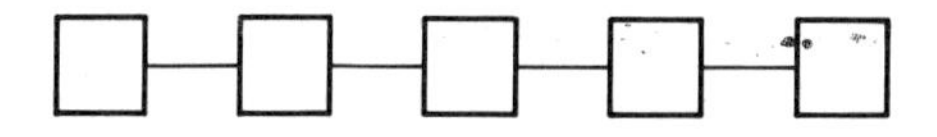

Fig. 1.6.1. Schematic diagram of a computer system whose individual computers are connected with each other and determine the distribution of tasks by themselves

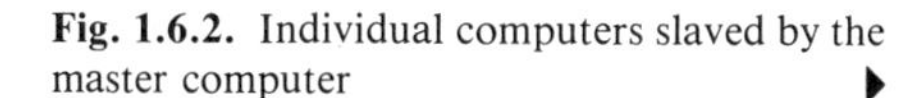

Fig. 1.6.2. Individual computers slaved by the master computer

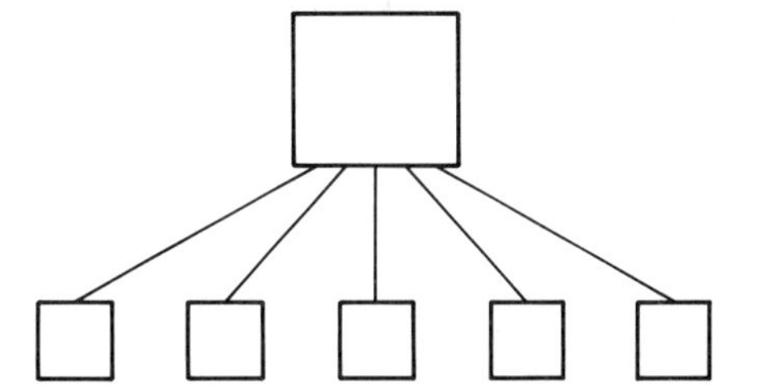

1.6.2 Pattern Recognition by Machines

For sake of completeness, pattern recognition is mentioned as a typical synergetic process. Its mathematical treatment using methods of synergetics is still in its infancy, however, so no further details are given here.

1.6.3 Reliable Systems from Unreliable Elements

Usually the individual elements of a system, especially those at the molecular level, may be unreliable due to imperfections, thermal fluctuations and other

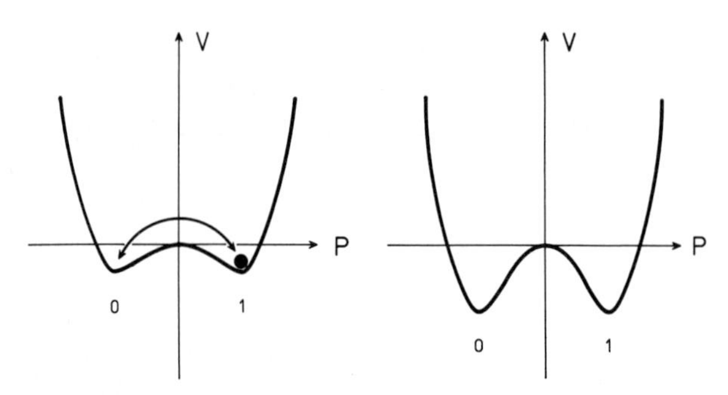

Fig. 1.6.3. How to visualize the buildup of a reliable system from unreliable elements. *(Left side)* The potential function V of an individual element being very flat allows for quick jumps of the system between the holding states 0 and 1. *(Right side)* By coupling elements together the effective potential can be appreciably deepened so that a jump from one holding state to another becomes very improbable

causes. It is suspected that the elements of our brain, such as neurons, are of such a type. Nature has mastered this problem to construct reliable systems out of these unreliable elements. When computer elements are made smaller and smaller, they become less and less reliable. In which way can one put computer elements together so that the total system works reliably? The methods of synergetics indicated below will allow us to devise systems which fulfill this task. Let us exemplify here how a reliable memory can be constructed out of unreliable elements. In order to describe the behavior of a single element, the concept of the order parameter is used (to be explained later). Here it suffices to interpret the order parameter by means of a particle moving in a potential (in an overdamped fashion) (Fig. 1.6.3). The two holding states 1 and 2 of the memory element are identified with the two minima of the potential. Clearly, if the "valleys" are flat, noise will drive the "particle" back and forth and holding a state is not possible. However, by coupling several elements together, a potential with deeper values can be obtained, which is obviously more reliable. Coupling elements in various ways, several reliable holding states can be obtained. Let us now turn to phenomena treated by disciplines other than the natural sciences.

1.7 Economy

In economy dramatic macroscopic changes can be observed. A typical example is the switching from full employment to underemployment. A change of certain control parameters, such as the kind of investment from production-increasing into rationalizing investments, may lead to a new state of economy, i.e., that of underemployment. Oscillations between these two states have been observed and are explainable by the methods of synergetics. Another example of the development of macroscopic structures is the evolution of an agricultural to an industrialized society.

1.8 Ecology

Dramatic changes on macroscopic scales can be observed in ecology and related fields. For instance, in mountainous regions the change of climate with altitude acting as a control parameter may cause different belts of vegetation. Similar observations are made with respect to different zones of climate on the earth, giving rise to different kinds of vegetation. Further examples of macroscopic changes are provided by pollution where the increase of only very few percent of pollution may cause the dying out of whole populations, e.g., of fish in a lake.

1.9 Sociology

Studies by sociologists strongly indicate that the formation of "public opinion" (which may be defined in various ways) is a collective phenomenon. One mechanism which might be fundamental was elucidated by experiments by S. Ash which, in principle, are as follows. The following task is given to about 10 "test persons": They have to say which reference line agrees in length with three other lines of different lengths (Fig. 1.9.1). Except for one genuine test person, all others were helpers of the experimenter, but the single test person did not know that. In a first run, the helpers gave the correct answer (and so did the test person). In the following runs, all helpers gave a wrong answer, and now about 60% of the test persons gave the wrong answer, too. This is a clear indication how individuals are influenced by the opinion of others. These results are supported by field experiments by E. Noelle-Neumann. Since individuals mutually influence each other during opinion-making, the field of public opinion formation may be analyzed by synergetics. In particular, under certain external conditions (state of economy, high taxes), public opinion can undergo dramatic changes, which can manifest itself even in revolutions. For further details, the reader is referred to the references.

Fig. 1.9.1. A typical arrangement of the experiment by Ash as described in the text

1.10 What are the Common Features of the Above Examples?

In all cases the systems consist of very many subsystems. When certain conditions ("controls") are changed, even in a very unspecific way, the system can develop new kinds of patterns on macroscopic scales. A system may go from a homogeneous, undifferentiated, quiescent state to an inhomogeneous, but well-ordered state or even into one out of several possible ordered states. Such systems can be operated in different stable states (bistability or multistability). These systems can be used, e.g., in computers as a memory (each stable state represents a number, e.g., 0 or 1). In the ordered state, oscillations of various types can also occur with a single frequency, or with different frequencies (quasiperiodic oscillations). The system may also undergo random motions (chaos). Furthermore, spatial patterns can be formed, e.g., honeycomb structures, waves or spirals. Such structures may be maintained in a dynamic way by a continuous flux of energy (and matter) through the systems, e.g., in fluid dynamics. In other cases, the structures are first generated dynamically but eventually a "solidification" occurs, e.g., in crystal growth or morphogenesis. In a more abstract sense, in social, cultural or scientific "systems", new patterns can evolve, such as ideas, concepts, paradigms. Thus in all cases we deal with processes of self-organization leading to qualitatively new structures on macroscopic scales. What are the mechanisms by which these new structures evolve? In which way can we describe these transitions from one state to another? Because the systems may be composed of quite different kinds of parts, such as atoms, molecules, cells, or animals, at first sight it may seem hopeless to search for general concepts and mathematical methods. But this is precisely the aim of the present book.

1.11 The Kind of Equations We Want to Study

In this section we shall discuss how we can deal mathematically with the phenomena described in the preceding sections and with many others. We shall keep in mind that our approach will be applicable to problems in physics, chemistry, and biology, and also in electrical and mechanical engineering. Further areas are economy, ecology, and sociology. In all these cases we have to deal with systems composed of very many subsystems for which not all information may be available. To cope with these systems, approaches based on thermodynamics or information theory are frequently used. But during the last decade at least, it became clear that such approaches (including some generalizations such as irreversible thermodynamics) are inadequate to cope with physical systems driven away from thermal equilibrium or with, say, economic processes. The reason lies in the fact that these approaches are basically static, ultimately based on information theory which makes guesses on the *numbers* of possible states. In [1], I have demonstrated how that formalism works and where its limitations lie. In all the systems considered here, dynamics plays the crucial role. As shall be demonstrated with mathematical rigor, it is the growth (or decay) rates of collective "modes" that determine which and how macroscopic states are formed. In a way, we are led to some kind of generalized Darwinism which even acts in the inanimate world, namely, the generation of collective modes by fluctuations, their competition and finally the selection of the "fittest" collective mode or a combination thereof, leading to macroscopic structures.

Clearly, the parameter "time" plays a crucial role. Therefore, we have to study the evolution of systems in time, whereby equations which are sometimes called "evolution equations" are used. Let us study the structure of such equations.

1.11.1 Differential Equations

We start with the example of a single variable q which changes in time. Such a variable can be the number of cells in a tissue, the number of molecules, or the coordinate of a particle. The temporal change of q will be denoted by $dq/dt = \dot{q}$. In many cases $\dot{q}$ depends on the present state of the system, e.g., on the number of cells present. The simplest case of the corresponding equation is

$$\dot{q} = \alpha q . \tag{1.11.1}$$

Such equations are met, for instance, in chemistry, where the production rate $\dot{q}$ of a chemical is proportional to its concentration q ("autocatalytic reaction"), or in population dynamics, where q corresponds to the number of individuals. In wide classes of problems one has to deal with oscillators which in their simplest form are described by

$$\ddot{q}_1 + \omega^2 q_1 = 0 , \tag{1.11.2}$$

where ω is the frequency of the oscillation.

1.11.2 First-Order Differential Equations

In contrast to (1.11.1), where a first-order derivative occurs, (1.11.2) contains a second-order derivative $\ddot{q}_1$. However, by introducing an additional variable q_2 by means of

$$\dot{q}_1 = q_2 , \tag{1.11.3}$$

we may split (1.11.2) into two equations, namely (1.11.3) and

$$\dot{q}_2 = -\omega^2 q_1 . \tag{1.11.4}$$

These two equations are equivalent to (1.11.2). By means of the trick of introducing additional variables we may replace equations containing higher-order derivatives by a set of differential equations which contain first-order derivatives only. For a long time, equations of the form (11.1.1 – 4) or generalizations to many variables were used predominantly in many fields, because these equations are linear and can be solved by standard methods. We now wish to discuss those additional features of equations characteristic of synergetic systems.

1.11.3 Nonlinearity

All equations of synergetics are *nonlinear*. Let us consider an example from chemistry, where a chemical with concentration q_1 is produced in an autocatalytic way by its interaction with a second chemical with a concentration q_2. The increase of concentration of chemical 1 is described by

$$\dot{q}_1 = \beta q_1 q_2 \,. \tag{1.11.5}$$

Clearly, the cooperation of parts (i.e., synergetics), in our case of molecules, is expressed by a nonlinear term. Quite generally speaking, we shall consider equations in which the right-hand side of equations of the type (1.11.5) is a nonlinear function of the variables of the system. In general, a set of equations for several variables q_j must be considered.

1.11.4 Control Parameters

The next important feature of synergetic systems consists in the control outside parameters may exercise over them. In synergetics we deal mainly with open systems. In physics, chemistry, or biology, systems are driven away from states of thermal equilibrium by an influx of energy and/or matter from the outside. We can also manipulate systems from the outside by changing temperature, irradiation, etc. When these external controls are kept constant for a while, we may take their effects into account in the equations by certain constant parameters, called control parameters. An example for such a parameter may be α which occurs in (1.11.1). For instance, we may manipulate the growth rate of cells by chemicals from the ouside. Let us consider α as the difference between a production rate p and a decay rate d, i.e., $\alpha = p - d$. We readily see that by control of the production rate quite different kinds of behavior of the population may occur, namely exponential growth, a steady state, or exponential decay. Such control parameters may enter the evolution equations at various places. For instance in

$$\dot{q}_1 = \alpha q_1 + \beta q_1 q_2 \,, \tag{1.11.6}$$

the constant β describes the coupling between the two systems q_1 and q_2. When we manipulate the strength of the coupling from the outside, β plays the role of a control parameter.

1.11.5 Stochasticity

A further important feature of synergetic systems is stochasticity. In other words, the temporal evolution of these systems depends on causes which cannot be predicted with absolute precision. These causes can be taken care of by "fluctuating" forces $f(t)$ which in their simplest form transform (1.11.1) into

$$\dot{q} = \alpha q + f(t) \,. \tag{1.11.7}$$

Sometimes the introduction of these forces causes deep philosophical problems which shall be briefly discussed, though later on a more pragmatic point of view will be taken, i.e., we shall assume that the corresponding fluctuating forces are given for each system under consideration. Before the advent of quantum theory, thinking not only in physics but also in many other fields was dominated by a purely mechanistic point of view. Namely, once the initial state of a system is known it will be possible to predict its further evolution in time precisely. This idea was characterized by Laplace's catch phrase, "spirit of the world". If such a creature knows the initial states of all the individual parts of a system (in particular all positions and velocities of its particles) and their interactions, it (or he) can predict the future for ever. Three important ideas have evolved since then.

a) Statistical Mechanics. Though it is still in principle possible to predict, for instance, the further positions and velocities of particles in a gas, it is either undesirable or in practice impossible to do this explicitly. Rather, for any purpose it will be sufficient to describe a gas by statistics, i.e., to make predictions in the sense of probability. For instance, how probable it will be to find n particles with a velocity v' between v and $v + dv$. Once such a probabilistic point of view is adopted, fluctuations will be present. The most famous example is Brownian motion in which $f(t)$ in (1.11.7) represents the impact of all particles of a liquid on a bigger particle. Such fluctuations occur whenever we pass from a microscopic description to one which uses more or less macroscopic variables, for instance when we describe a liquid not by the positions of its individual molecules but rather by the local density and velocity of its molecules.

b) Quantum Fluctuations. With the advent of quantum theory in the 1920s it became clear that it is impossible even in principle to make predictions with absolute certainty. This is made manifest by Heisenberg's uncertainty principle which states that it is impossible to measure at the same time the velocity and the position of a particle with absolute precision. Thus the main assumption of Laplace's "spirit of the world" is no longer valid. This impossibility is cast into its most precise form by Born's probability interpretation of the wave function in quantum theory. Since quantum theory lies at the root of all events in the material world, uncertainties caused by quantum fluctuations are unavoidable. This will be particularly important where microscopic events are amplified to acquire macroscopic dimensions. (For instance, in biology mutations may be caused by quantum fluctuations.)

c) Chaos. There is a third, more recent, development which shows that even without quantum fluctuations the future path of systems cannot be predicted. Though the equations describing temporal evolution are entirely deterministic, future development can proceed along quite different routes. This rests on the fact that some systems are extremely sensitive in their further development to initial conditions. This can be most easily visualized by a simple mechanical example. When a steel ball falls on a vertical razor-blade, its future trajectory will depend extremely sensitively on its relative position before it hits the razor-blade. Actually, a whole industry of gambling machines is based on this and similar phenomena.

If the impact of fluctuations on a system is taken care of by a fluctuating force, such as in (1.11.7), we shall speak of additive noise. A randomly fluctuating environment may cause other types of noise, too. For instance, the growth rate in (1.11.1) may fluctuate. In such a case, we have

$$\dot{q} = \alpha(t)q \,, \tag{1.11.8}$$

where we speak of multiplicative noise.

1.11.6 Many Components and the Mezoscopic Approach

So far we have discussed the main building blocks of the equations we want to study. Finally, we should take into consideration that synergetic system are generally composed of very many subsystems. Accordingly, these systems must be described by many variables, denoted by $q_1, \ldots, q_n$. Because the values of these variables at a given time t describe the state of the system, we shall call these variables "state variables". All these variables can be lumped together into a state vector $\boldsymbol{q}$,

$$(q_1, \ldots, q_n) = \boldsymbol{q} \,. \tag{1.11.9}$$

For practical purposes it is important to choose the variables q in an adequate fashion. To this end we distinguish between the microscopic, the mezoscopic and the macroscopic levels of approach. Let us take a liquid as an example. According to our understanding, at the *microscopic level* we are dealing with its individual atoms or molecules described by their positions, velocities and mutual interactions. At the *mezoscopic level* we describe the liquid by means of *ensembles* of many atoms or molecules. The extension of such an ensemble is assumed large compared to interatomic distances but small compared to the evolving macroscopic pattern, e.g., the dimension of hexagons in the Bénard instability. In this case, the variables q_i refer to the ensembles of atoms or molecules. In the case of a fluid we may identify q_i with the density and the mean local velocity. When macroscopic patterns are formed, the density and velocity may change locally. In other words, q_i becomes a time- *and space*-dependent variable. Finally, at the *macroscopic level* we wish to study the corresponding spatial patterns. When we treat continuously extended systems (fluids, chemical reactions, etc.), we shall start from the *mezoscopic level* and devise methods to predict the evolving macroscopic patterns.

The mezoscopic level allows us to introduce concepts which refer to ensembles of atoms but cannot be defined with respect to an individual atom. Such a concept is, e.g., temperature. Another one is that of phases, such as liquid or solid. Correspondingly, we may introduce two kinds of variables, say $q_1(\boldsymbol{x}, t)$ and $q_2(\boldsymbol{x}, t)$, where q_1 refers to the density of molecules in the liquid and q_2 to their density in the solid phase. In this way, e.g., crystal growth can be mathematically described by evolution equations.

In other fields of science the microscopic level need not be identified with atoms or molecules. For instance, when mathematically treating a cell tissue, it may be sufficient to consider the cells as the individual elements of the microscopic level and their *density* (or *type*) as the appropriate variable at the *mezoscopic* level.

In all such cases q_j or the state vector $\boldsymbol{q}$ become functions of space and time

$$(q_1(\boldsymbol{x}, t), \ldots, q_n(\boldsymbol{x}, t)) = \boldsymbol{q}(\boldsymbol{x}, t) . \tag{1.11.10}$$

In addition to the temporal change we now have to consider spatial changes which will be mostly taken care of by spatial derivatives. An example is the equation for the diffusion of a substance, e.g.,

$$\dot{q} = D \Delta q , \tag{1.11.11}$$

where Δ is the Laplace operator which reads in Cartesian coordinates

$$\Delta = \frac{\partial^2}{\partial x^2} + \frac{\partial^2}{\partial y^2} + \frac{\partial^2}{\partial z^2} , \tag{1.11.12}$$

and D the diffusion constant.

Putting all these individual aspects together, we are led to *nonlinear stochastic partial differential equations* of the general type

$$\dot{\boldsymbol{q}} = \boldsymbol{N}(\boldsymbol{\alpha}, \boldsymbol{q}, \nabla, \boldsymbol{x}, t) , \tag{1.11.13}$$

where ∇ is the operator $(\partial/\partial x, \partial/\partial y, \partial/\partial z)$. The study of such equations is an enormous task and we shall proceed in two steps. First we shall outline how at least in some simple cases the solutions of these equations can be visualized, and then we shall focus our attention on those phenomena which are of a general nature.

1.12 How to Visualize the Solutions

At least in principle it is possible to represent the solutions $\boldsymbol{q}(t)$ or $\boldsymbol{q}(\boldsymbol{x}, t)$ by means of graphs. Let us first deal with q_j's, i.e., $q_1, \ldots, q_n$, which are independent of $\boldsymbol{x}$ but still dependent on time. The temporal evolution of each $q_j(t)$ can be represented by the graphs of Fig. 1.12.1. In many cases it is desirable to obtain an overview of all variables. This may be achieved by taking $q_1, \ldots, q_n$ as coordinates of an n-dimensional space and attaching to each time t_1, t_2, etc., the corresponding point $(q_1(t), q_2(t), \ldots)$ (Fig. 1.12.2). When we follow up the continuous sequence of points to $t \to +\infty$ and $t \to -\infty$ we obtain a trajectory (Fig. 1.12.3). When we choose a different starting point we find a different trajectory (Figs. 1.12.1 – 3). Plotting neighboring trajectories we find a whole set of trajectories (Fig. 1.12.4). Since these trajectories are reminiscent of streamlines of a fluid they are sometimes called "streamlines" and their entity the "flow".

It is well known (see, for instance, [1]) that such trajectories need not always go (in one dimension) from $q = -\infty$ to $q = +\infty$, but they may terminate in dif-

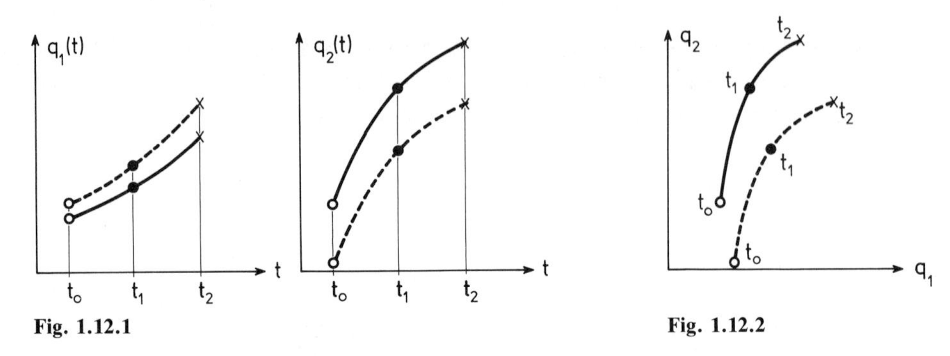

Fig. 1.12.1 **Fig. 1.12.2**

Fig. 1.12.1. Temporal evolution of the variables q_1 and q_2, respectively, in the case of time t. At an initial time t_0 a specific value of q_1 and q_2 was prescribed and then the solid curves evolved. If different initial values of q_1, q_2 are given, other curves (– – –) evolve. If a system possesses many variables, for each variable such plots must be drawn

Fig. 1.12.2. Instead of plotting the trajectories as a function of time as in Fig. 1.12.1, we may plot the trajectory also in the q_1, q_2 plane where for each time t_j the corresponding point $q_1(t_j)$, $q_2(t_j)$ is plotted. If the system starts at an initial time t_0 from different points, different trajectories evolve. In the case of n variables the trajectories must be drawn in an n-dimensional space

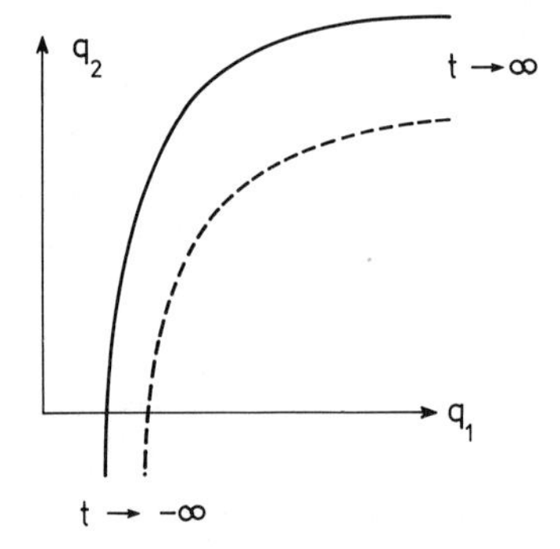

Fig. 1.12.3. In general, trajectories are plotted for $t \to +\infty$ and $t \to -\infty$ if one follows the trajectory

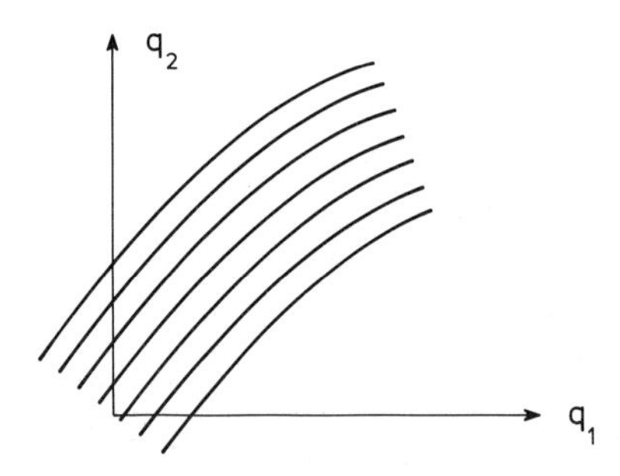

Fig. 1.12.4. Example of a set of trajectories

ferent ways. For instance, in two dimensions they may terminate in a node (Fig. 1.12.5) or in a focus (Fig. 1.12.6). Because the streamlines are attracted by their endpoints, these endpoints are called attractors. In the case of a node, the temporal behavior of $\boldsymbol{q}$ is given by a graph similar to Fig. 1.12.7, whereas in the case of a focus the corresponding graph is given in Fig. 1.12.8. In the plane the only other singular behavior of trajectories besides nodes, foci and saddle points, is a limit cycle (Fig. 1.12.9). In the case shown in Fig. 1.12.9 the limit cycle is stable because it attracts the neighboring trajectories. It represents an "attractor". The corresponding temporal evolution of q_1, which moves along the limit cycle, is shown in Fig. 1.12.10 and presents an undamped oscillation. In dimensions greater than two other kinds of attractors may also occur. An important class consists of attractors which lie on *manifolds* or form manifolds. Let us explain the concept of a manifold in some detail. The cycle along which the motion occurs in Fig. 1.12.9 is a simple example of a manifold. Here each point on the manifold can be mapped to a point on an interval and vice versa (Fig. 1.12.11).

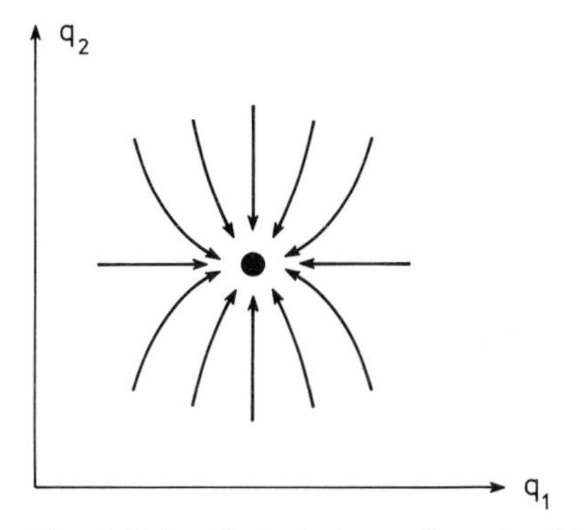

Fig. 1.12.5. Trajectories ending at a (stable) node. If time is reversed, the trajectories start from that node and the node is now unstable

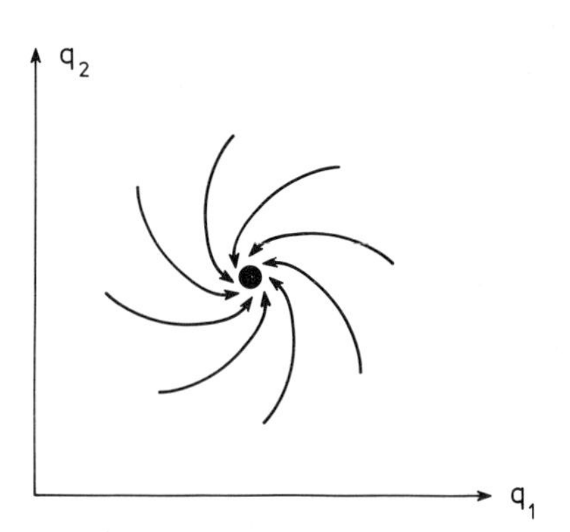

Fig. 1.12.6. Trajectories ending at a (stable) focus. If time is reversed the trajectories start from that focus and it has become an unstable focus

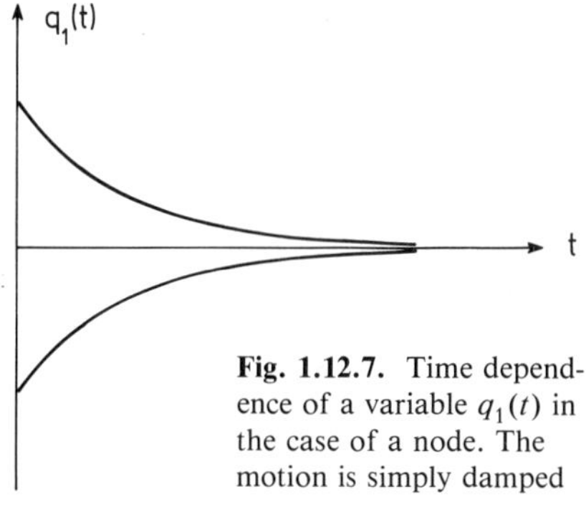

Fig. 1.12.7. Time dependence of a variable $q_1(t)$ in the case of a node. The motion is simply damped

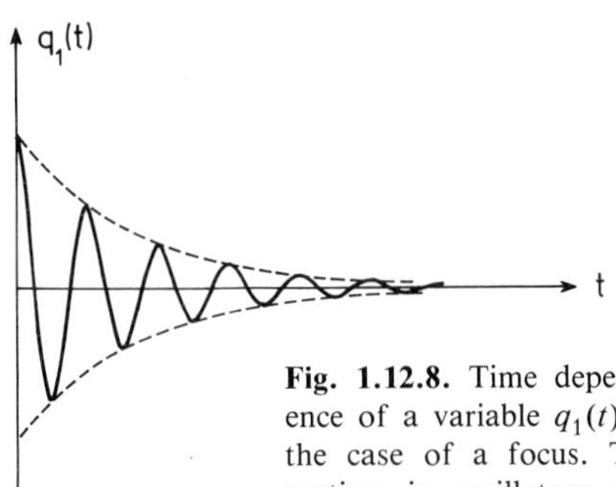

Fig. 1.12.8. Time dependence of a variable $q_1(t)$ in the case of a focus. The motion is oscillatory but damped

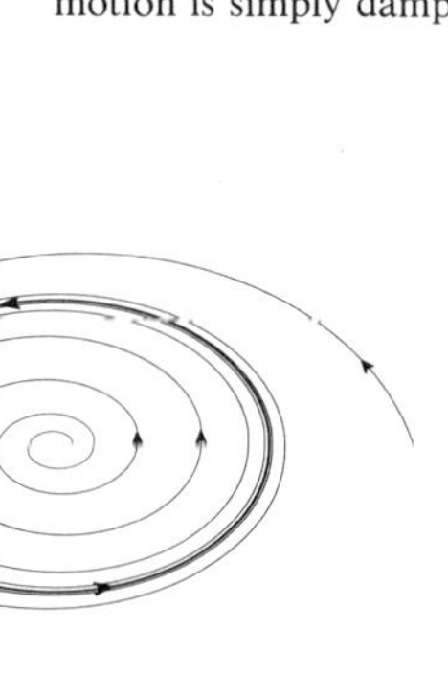

Fig. 1.12.9. Stable limit cycle in a plane. The trajectories approach the limit cycle from the outside and inside of the limit cycle

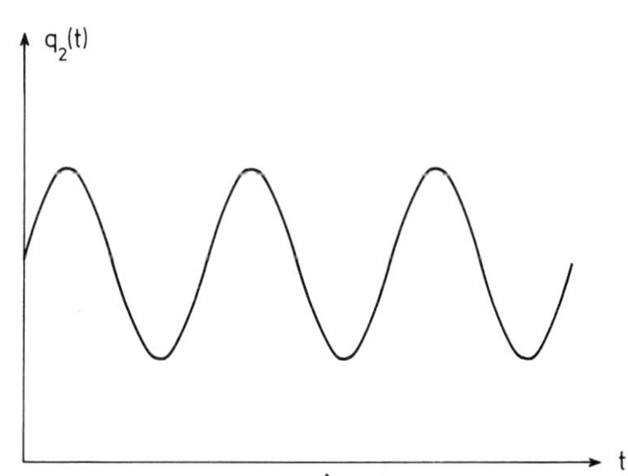

Fig. 1.12.10. Temporal behavior of a variable, e.g., $q_2(t)$, in the case of a limit cycle

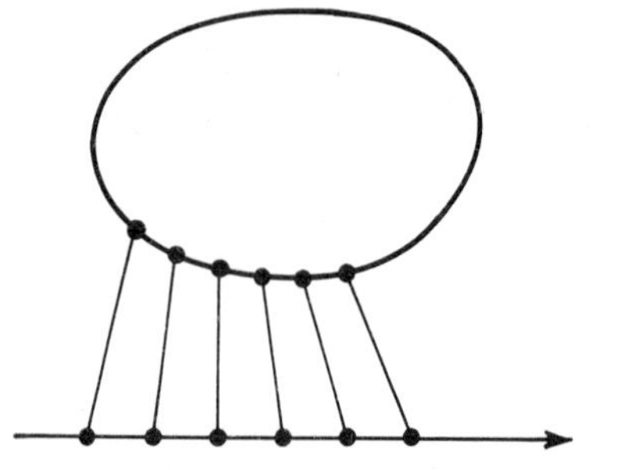

Fig. 1.12.11. One-to-one mapping of the individual points of a limit cycle onto points on a line

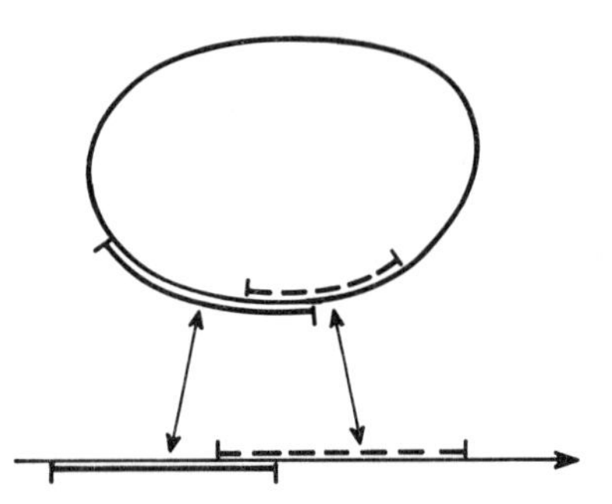

Fig. 1.12.12. One-to-one mapping of overlapping pieces on a limit cycle onto overlapping pieces on a line

The total manifold, especially, can be covered by pieces which can be mapped on overlapping pieces along the line and vice versa (Fig. 1.12.12). Each interval on the circle corresponds to a definite interval along the line. If we take as the total interval 0 till 2π we may describe each point along the circle by a point on the φ axis from 0 till 2π. Because there is one-to-one mapping between points on the limit cycle and on the interval $0 \dots 2\pi$, we can directly use such a coordinate (system) on the limit cycle itself. This coordinate system is independent of any coordinate system of the plane, in which the limit cycle is embedded.

The limit cycle is a *differentiable* manifold because when we use, e. g., time as a parameter, we assume $\dot{q}$ exists – or geometrically expressed – the limit cycle possesses everywhere a tangent.

Of course, in general, limit cycles need not be circles but may be other closed orbits which are repeated after a period $T = 2\pi/\omega$ (ω: frequency). If $\dot{q}$ exists, this orbit forms a differentiable manifold again. In one dimension such a periodic motion is described by

$$q_1(\omega t) = \sum_n c_n \mathrm{e}^{\mathrm{i}n\omega t}, \quad n \text{ integers}, \tag{1.12.1}$$

or in m dimensions by

$$\boldsymbol{q}(\omega t) = \sum_n \boldsymbol{c}_n \mathrm{e}^{\mathrm{i}n\omega t}, \quad n \text{ integers}, \tag{1.12.2}$$

where the vectors $\boldsymbol{c}_n$ have m components.

Another example of a limit cycle is provided by Fig. 1.12.13 which must be looked at at least in 3 dimensions. The next example of a manifold is a torus which is presented in Fig. 1.12.14. In this case we can find a one-to-one correspondence between each surface element on the torus and an element in the plane where we may attach a point of the plane element to each point of the torus element and vice versa. Furthermore, the torus can be fully covered by overlapping pieces of such elements.

When we put the individual pieces on the plane together in an adequate way we find the square of Fig. 1.12.15. From it we can construct the torus by folding the upper and lower edge together to form a tube, bending the tube and glueing the ends together. This makes it clear that we can describe each point of the torus

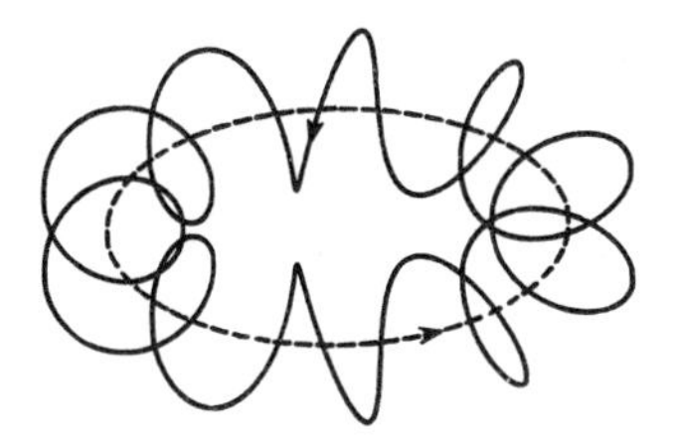

Fig. 1.12.13. Limit cycle in three dimensions

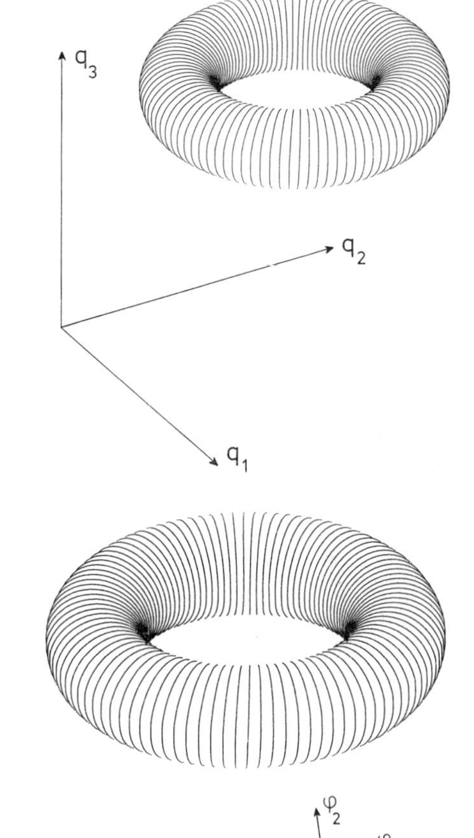

Fig. 1.12.14. Two-dimensional torus in three-dimensional space

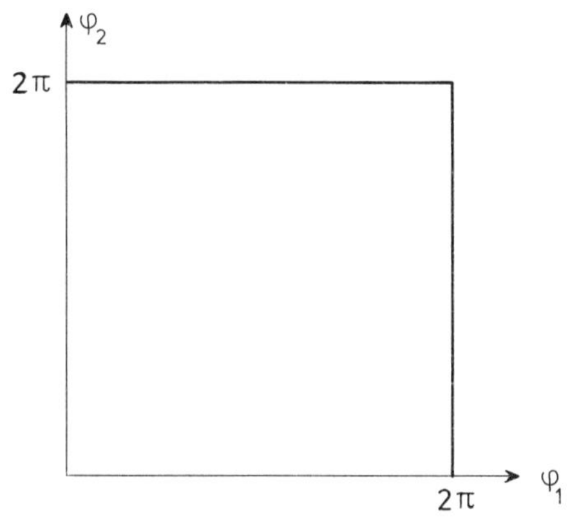

Fig. 1.12.15. The two-dimensional torus with local coordinates φ_1 and φ_2 can be mapped one-to-one on a square (*left side of figure*)

by the coordinate system φ_1, φ_2. Because in each point φ_1, φ_2 a tangent plane exists, the torus forms a *differentiable* manifold. Such two- and higher-dimensional tori are adequate means to visualize quasiperiodic motion, which takes place at several frequencies. An example is provided by

$$q = \sin(\underbrace{\omega_1 t}_{\varphi_1}) \sin(\underbrace{\omega_2 t}_{\varphi_2}) \tag{1.12.3}$$

or, more generally, by

$$\boldsymbol{q}(\underbrace{\omega_1 t}_{\varphi_1}, \underbrace{\omega_2 t}_{\varphi_2}) . \tag{1.12.4}$$

$\boldsymbol{q}$ can be represented by multiple Fourier series in the form

$$q = \sum_n c_n e^{i n \cdot \omega t} , \tag{1.12.5}$$

where $\boldsymbol{n} \cdot \boldsymbol{\omega} = n_1 \omega_1 + n_2 \omega_2 + \ldots + n_N \omega_N$, n_j integer.

Whether a vector function (1.12.4) fills the torus entirely or not depends on the ratio between ω's. If the ω's are rational, only lines on the torus are covered. This is easily seen by looking at the example in Fig. 1.12.16 in which we have chosen $\omega_2 : \omega_1 = 3:2$. Starting at an arbitrary point we find (compare also the legend of Fig. 1.12.16) only one closed trajectory which then occurs as a closed trajectory on a torus.

Another example is provided by $\omega_2 : \omega_1 = 1:4$ and $\omega_2 : \omega_1 = 5:1$, respectively (Figs. 1.12.17, 18). On the other hand, if the frequencies are irrational, e.g., $\omega_2 : \omega_1 = \pi : 1$ the trajectory fills up the total square, or in other words, it fills up the whole torus (Fig. 1.12.19) because in the course of time the trajectory comes arbitrarily close to any given point. (The trajectory is dense on the torus.) These

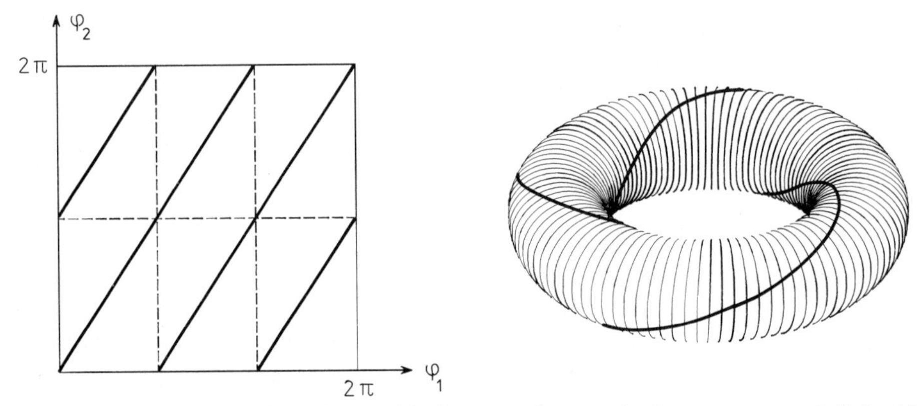

Fig. 1.12.16. A trajectory in a plane and its image on the torus in the case $\omega_2 : \omega_1 = 3:2$. In this figure we have started the trajectory at $\varphi_1 = 0$, $\varphi_2 = 0$. Because of the periodicity condition we continue that trajectory by projecting its cross section with the axis $\varphi_2 = 2\pi$ down to the axis $\varphi_2 = 0$ (– – –). It then continues till the point $\varphi_1 = 2\pi$, $\varphi_2 = \pi$ from where we projected it on account of the periodicity condition on $\varphi_1 = 0$, $\varphi_2 = \pi$ (– – –). From there it continues by means of the solid line, etc. Evidently by this construction we eventually find a closed line

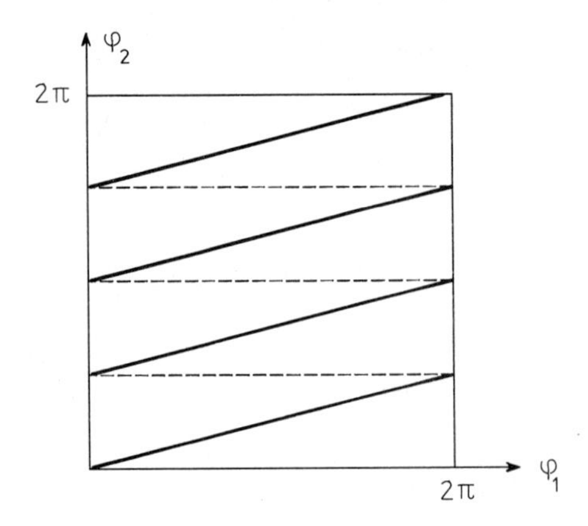

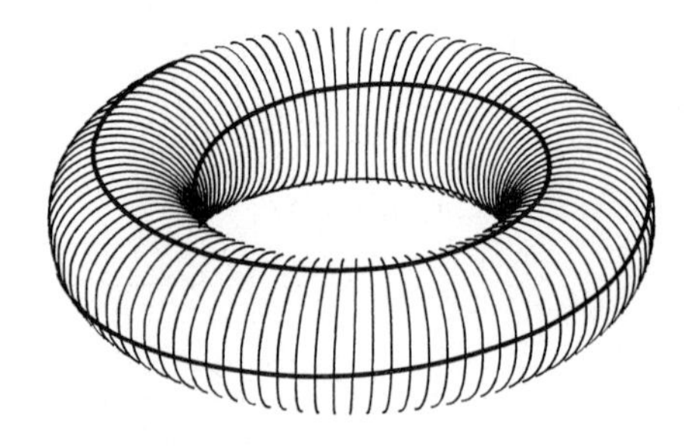

Fig. 1.12.17. The case $\omega_1 : \omega_2 = 1:4$

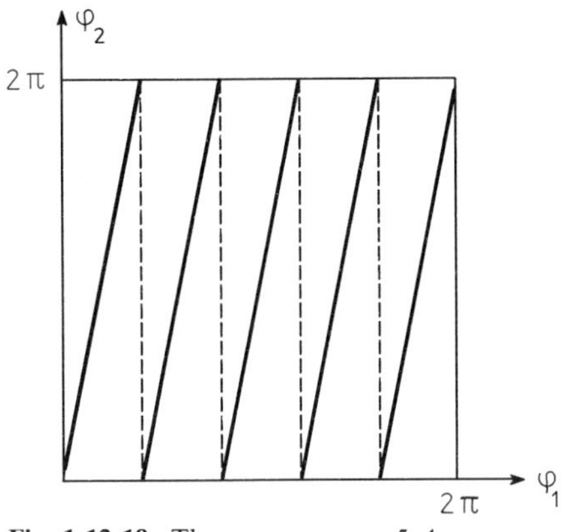

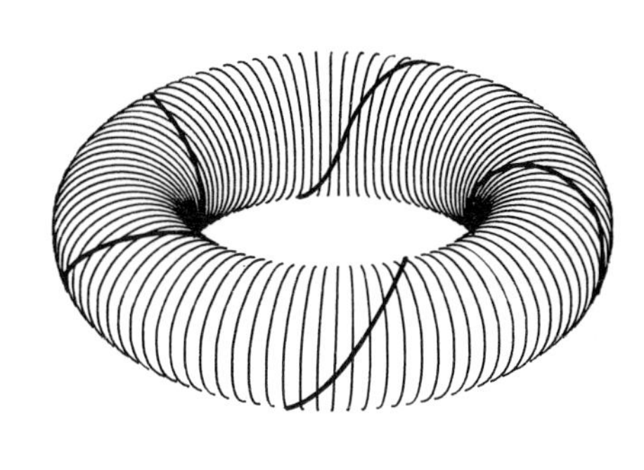

Fig. 1.12.18. The case $\omega_2 : \omega_1 = 5:1$

considerations can be easily generalized to tori in multidimensional spaces which can be mapped on cubes with coordinates $\varphi_1, \ldots, \varphi_N$, $0 \leq \varphi_j \leq 2\pi$. After these two explicit examples of (differentiable) manifolds (circle and torus), it is sufficiently clear what we understand by a (differentiable) manifold. (For an abstract definition the reader is referred to the references.) In the above example of a stable limit cycle, all trajectories which started in the neighborhood of the limit cycle eventually ended on that manifold. Manifolds which have such a property will be called attracting manifolds.

Once $\boldsymbol{q}(t)$ of a system is on a stable limit cycle it will stay there for ever. In such a case we call the limit cycle an *invariant manifold* because the manifold (the limit cycle) is not changed during the motion. It is invariant against temporal evolution. Such a definition applies equally well to all other kinds of manifolds.

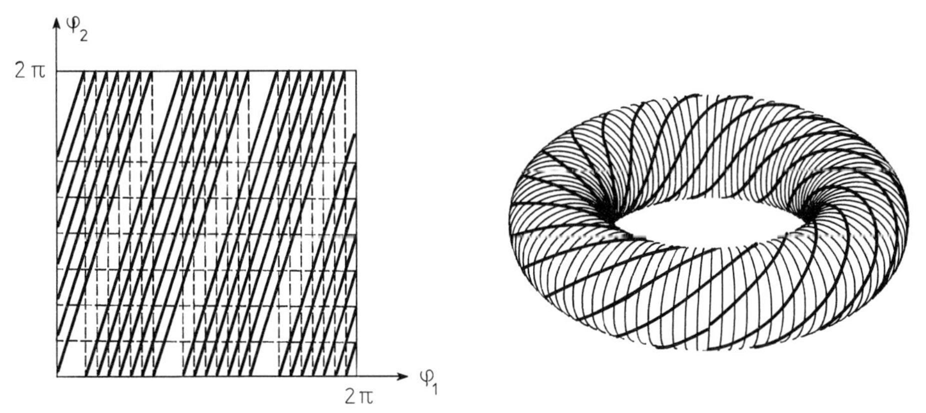

Fig. 1.12.19. The case $\omega_2 : \omega_1 = \pi : 1$, the case of an irrational number. In this computer plot neither the trajectories in the φ_1, φ_2 plane nor on the torus ever close and indeed the trajectories fill the whole plane or the whole torus, respectively. Note that in this plot the computer ran only a finite time, otherwise the square or the torus would appear entirely black

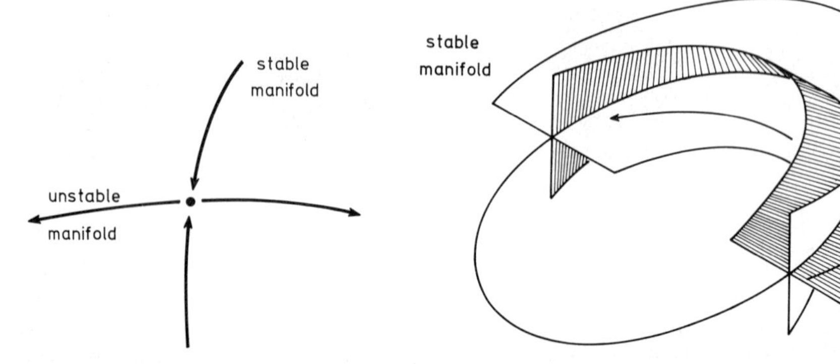

Fig. 1.12.20. The stable and unstable manifold of a saddle point (compare text)

Fig. 1.12.21. The unstable and stable manifold connected with a limit cycle

An important concept refers to stable and unstable manifolds. To exhibit the ideas behind this concept we again shall discuss examples. Figure 1.12.20 shows the stable and unstable manifold of a fixed point of the saddle type in the plane. The stable manifold (of a fixed point) is defined as the set of all points which are the initial points of trajectories which in the limit $t \to +\infty$ terminate at the fixed point. Obviously in our present example the stable manifold has the same dimension as the real line. This is so because all of the trajectories starting close to the stable manifold will pass the saddle in a finite distance but then tend away from it. The unstable manifold (of a fixed point) is defined as the set of initial points of trajectories which end in the limit $t \to -\infty$ at the fixed point. We note that both types of manifolds fulfill the properties of an invariant manifold. A further example is shown in Fig. 1.12.21, where the stable and unstable manifolds of a limit cycle, which is embedded in a three-dimensional Euclidian space, are drawn. These manifolds can be constructed locally from the linearized equations of motion. Consider again as an example the saddle of Fig. 1.12.20. If we denote the deviation from the saddle by $\boldsymbol{q} = (q_1, q_2)$, we have the following set of equations

$$\dot{q}_1 = \alpha q_1 + N_1(q_1, q_2) , \tag{1.12.6}$$

$$\dot{q}_2 = -\gamma q_2 + N_2(q_1, q_2) , \tag{1.12.7}$$

where $\alpha, \gamma > 0$ and the N_j's $(j = 1, 2)$ are nonlinear functions of q_j. Confining ourselves to small deviations q_j, we may safely neglect the nonlinear terms N_j which are assumed to be of the order $O(|q|^2)$. We then notice that small deviations q_1 are exponentially enhanced in time, meaning that q_1 is tangent to the unstable manifold of the saddle. Conversely, small perturbations q_2 are exponentially damped in time and we may therefore conclude that direction q_2 is tangent to the stable manifold at the saddle.

But in addition, generally a third class of directions might exist along which a perturbation is neither enhanced nor damped, i.e., the behavior along these

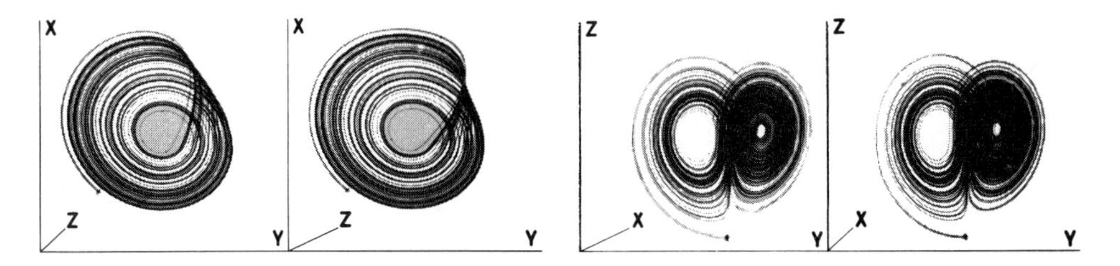

Fig. 1.12.22 **Fig. 1.12.23**

Fig. 1.12.22. Stereoplot of the Rössler attractor. To obtain a stereoscopic impression, put a sheet of paper between your nose and the vertical middle line. Wait until the two pictures merge to one. The parameter values are: $a = b = 0.2$, $c = 5.7$, $x(0) = y(0) = -0.7$, $z(0) = 1$. Axes: $-14 \ldots 14$ for x and y, $0 \ldots 28$ for z. [After O. E. Rössler: In *Synergetics, A Workshop*, ed. by H. Haken, Springer Ser. Synergetics, Vol. 2 (Springer, Berlin, Heidelberg, New York 1977) p. 184]

Fig. 1.12.23. Stereoplot of the (modified) Lorenz attractor. The parameters are: $a = 2.2$, $\sigma = 4$, $r = 80$, $b = 8/3$, $x(0) = 5.1$, $y(0) = -13.72$, $z(0) = 52$. Axes: $-50 \ldots 50$ for x, $-70 \ldots 70$ for y, $0 \ldots 14$ for z. [After O. E. Rössler: In *Synergetics, A Workshop*, ed. by H. Haken, Springer Ser. Synergetics, Vol. 2 (Springer, Berlin, Heidelberg, New York 1977) p. 184]

directions is neutral. These directions are tangent to the so-called center manifold. An example is provided by the limit cycle of Fig. 1.12.21. Obviously a perturbation tangent to the cycle can be neither enhanced nor damped in time. Furthermore, we observe that in the case of the saddle (Fig. 1.12.20) the center manifold is reduced to the point itself.

From recent work it appears that there may be attractors which are not manifolds. Such attractors have been termed "strange attractors" or "chaotic attractors". The reader should be warned of some mathematical subtleties. The notion "strange attractor" is nowadays mostly used if certain mathematical axioms are fulfilled and it is not known (at least at present) whether systems in nature fulfill these axioms. Therefore we shall use the more loosely defined concept of a chaotic attractor. Once the vector $\boldsymbol{q}(t)$ has entered such a region it will stay there forever. But the trajectory does not lie on a manifold. Rather, the vector $\boldsymbol{q}(t)$ will go on like a needle which we push again and again through a ball of thread. Chaotic attractors may occur in three and higher dimensions. Examples are presented in Figs. 1.12.22 and 23.

The trajectories of a chaotic attractor may be generated by rather simple differential equations (if the parameters are adequately chosen). The simplest example known is the Rössler attractor. Its differential equations possess only one nonlinearity and read

$$\dot{x} = -y - z \,, \tag{1.12.8}$$

$$\dot{y} = x + ay \,, \tag{1.12.9}$$

$$\dot{z} = b + z(x - c) \,, \tag{1.12.10}$$

where a, b, c are constant parameters. A plot of this attractor is presented in Fig. 1.12.22.

The next simple (and historically earlier) example is the Lorenz attractor. The corresponding differential equations are

$$\dot{x} = \sigma(y - x) , \tag{1.12.11}$$

$$\dot{y} = x(r - z) - y , \tag{1.12.12}$$

$$\dot{z} = xy - bz , \tag{1.12.13}$$

where σ, b, r are constant parameters. This model was derived for the convection instability in fluid dynamics. The single mode laser is described by equations equivalent to the Lorenz equations. A plot of a (modified) Lorenz attractor is provided by Fig. 1.12.23. The modification consists in adding a constant a to the rhs of (1.12.11).

1.13 Qualitative Changes: General Approach

A general discussion of the nonlinear partial stochastic differential equations (1.11.13) seems rather hopeless, because they cover an enormous range of phenomena with which nobody is able to deal. On the other hand, in the realm of synergetics we wish to find out general features of complex systems. We can take a considerable step towards that goal by focusing attention on those situations in which the macroscopic behavior of a system changes dramatically. We wish to cast this idea into a mathematical form. To this end we first discuss the concept of structural stability, taking an example from biology.

Figure 1.13.1 shows two different kinds of fish, namely porcupine fish and sun fish. According to studies by d'Arcy Wentworth Thompson at the beginning of the twentieth century, the two kinds of fish can be transformed into each other by a simple grid transformation. While from the biological point of view such a grid transformation is a highly interesting phenomenon, from the mathematical point of view we are dealing here with an example of structural stability[1]. In a mathematician's interpretation the two kinds of fish are the same. They are just deformed copies of each other. A fin is transformed into a fin, an eye into an eye, etc. In other words, no new qualitative features, such as a new fin, occur. In the following, we shall have structural changes (in the widest sense of this word) in

[1] The concept of structural stability seems to play a fundamental role in biology in a still deeper sense than in the formation of different species by way of deformation (Fig. 1.13.1). Namely, it seems that, say within a species, organisms exhibit a pronounced invariance of their functions against spatial or temporal deformations. This makes it sometimes difficult to perform precise (and reproducible) physical measurements on biological objects. Most probably, in such a case we have to look for transformation groups under which the function of an organ (or animal) is left invariant. This invariance property seems to hold for the most complicated organ, the human brain. For example, this property enables us to recognize the letter *a* even if it is strongly deformed. From this ability an art out of writing letters (in the double sense of the word) developed in China (and in old Europe).

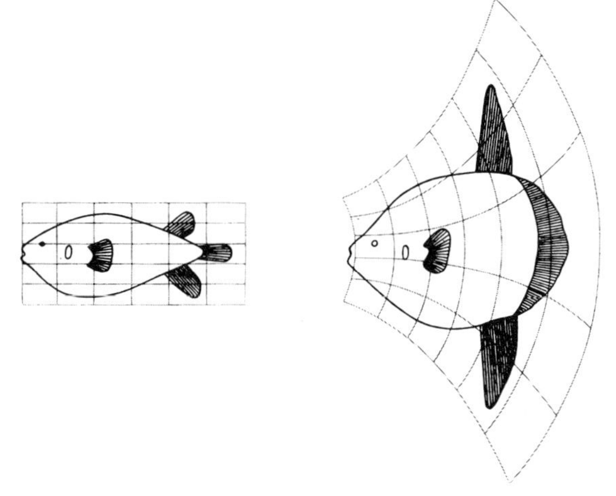

Fig. 1.13.1. The porcupine fish (left) and the sun fish (right) can be transformed into each other by a simple grid transformation. [After D'Arcy Thompson: In *On Growth and Form*, ed. by J. T. Bonner (University Press, Cambridge 1981)]

mind. In contrast to our example of the two kinds of fish (Fig. 1.13.1), we shall not be concerned with static patterns but rather with patterns of trajectories, i. e., in other words, with the flows we treated in the foregoing section. As we know, we may manipulate a system from the outside, which in mathematical form will be done by changing certain control parameters. We want to show how the properties of a system may change dramatically even if we alter a control parameter only slightly. In [1] a simple example for this kind of behavior was presented.

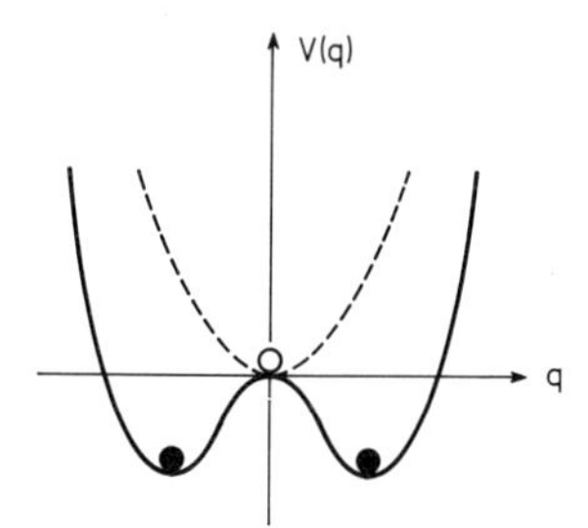

Fig. 1.13.2. Equilibrium position of a ball in two kinds of vase (*dashed line* or *solid line*)

Consider a ball which slides down along the walls of a vase. If the vase has the form of the dashed line of Fig. 1.13.2, the ball comes to rest at $q = 0$. If the vase, however, is deformed as indicated by the solid line, the ball comes to rest at $q = +a$ or $q = -a$. The flow diagram of this ball is easily drawn. In the case of the potential curve with a single minimum, we obtain Fig. 1.13.3, whereas in the case of two minima, the flow diagram is given by Fig. 1.13.4. A related transition from a single attractor to two attractors is provided by the self-explaining Figs.

Fig. 1.13.3. One-dimensional motion of a ball with coordinate q ending at one stable point

Fig. 1.13.4. One-dimensional motion of a ball with two points of stable equilibrium (●) and one unstable point (○)

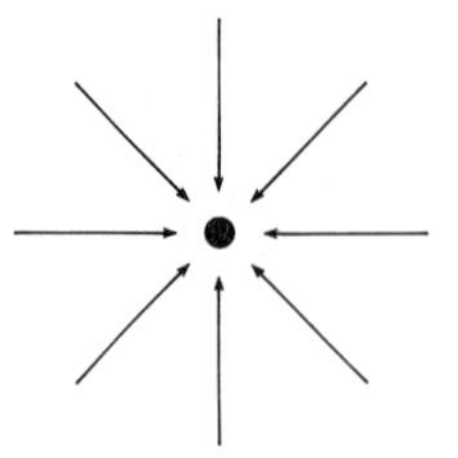

Fig. 1.13.5. Trajectories ending at a node

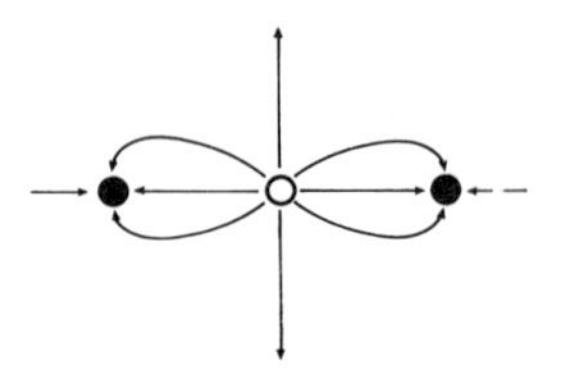

Fig. 1.13.6. Two stable and an unstable node

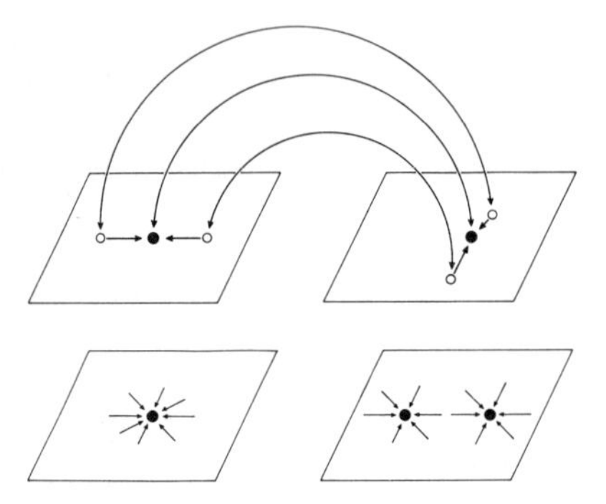

Fig. 1.13.7. One-to-one mapping between these two flow diagrams is possible

Fig. 1.13.8. For these two flow diagrams a one-to-one mapping is impossible

1.13.5, 6. By the way, the example of Fig. 1.13.2 reveals the important role played by fluctuations. If the ball is initially at $q = 0$, to which valley it will eventually go entirely depends on fluctuations.

What makes the transition from Figs. 1.13.3 to 4 or from Figs. 1.13.5 to 6 so different from the example of the two kinds of fish is the following. Let us draw one of the two fishes on a rubber sheet. Then by mere stretching or pushing together the rubber sheet, we may continuously proceed from one picture to the other. However, there is no way to proceed from Fig. 1.13.5 to 6 by merely stretching or deforming a rubber sheet continuously as it is possible in Fig. 1.13.7. In other words, there is no longer any one-to-one mapping between the stream lines of one flow to the stream lines of the other (Fig. 1.13.8). In the mathematical sense we shall understand by "structural instability" or "structural changes" such situations in which a one-to-one mapping becomes impossible.

We now wish to discuss briefly how to check whether the change of a control parameter causes structural instability. We shall come to this problem in much greater detail later. Here the main method is illustrated by means of a most simple example. The equation which describes the sliding motion of a ball in a vase reads:

$$\dot{q} = -\alpha q - |k| q^3 . \tag{1.13.1}$$

For $\alpha > 0$ the solution reads $q = 0$, although $q = 0$ still remains a solution for $\alpha < 0$. Of course, by looking at Fig. 1.13.2 we immediately recognize that the position $q = 0$ is now unstable. However, in many cases of practical interest we may not invoke the existence of a potential curve as drawn in Fig. 1.13.2. Rather, we must resort to another approach, namely *linear stability analysis*. To this end a small time-dependent deviation u is introduced so that we write the solution q of (1.13.1) as

$$q = q_0 + u = u\,. \tag{1.13.2}$$

Inserting (1.13.2) into (1.13.1) and keeping only linear terms, we find the equation (for $\alpha < 0$)

$$\dot{u} = |\alpha| u\,, \tag{1.13.3}$$

which gives the solution

$$u(t) = u(0)\,\mathrm{e}^{|\alpha| t}\,. \tag{1.13.4}$$

Because $|\alpha| = -\alpha > 0$, $u(t)$ grows exponentially. This indicates that the state $q_0 = 0$ is unstable. In Chaps. 2, 3 linear stability analysis will be quite generally presented. In particular, we shall study not only the case in which a constant $\boldsymbol{q}_0$ becomes unstable, but also the case in which motions on a limit cycle or on a torus become unstable. The latter problem leads to the rather strange country of quasiperiodic motions where still a great many discoveries can be made, to which this book has made some contribution. After having performed the stability analysis we ask the question to which new states the system will go. In this context two concepts are central to synergetics, namely the concept of the *order parameter* and that of *slaving*. Consider to this end the following two differential equations

$$\dot{q}_1 = \lambda_1 q_1 - q_1 q_2\,, \tag{1.13.5}$$

$$\dot{q}_2 = -\lambda_2 q_2 + q_1^2\,, \tag{1.13.6}$$

which may occur in a number of fields, for instance, chemistry. Equation (1.13.5) describes the autocatalytic generation of a concentration q_1 of a chemical 1 by $\lambda_1 q_1$ and the decay of the molecules of 1, due to their interaction with a different kind of molecules with concentration q_2, by $-q_1 q_2$. Equation (1.13.6) describes by its first term the spontaneous decay of the molecules q_2 and their generation by a bimolecular process from q_1. Of course, for a mathematical treatment these interpretations are of no importance at all and we rather focus our attention on the mathematical features. Let us assume that λ_1 is very small or slightly positive. In such a case q_1 will change very slowly provided q_1 and q_2 are small quantities (which allows us to neglect the quadratic term in a first approximation). According to (1.13.6), q_2 is driven by q_1^2. But because q_1 changes very slowly, we may expect that q_2 changes very slowly, too. If λ_2 is positive and much bigger than λ_1 we may neglect $\dot{q}_2$ compared to $\lambda_2 q_2$. This result was deduced still more explicitly in [1]. Putting

$$\dot{q}_2 \approx 0 \tag{1.13.7}$$

we may immediately solve (1.13.6) by means of

$$q_2 \approx q_1^2/\lambda_2 . \tag{1.13.8}$$

This approach, which is often called the *adiabatic approximation*, allows us to express q_2 explicitly by q_1. Or, in other words, q_2 is *slaved* by q_1. When we are dealing with very many variables, which are slaved by q_1, we may reduce a complex problem quite considerably. Instead of very many equations for the q's we need to consider only a single equation for q_1, and then we may express all other q's by q_1 according to the *slaving principle*. We shall call such a q_1 an *order parameter*. Of course, in reality the situation may be more difficult. Then (1.13.5, 6) must be replaced by equations which are much more involved. They may depend on time-dependent coefficients and they may contain fluctuating forces. Furthermore (1.13.7) and therefore (1.13.8) are just approximations. Therefore it will be an important task to devise a general procedure by which one may express q_2 by q_1. We shall show in Chap. 7 that it is indeed possible to express $q_2(t)$ by $q_1(t)$ at the same time t, and an explicit procedure will be devised to find the function $q_2(t) = f(q_1(t))$ explicitly for a large class of stochastic nonlinear partial differential equations. There is a most *important internal relation* between the *loss of linear stability*, the occurrence of *order parameters* and the validity of the *slaving principle*. When we change control parameters, a system may suffer loss of linear stability. As can be seen from (1.13.5, 6), in such a case $\mathrm{Re}\{\lambda_1\}$ changes its sign, which means that it becomes very small. In such a situation the slaving principle applies. Thus we may expect that at points where structural changes occur, the behavior of a system is governed by the order parameters alone. As we shall see, this connection between the three concepts allows us to establish far-reaching analogies between the behavior of quite different systems when macroscopic changes occur.

1.14 Qualitative Changes: Typical Phenomena

In this section we want to give a survey on qualitative changes caused by instabilities. There are wide classes of systems which fall into the categories discussed below. To get an overview we shall first neglect the impact of noise. Let us start from equations of the form (1.11.13)

$$\dot{q} = N(q, \alpha, \nabla, x, t) . \tag{1.14.1}$$

We assume that N does not explicitly depend on time, i.e., we are dealing with *autonomous systems*. Ignoring any spatial dependence, we start with equations of the form

$$\dot{q}(t) = N(q(t), \alpha) . \tag{1.14.2}$$

We assume that for a certain value (or a range of values) of the control parameter α, a stable solution $q_0(t, \alpha)$ of (1.14.2) exists. To study the stability of this solution when α is changed, we make the hypothesis

$$q = q_0 + w(t), \tag{1.14.3}$$

where $w(t)$ is assumed to be small. Inserting (1.14.3) into (1.14.2) and keeping only terms linear in w, we arrive at equations of the form

$$\dot{w} = Lw, \tag{1.14.4}$$

where L depends on q_0. In Sect. 1.14.6 it will be shown explicitly how L is connected with N. For the moment it suffices to know that L bears the same time dependence as q_0.

Let us assume that q_0 is time independent. To solve (1.14.4) we make the hypothesis

$$w(t) = e^{\lambda t} v, \quad v \text{ constant vector}, \tag{1.14.5}$$

which transforms (1.14.4) into

$$Lv = \lambda v, \tag{1.14.6}$$

which is a linear algebraic equation for the constant vector v and the eigenvalue λ.

To elucidate the essential features of the general approach, let us assume that the matrix L is of finite dimension n. Then there are at most n different eigenvalues λ_j, which in general depend on the control parameter α. An instability occurs if at least one eigenvalue acquires a nonnegative real part.

1.14.1 Bifurcation from One Node (or Focus) into Two Nodes (or Foci)

Let us treat the case in which $\mathrm{Re}\{\lambda_j\} \geq 0$ just for *one* j, say $j = 1$, and $\mathrm{Re}\{\lambda_j\} < 0$ for all other j. This example illustrates some of the essential features of our approach, an overview of which will be presented later. Since our final goal is to solve the fully nonlinear equation (1.14.2), we must make a suitable hypothesis for its solution $q(t)$. To this end we repesent $q(t)$ in the following form

$$q(t) - q_0 + \sum \xi_j(t) v_j, \tag{1.14.7}$$

where v_j are the solutions of (1.14.6), while ξ_j are still unknown time-dependent coefficients. Inserting (1.14.7) into (1.14.2), we may find after some manipulations (whose explanation is not important here, cf. Chap. 8) equations for ξ_j. Let us take for illustration just $j = 1, 2$. The corresponding equations read

$$\dot{\xi}_1 = \lambda_1 \xi_1 + \hat{N}_1(\xi_1, \xi_2), \tag{1.14.8}$$

$$\dot{\xi}_2 = \lambda_2 \xi_2 + \hat{N}_2(\xi_1, \xi_2). \tag{1.14.9}$$

Here λ_1, λ_2 are the eigenvalues of (1.14.6), while $\hat{N}_j$ are nonlinear functions of ξ_1,

ξ_2 which start at least with terms quadratic (or bilinear) in ξ_1, ξ_2. Now remember that we are close to a control parameter value α where the system loses linear stability, i.e., where $\mathrm{Re}\{\lambda_1\}$ changes sign. But this means that $|\lambda_1| \ll |\lambda_2|$ so that the slaving principle applies. Therefore we may express ξ_2 by ξ_1, $\xi_2 = f(\xi_1)$, and need solve only a single equation of the form

$$\dot{\xi}_1 = \lambda_1 \xi_1 + \tilde{N}_1(\xi_1)\,, \quad \text{where} \tag{1.14.10}$$

$$\tilde{N}_1(\xi_1) = \hat{N}_1(\xi_1, f(\xi_1))\,. \tag{1.14.11}$$

Now consider a small surrounding of $\xi_1 = 0$, in which we may approximate $\tilde{N}_1$ by a polynomial whose leading terms, only, need to be kept for small enough ξ_1. If the leading term of $\tilde{N}_1$ reads $-\beta\xi_1^3$, we obtain as the order parameter equation just

$$\dot{\xi}_1 = \lambda_1 \xi_1 - \beta\xi_1^3\,, \quad \beta > 0\,. \tag{1.14.12}$$

But when confining to real λ_1 this is precisely our former equation (1.13.1) which describes the sliding motion of a ball in a vase with one valley ($\lambda_1 < 0$) or two valleys ($\lambda_1 > 0$). Therefore the single node present for $\lambda_1 < 0$ is replaced by two nodes for $\lambda_1 > 0$ (Figs. 1.13.5 and 6). Or, in other words, the single node *bifurcates* into two nodes (Fig. 1.14.1). It is worth mentioning that (1.14.12) not only describes the new equilibrium position, but also the relaxation of the system into these positions and thus allows their stability to be verified. This example may suffice here to give the reader a feeling how the nonlinear equation (1.14.6) can be solved and what kind of qualitative change occurs in this case.

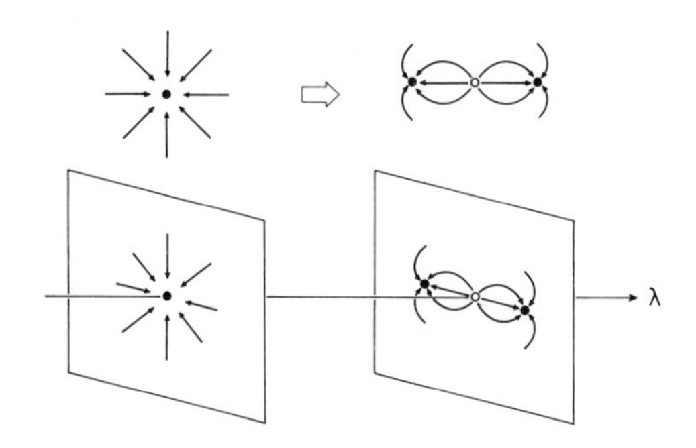

Fig. 1.14.1. Two ways of representing the bifurcation of a stable node into two stable nodes. In the upper part the two flow diagrams are plotted in the same plane. In the lower part the control parameter λ is plotted along the abscissa and the planes representing the flow diagrams corresponding to the upper part of the figure are shown on individual planes perpendicular to the λ axis

From now on this section will deal with classification and qualitative description of the phenomena at instability points. The detailed mathematical approaches for their adequate treatment will be presented in later chapters. For instance, in Sect. 8.3 we shall discuss more complicated problems in which several eigenvalues, which are real, become positive.

1.14.2 Bifurcation from a Focus into a Limit Cycle (Hopf Bifurcation)

A famous example of bifurcation is the *Hopf bifurcation*. In this case two complex conjugate eigenvalues

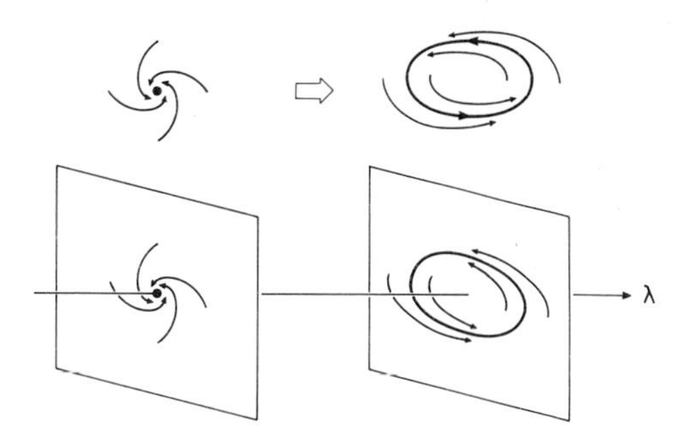

Fig. 1.14.2. Two ways of representing the bifurcation of a stable focus into a stable limit cycle. The representation technique is the same as in Fig. 1.14.1

$$\lambda_1 = \lambda' + \mathrm{i}\omega\,, \quad \lambda_2 = \lambda' - \mathrm{i}\omega\,, \tag{1.14.13}$$

where λ', ω real, $\neq 0$, cross the imaginary axis so that $\lambda' \geq 0$. In such a case an *oscillation* sets in and the originally stable focus bifurcates into a limit cycle (Fig. 1.14.2). At this moment a general remark should be made. Loss of linear stability does not guarantee that the *newly evolving* states are stable. Rather, we have to devise methods which allow us to check the stability of the new solutions explicitly. Our approach, discussed later in this book, will make this stability directly evident.

1.14.3 Bifurcations from a Limit Cycle

In a number of realistic cases a further change of the control parameter can cause an instability of a limit cycle. This requires an extension of the linear stability analysis by means of (1.14.5), because the motion on the limit cycle is described by a time-dependent $q_0(t)$. Therefore L of (1.14.4) becomes a function of t which is periodic. In such a case we may again study the stability by means of the Floquet exponents λ occurring in (1.14.5), using the results derived in Sect. 2.7.

If a single real λ becomes positive, the old limit cycle may split into new limit cycles (Figs. 1.14.3 and 4). If, on the other hand, a complex eigenvalue (1.14.13) acquires a positive real part, a new limit cycle is superimposed on the old one.

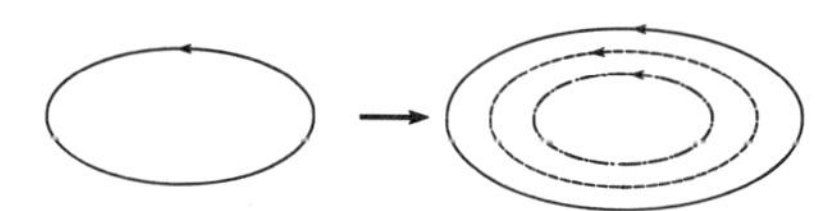

Fig. 1.14.3. Bifurcation of a limit cycle in the plane into two limit cycles in the same plane. The old limit cycle, which has become unstable, is still represented as a dashed line on the right-hand side of this figure

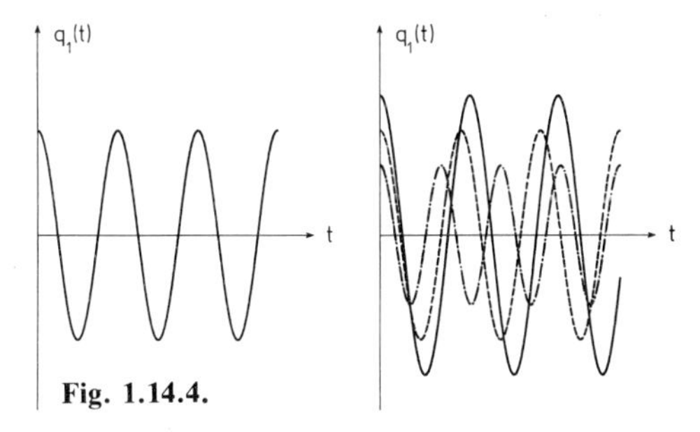

Fig. 1.14.4. The temporal behavior of a variable $q_1(t)$ of the limit cycle before bifurcation (*left-hand side*) and after bifurcation (*right-hand side*). The variables q_1 belonging to the new stable limit cycles are shown as solid and dashed-dotted lines, respectively. The unstable limit cycle is represented by a dashed line

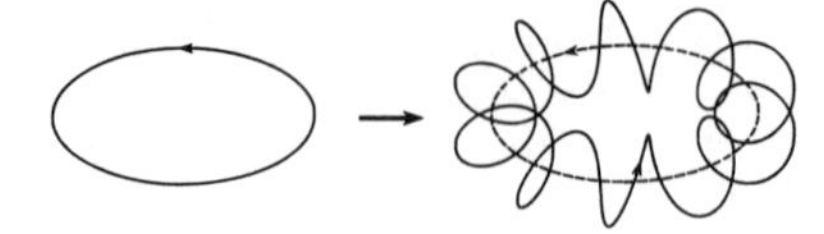

Fig. 1.14.5. Bifurcation of a limit cycle in two dimensions to a limit cycle in three dimensions. Depending on the frequency of rotation along and perpendicular to the dashed line, closed or unclosed orbits may evolve. In the case of a closed orbit, again a new limit cycle arises, whereas in the other case the trajectory fills a torus

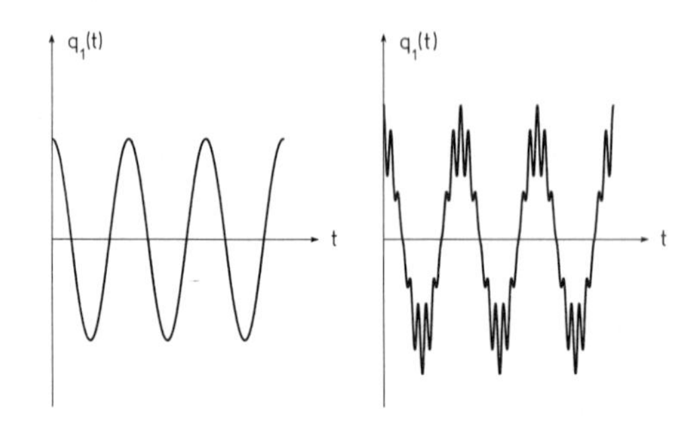

Fig. 1.14.6. The temporal evolution of $q_1(t)$ of a limit cycle before bifurcation (*left-hand side*) and after bifurcation (*right-hand side*)

This new motion of the solution vector $\boldsymbol{q}(t)$ can be visualized as that of a motion on a torus (Figs. 1.14.5 and 6). In other words, we are now dealing with a quasi-periodic motion. When a limit cycle becomes unstable, other phenomena may occur also. The limit cycle can be replaced by another one where the system needs twice the time to return to its original state or, to put it differently, the *period* has *doubled*, or a *subharmonic* is generated. There are a number of systems known, ranging from fluid dynamics to electronic systems, which pass through a hierarchy of subsequent period doublings when the control parameter is changed. Figures 1.14.7 – 12 survey some typical results.

The phenomenon of period doubling or, in other words, of subharmonic generation, has been known for a long time in electronics. For instance, a number of electronic circuits can be described by the Duffing equation

$$\ddot{q} + \beta\dot{q} + \gamma q + \delta q^3 = \alpha_1 + \alpha_2 \sin(\omega_0 t) , \tag{1.14.14}$$

which describes the response of a nonlinear oscillator to a periodic driving force.

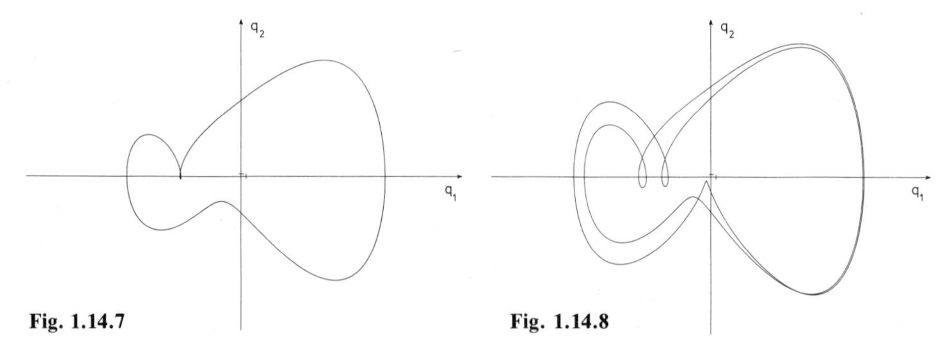

Fig. 1.14.7

Fig. 1.14.8

Fig. 1.14.7. Projection (from the q_1, q_2, t space) of a trajectory on the q_1, q_2 plane of a trajectory of the Duffing equation $\ddot{x} + k\dot{x} + x^3 = A \cos t$, where $q_1 \equiv x$, $q_2 \equiv \dot{x}$. The control parameter values are fixed at $k = 0.35$, $A = 6.6$. In this case a limit cycle occurs

Fig. 1.14.8. Solution of the same equation as in Fig. 1.14.7 but with $k = 0.35$, $A = 8.0$. The period of evolution is now twice as large as that of Fig. 1.14.7 but again a closed orbit occurs (period doubling)

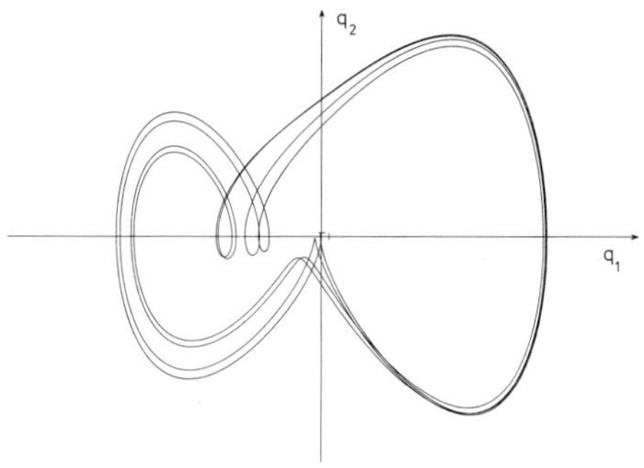

Fig. 1.14.9. Same equation as in Figs. 1.14.7 and 8 with $k = 0.35$, $A = 8.5$. The period is 4 times as large as that of the original limit cycle of Fig. 1.14.7

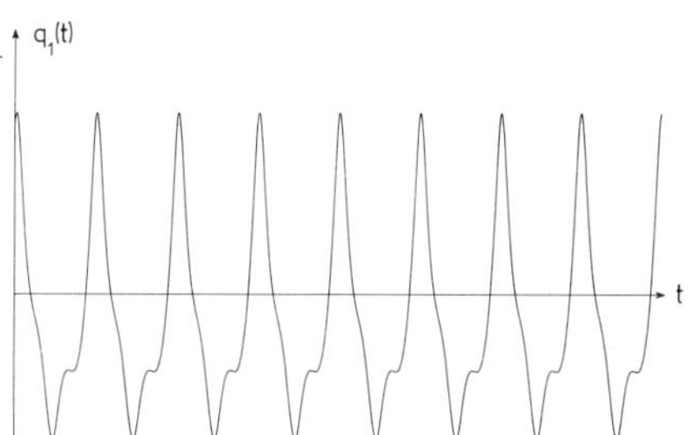

Fig. 1.14.10

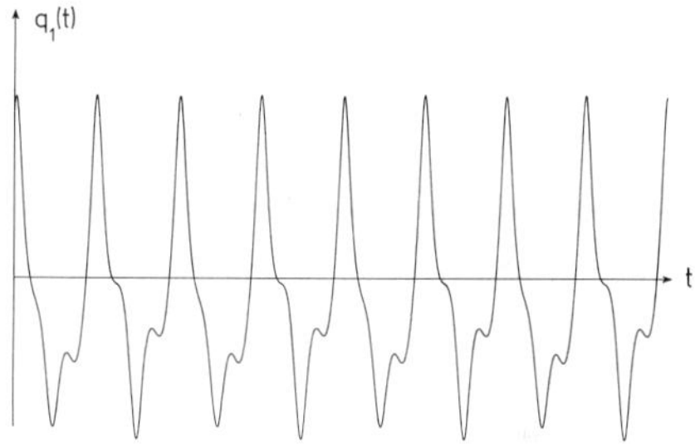

Fig. 1.14.11

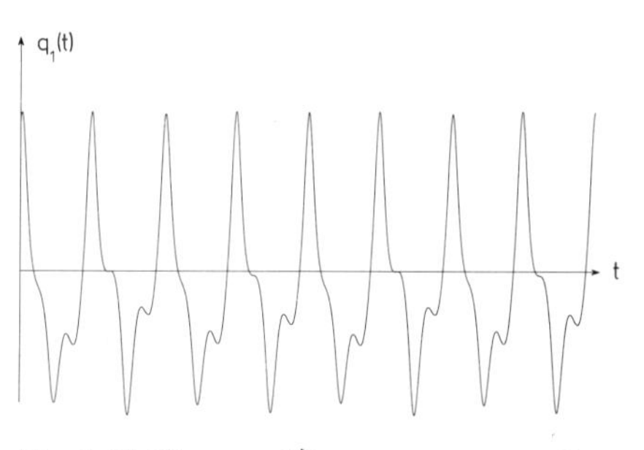

Fig. 1.14.12

Fig. 1.14.10. Temporal evolution of coordinate q_1 belonging to the case of Fig. 1.14.7

Fig. 1.14.11. Temporal evolution of the coordinate q_1 after period doubling has occurred. Note that two subsequent minima have different depths

Fig. 1.14.12. Temporal evolution of q_1 belonging to Fig. 1.14.9. Close inspection of the lower minima reveals that the cycle is repeated only after a time 4 times as big as in Fig. 1.14.10

Even this still rather simple equation describes a variety of phenomena including period doubling and tripling. But other subharmonics may also be present. A whole sequence of period doublings or triplings may occur, too. It seems that here we are just at the beginning of a new development, which allows us to study not only one or a few subsequent bifurcations, but a whole hierarchy. A warning seems to be necessary here. While some classes of equations (certain "discrete maps", cf. Sect. 1.17) indicate a complete sequence of period doubling, real systems may be more complicated allowing, e.g., both for doubling and tripling sequences or even for mixtures of them.

1.14.4 Bifurcations from a Torus to Other Tori

We have seen above that a limit cycle may bifurcate into a torus. Quite recently the bifurcation of a torus into other tori of the same dimension or of higher dimensions has been studied. Here quite peculiar difficulties arise. The mathematical analysis shows that a rather strange condition on the relative irrationality of the involved basic frequencies $\omega_1, \omega_2, \ldots$ of the system plays a crucial role. Many later sections will be devoted to this kind of problem. Some aspects of this problem are known in celestial mechanics, where the difficulties could be overcome only about one or two decades ago. While in celestial mechanics scientists deal with Hamiltonian, i.e., nondissipative, systems and are concerned with the stability of motion, here we have to be concerned with the still more complicated problem of dissipative, pumped systems, so that we shall deal in particular with qualitative changes of macroscopic properties (in particular the bifurcation from a torus into other tori).

1.14.5 Chaotic Attractors

When the motion on a torus becomes unstable due to a change of control parameter and if the specific conditions on the irrationality of the frequencies are not fulfilled, then quite different things may happen to the flow $\boldsymbol{q}(t)$. A torus may again collapse into a limit cycle, i.e., one or several frequencies may lock together (Fig. 1.14.13). A large class of phenomena, in the focus of present research, concerns irregular motions, i.e., the "chaotic attractors" briefly mentioned above.

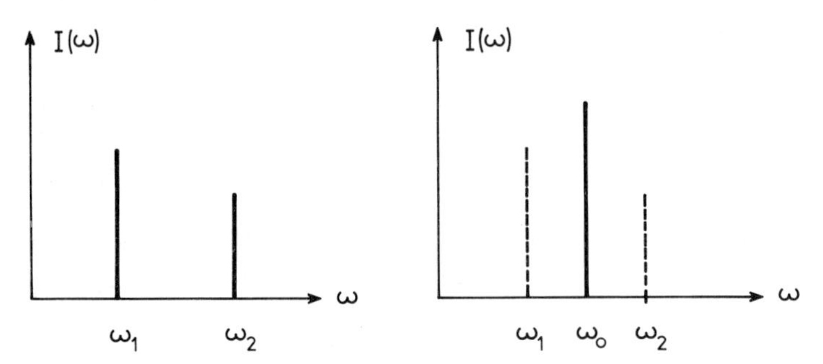

Fig. 1.14.13. How the power spectrum $I(\omega)$ reveals frequency locking. (*Left-hand side*) In the unlocked state the system oscillates at two fundamental frequencies ω_1 and ω_2. (*Right-hand side*) The power spectrum in case of a locked state. The two former frequencies ω_1 and ω_2 (– – –) have disappeared and are replaced by a single line at frequency ω_0

The results of all these studies are of great importance for many problems in the natural sciences and other fields, because "motion on a torus" means for a concrete system that it exerts a motion at several fundamental frequencies and their suitable linear combinations. In many cases such a motion is caused by a system of nonlinear oscillators which occur in nature and in technical devices quite frequently. It is then important to know in which way such a system changes its behavior if a control parameter is changed.

1.14.6 Lyapunov Exponents*

As we have seen in the preceding sections, there may be quite different kinds of attractors, such as a stable focus, a limit cycle, a torus, or finally a chaotic attractor. It is therefore highly desirable to develop criteria which allow us to distinguish between different attractors. Such a criterium is provided by the Lyapunov exponents which we are going to explain now.

Let us consider the simplest example, namely one variable which obeys the nonlinear differential equation

$$\dot{q} = N(q) . \tag{1.14.15}$$

In this case the only attractor possible is a stable fixed point (a "one-dimensional node") and the "trajectory" of this attractor is just a constant, $q = q_0$, denoting the position of that fixed point. In order to prove the stability of this point, we make a linear stability analysis introduced in Sect. 1.13. To this end we insert

$$q(t) = q_0 + \delta q(t) \tag{1.14.16}$$

into (1.14.15), linearize this equation with respect to δq, and obtain

$$\frac{d}{dt} \delta q = L \delta q , \tag{1.14.17}$$

where $L = \partial N / \partial q \,|_{q=q_0}$ is a constant.

The solution of (1.14.17) reads, of course,

$$\delta q(t) = \delta q(0) \, \mathrm{e}^{Lt} . \tag{1.14.18}$$

If L is negative, the fixed point is stable. In this trivial example, we can directly read off L from (1.14.18). But in more complicated cases to be discussed below, only a computer solution might be available. But even in such a case, we may derive L from (1.14.18) by a simple prescription. Namely, we form

$$\frac{1}{t} \ln |\delta q(t)|$$

and take the limit of $t \to \infty$. Then, clearly

$$L = \lim_{t \to \infty} \frac{1}{t} \ln |\delta q(t)| . \tag{1.14.19}$$

The concept of the Lyapunov exponent is a generalization of (1.14.19) in two ways:

1) One admits trajectories in a multi-dimensional space so that $\boldsymbol{q}(t)$ is the corresponding vector which moves along its trajectory when time t elapses.

2) One tests the behavior of the system in the neighborhood of the trajectory $\boldsymbol{q}_0(t)$ under consideration (which may, in particular, belong to an attractor). In analogy to (1.14.16) we put

$$\boldsymbol{q}(t) = \boldsymbol{q}_0(t) + \delta \boldsymbol{q}(t) \tag{1.14.20}$$

so that $\delta \boldsymbol{q}(t)$ "tests" in which way the neighboring trajectory $\boldsymbol{q}(t)$ behaves, i.e. whether it approaches $\boldsymbol{q}_0(t)$ or departs from it. In order to determine $\delta \boldsymbol{q}(t)$ we insert (1.14.20) into the nonlinear equation

$$\dot{\boldsymbol{q}}(t) = \boldsymbol{N}(\boldsymbol{q}(t)) \tag{1.14.21}$$

(of which $\boldsymbol{q}_0$ is a solution), and linearize $\boldsymbol{N}$ with respect to $\delta \boldsymbol{q}(t)$. Writing (1.14.21) in components we readily obtain

$$\frac{d}{dt}\delta q_j(t) = \sum_k \partial N_j(\boldsymbol{q}(t))/\partial q_k|_{\boldsymbol{q}=\boldsymbol{q}_0}\delta q_k(t)\,. \tag{1.14.22}$$

This is a set of *linear* differential equations with, in general, time-dependent coefficients $(\partial N_j/\partial q_k)$.

In generalization of (1.14.19) one is tempted to define the Lyapunov exponents by

$$\lambda = \lim_{t\to\infty}\frac{1}{t}\ln|\delta \boldsymbol{q}(t)|. \tag{1.14.23}$$

This definition, though quite often used in scientific literature, is premature, however, because the limit needs not to exist. Consider as an example

$$\delta q = \mathrm{e}^{\lambda_1 t}\sin(\omega t) + \mathrm{e}^{\lambda_2 t}\cos(\omega t)\,. \tag{1.14.24}$$

If we choose, in (1.14.23), $t = t_n = 2\pi n/\omega$, n being an integer, so that $\sin\omega t$ vanishes, δq behaves like $\exp(\lambda_2 t_n)$ and (1.14.23) yields $\lambda = \lambda_2$. If, on the other hand, we put $t = t'_n = 2\pi(n + 1/2)/\omega$, δq behaves like $\exp(\lambda_1 t'_n)$ and (1.14.23) yields

$$\lambda = \lambda_1\,.$$

Clearly, if $\lambda_1 \neq \lambda_2$, the limit (1.14.23) does not exist. Therefore one has to refine the definition of the Lyapunov exponent. Roughly speaking, one whishes to select the biggest rate λ, and one therefore replaces "lim" in (1.14.23) by the "limes superior", or in short lim sup, so that (1.14.23) is replaced by

$$\lambda = \limsup_{t\to\infty}\frac{1}{t}\ln|\delta \boldsymbol{q}(t)|. \tag{1.14.25}$$

We shall give the precise definition of lim sup in Chap. 2 where we shall present a theorem on the existence of Lyapunov exponents, too. Depending on different initial values of $\delta \boldsymbol{q}$ at $t = t_0$, different Lyapunov exponents may exist, but not more than m different ones, if m is the dimension of the vector space of $\boldsymbol{N}$ (or $\boldsymbol{q}$).

We are now able to present the criterium anounced above, which helps us to distinguish between different kinds of *attractors*.

In one dimension, there are only stable fixed points, for which the Lyapunov exponents λ are negative $(-)$. In two dimensions, the only two possible classes of attractors are stable fixed points, or limit cycles, as is proven rigorously in mathematics. In the case of a stable fixed point (focus) the two Lyapunov exponents (λ_1, λ_2) (which may coincide) are negative $(-,-)$. In the case of a stable limit cycle, the Lyapunov exponents belonging to a motion $\delta \boldsymbol{q}$ transversal to the limit cycle $\boldsymbol{q}_0(t)$ is negative (stability!), whereas the Lyapunov exponent belonging to $\delta \boldsymbol{q}$ in tangential direction vanishes, as we shall demonstrate in a later section. Therefore $(\lambda_1, \lambda_2) = (-,0)$. There may be a "pathological" case, for which $(\lambda_1, \lambda_2) = (-,0)$, but no limit cycle present, namely if there is a line of fixed points.

Finally we discuss typical cases in three dimensions. In each case we assume an *attractor* $\boldsymbol{q}_0$, i.e. $|\boldsymbol{q}_0|$ remains bounded for $t \to \infty$.

$$(\lambda_1, \lambda_2, \lambda_3) = (-,-,-) \quad \text{stable focus (fixed point)}$$

$$(\lambda_1, \lambda_2, \lambda_3) = (-,-,0) \quad \text{stable limit cycle .}$$

Neighboring trajectories of the limit cycle can approach the limit cycle from two linearly independent directions transversal to the limit cycle so that $(\lambda_1, \lambda_2) = (-,-)$, whereas the third Lyapunov exponent corresponding to a shift of the trajectory in tangential direction is equal to zero.

$$(\lambda_1, \lambda_2, \lambda_3) = (-,0,0) \quad \text{stable torus .}$$

The discussion is similar to the case of the limit cycle. (There are still some subtleties analogous to the "pathological" case just mentioned.)

Chaos may occur, if one Lyapunov exponent becomes positive. But at any rate, some more discussion is needed. For instance $(\lambda_1, \lambda_2, \lambda_3) = (+,0,0)$ may mean we are dealing with an unstable torus (i.e. *no attractor*). If an *attractor* possesses the exponents $(\lambda_1, \lambda_2, \lambda_3) = (+,0,-)$, it is considered as a *chaotic* attractor, $(\lambda_1, \lambda_2, \lambda_3) = (+,+,0)$ may mean an unstable limit cycle (i.e. no attractor) etc. Because in a chaotic attractor at least one Lyapunov exponent is positive, neighboring trajectories depart very quickly from each other. But since neighboring trajectories stem from initial conditions which differ but slightly we recognize that the phenomena described by a chaotic attractor *sensitively depend on the initial conditions.* It might be noteworthy that the research on Lyapunov exponents – both what they mean for attractors and how they can be determined – is still under its way.

In conclusion we mention the following useful theorem: If $\boldsymbol{q}(t)$ is a trajectory which remains in a bounded region (e.g. the trajectory of an attractor) and if it does not terminate at a fixed point, at least one of its Lyapunov exponents vanishes. We shall present a detailed formulation of this theorem and its proof in Sect. 2.4.

1.15 The Impact of Fluctuations (Noise). Nonequilibrium Phase Transitions

So far we have discussed a number of typical phenomena neglecting noise, i.e., the impact of fluctuations on the system. Over the past years it has become more and more evident that just at critical points, where the system changes its macroscopic behavior, fluctuations play a decisive role. According to fundamental laws of theoretical physics, whenever dissipation occurs, fluctuations must be present. Therefore as long as we deal with physical, chemical, biological, mechanical or electrical systems, fluctuations must not be ignored, at least in systems close to critical points. For phase transitions of systems in thermal equilibrium, the adequate treatment of fluctuations had been a long standing problem and could be solved only recently by renormalization group techniques. In this book we are concerned with instabilities of physical and chemical systems *far from thermal equilibrium* and other systems. Here fluctuations play an at least equally important role and ask for new approaches. For instance, the slaving principle we got to know in Sect. 1.13 must be able to take into account fluctuations (cf. Chap. 7), and the order parameter equations must be solved with the adequate inclusion of fluctuations (cf. Chap. 10). In short, fluctuations render the phenomena and problems of *bifurcations* (which are difficult enough) into the still more complex phenomena and correspondingly more difficult problems of *nonequilibrium phase transitions.*

In order to get some insight into the role played by fluctuations we wish to study some relatively simple examples. Let us consider the transition of a stable node to two stable nodes and an unstable node which we came across in Figs. 1.13.5 and 6. In the presence of noise, even in the steady state, the representative point of a system, $\boldsymbol{q}(t)$, is pushed backwards and forwards in a random sequence all the time. Therefore we can tell only what the *probability* will be to find the system's vector $\boldsymbol{q}$ in a certain volume element $dV = dq_1 dq_2 \dots dq_n$. This probability is decribed by a probability distribution function $f(\boldsymbol{q}, t)$ multiplied by dV. This function may change its shape dramatically at transition points (Figs. 1.15.1, 2). Close to these points the fluctuations of the order parameters become

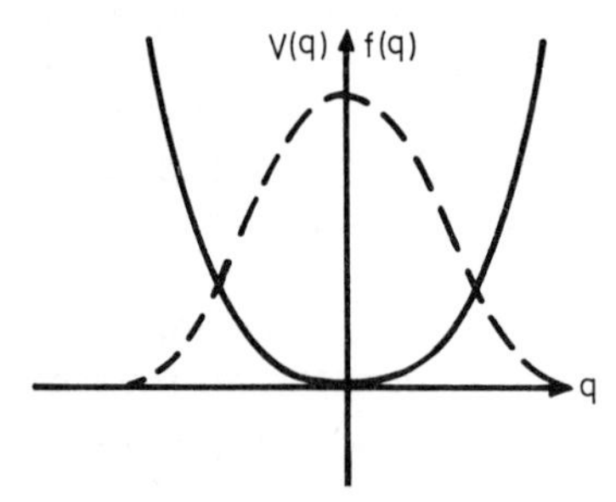

Fig. 1.15.1. The distribution function $f(q)$ (– – –) belonging to a node. The solid line represents the potential in which the "particle" with coordinate q moves

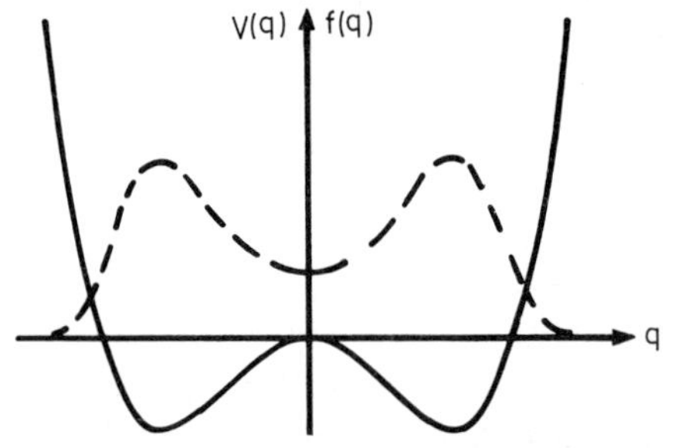

Fig. 1.15.2. The distribution function $f(q)$ (– – –) belonging to two nodes and a saddle point (in one dimension). The solid line represents the potential in which the "particle" moves

particularly large ("critical fluctuations"). When we retreat further from the transition points $\alpha = \alpha_0$, the new double peak may become very sharp, indicating that the system has a high probability to be found in either of the new states. Another feature of a system subjected to noise is the following. At an initial time we may find ("measure") the system in a state $\boldsymbol{q} = \boldsymbol{q}_0$ or within a neighborhood of such a state. But the fluctuations may drive the system away from this state $\boldsymbol{q}_0$. We then may ask how long it will take for the system to reach another given state $\boldsymbol{q}_1$ for the first time. Since we are dealing with stochastic processes, such a transition time can be determined only in a statistical sense. This problem is called the *first passage time problem.* As a special case the problem arises how to determine the time which a system needs to go from one maximum of the distribution function to the other one.

Similarly, effects typical for the impact of noise occur when a focus becomes unstable and is replaced by a limit cycle. Writing the newly evolving solution q in the form $\exp[\mathrm{i}\omega t + \mathrm{i}\varphi(t)]\, r(t)$ (ω, φ, r are assumed real), it turns out that the phase $\varphi(t)$ undergoes a diffusion process, while the amplitude r performs fluctuations around a stable value r_0. Important quantities to be determined are the relaxation time of r as well as the diffusion constant of the phase diffusion. Other phenomena connected with noise concern the destruction of a frequency-locked state. Noise may drive a system from time to time away from that state so that occasionally, instead of the one frequency, two frequencies may occur again. In later chapters we shall treat the most important aspects of noise in detail, illustrating the procedure by characteristic examples and presenting several general theorems which have turned out to be most useful in practical applications.

1.16 Evolution of Spatial Patterns

So far we have discussed qualitative changes of the temporal behavior of systems, such as onset of oscillations, occurrence of oscillations at several frequencies, subharmonic oscillations, etc. In many physical, chemical, and biological systems we must not ignore the spatial dependence of the system's variables. For instance, we have seen in Sect. 1.2.1 that spatial patterns can arise in fluids. In the simplest case we start from a spatially homogeneous state. At a certain control parameter value the homogenous solution may become unstable, as is revealed by linear stability analysis. Therefore we have to study linear equations of the form

$$\dot{w} = Lw\,. \tag{1.16.1}$$

Since N in (1.14.1) contains spatial derivatives, so does L. To illustrate the essential features, assume that L is of the form

$$L = L_0 + D\Delta\,, \tag{1.16.2}$$

where L_0 is a constant matrix and D a constant diagonal matrix. In general, L_0

depends on the control parameter α. By the hypothesis $\boldsymbol{w}(\boldsymbol{x}, t) = \exp(\lambda t)\boldsymbol{v}(\boldsymbol{x})$, we transform (1.16.1) into

$$(L_0 + D\Delta)\boldsymbol{v}(\boldsymbol{x}) = \lambda \boldsymbol{v}(\boldsymbol{x}), \tag{1.16.3}$$

which is of the form of an elliptic partial differential equation. Under given boundary conditions on $\boldsymbol{v}(\boldsymbol{x})$, (1.16.3) allows for a set of spatial modes $\boldsymbol{v}_j(\boldsymbol{x})$ with eigenvalues λ_j. When we change α, one or several of the λ_j's may cross the imaginary axis, i.e., the corresponding modes become unstable. In principle, the method of solution of the nonlinear equations (1.14.1) is the same as that of Sect. 1.14 in which we neglected any spatial dependence of $\boldsymbol{q}$ on space coordinate $\boldsymbol{x}$. We put

$$\boldsymbol{q}(\boldsymbol{x}, t) = \boldsymbol{q}_0 + \sum_j \xi_j(t)\, \boldsymbol{v}_j(\boldsymbol{x}). \tag{1.16.4}$$

Again we may identify order parameters ξ_j, for which $\mathrm{Re}\{\lambda_j\} \geqslant 0$. We may apply the slaving principle and establish order parameter equations. Confining (1.16.4) to the leading terms, i.e., the order parameters $\xi_j \equiv u_j$ alone, we obtain the skeleton of the evolving patterns. For instance, if only one $\mathrm{Re}\{\lambda_j\} \geqslant 0$, the "skeleton" reads

$$\boldsymbol{q}(\boldsymbol{x}, t) = \mathrm{const} + u_1(t)\, \boldsymbol{v}_1(\boldsymbol{x}). \tag{1.16.5}$$

Since in many cases $u_1(t)$ obeys the equation

$$\dot{u}_1 = \lambda_1 u_1 - \beta u_1^3, \tag{1.16.6}$$

we obtain the following result: while for $\lambda_1 < 0$ only the solution $u_1 = 0$ and therefore $\boldsymbol{q}(\boldsymbol{x}, t) = \mathrm{const}$ and such that the *homogeneous distribution* is stable, for $\lambda_1 > 0$ the solution $u_1 \neq 0$ and therefore the spatially *inhomogeneous* solution (1.16.5) arises. This approach allows us to study the growth of a *spatial pattern*. If several order parameters $u_j \neq 0$ are present, the "skeleton" is determined by

$$\boldsymbol{q}(\boldsymbol{x}, t) = \mathrm{const} + \sum_j u_j \boldsymbol{v}_j(\boldsymbol{x}). \tag{1.16.7}$$

But in constrast to linear theories the u_j's cannot be chosen arbitrarily. They are determined by certain nonlinear equations derived later and which thus determine the possible "skeletons" and their growths. In higher approximation the slaved modes contribute to the spatial pattern also. An important difference between these transitions and phase transitions of systems in thermal equilibrium, where long-range order is also achieved, should be mentioned. With few exceptions, current phase transition theory is concerned with infinitely extended media because only here do the singularities of certain thermodynamic functions (entropy, specific heat, etc.) become apparent. On the other hand, in nonequilibrium phase transition, i.e., in the transitions considered here, in general the finite geometry plays a decisive role. The evolving patterns depend on the shape of the boundaries, for instance square, rectangular, circular, etc., and in addition the patterns depend on the size of the system. In other words, the evolving patterns, at least in general, bring in their own length scales which must be matched with the given geometry.

With respect to applications to astrophysics and biology we have to study the evolution of patterns not only in the plane or in Euclidian space, but we have to consider that evolution also on spheres and still more complicated manifolds. Examples are the first stages of embryonic evolution or the formation of atmospheric patterns on planets, such as Jupiter. Of course, in a more model-like fashion we may also study infinitely extended media. In this case we find phenomena well known from phase transition theory and we may apply renormalization group techniques to these phenomena.

Within active media such as distributed chemical reactions, interaction of cells of a tissue, neuronal networks, etc., combined spatio-temporal patterns may evolve also. The methods described later allow a mathematical treatment of large classes of these phenomena.

1.17 Discrete Maps. The Poincaré Map

In the previous sections a brief outline was given of how to model many processes by means of evolution equations of the form

$$\dot{\boldsymbol{q}}(t) = \boldsymbol{N}(\boldsymbol{q}(t)) \, . \tag{1.17.1}$$

In order to know the temporal evolution of the system, which is described by $\boldsymbol{q}(t)$, we have to know $\boldsymbol{q}$ for all times. Since time is continuous, we need to know a continuous infinity of data! This is evidently an unsolvable task for any human being. There are several ways out of this difficulty. We may find steady states for which $\boldsymbol{q}$ is time independent and can be represented by a finite number of data. Or we may write down an exact or approximate closed form of $\boldsymbol{q}$, e.g., $\boldsymbol{q} = \boldsymbol{q}_0 \sin(\omega t)$.

The important thing is not so much that we can calculate $\sin(\omega t)$ for any time t to (at least in principle) any desired degree of accuracy. Rather, we can immediately visualize the function $\sin(\omega t)$ as a periodic oscillation, or $\exp(\lambda t)$ as an ever increasing or decreasing function, depending on the sign of λ.

There is yet another way we can do away with the problem of "infinite" information required. Namely, as in a digital computer, we may consider $\boldsymbol{q}$ at a discrete sequence of times t_n only. The differential equation (1.17.1) is then replaced by a suitable set of difference equations. A still more dramatic reduction is made by the Poincaré map. Consider as an example the trajectory in a plane (Fig. 1.17.1). Instead of following up the whole trajectory over all times, we consider only the crossing points with the q_1 axis. Let us denote these points by $q_1(n) \equiv x_n$. Because $(q_1 \equiv x_n, q_2 = 0)$ for a fixed n may serve as initial value for the half-trajectory which hits the q_1 axis again at x_{n+1}, it follows that x_{n+1} is uniquely determined by x_n, i.e.,

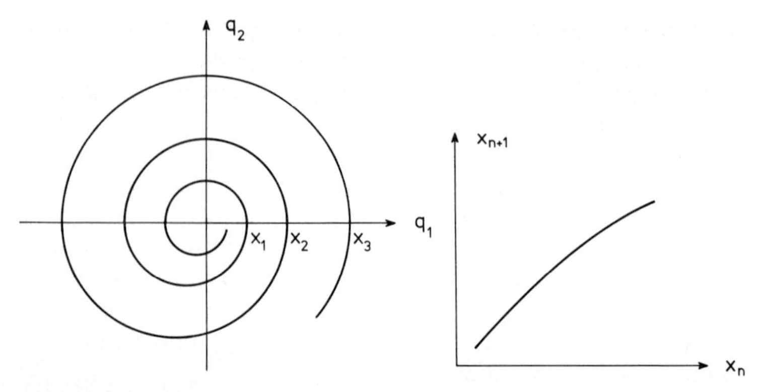

Fig. 1.17.1. A trajectory (———) hits the q_1 axis at intersection points x_1, x_2, x_3, etc.

Fig. 1.17.2. A plot of x_{n+1} versus x_n for the example of Fig. 1.17.1 represented as continuous graph

$$x_{n+1} = f(x_n) . \tag{1.17.2}$$

Of course, in order to find this connection, one has to integrate (1.17.1) over a time interval from t_n, where x_n is hit, to t_{n+1}, where x_{n+1} is hit. Thus no simplification seems to be reached at all. The following idea has turned out to be most useful. Let us consider (1.17.2) for all n and a fixed function f in terms of a model! Because we no longer need to integrate (1.17.1) over continuous times, we expect to gain more insight into the global behavior of x_n.

Equation (1.17.2) is called a map because each value of x_n is mapped onto a value of x_{n+1}. This can be expressed in form of a graph (Fig. 1.17.2). Probably the simplest graph which gives rise to nontrivial results is described by the "logistic" map which reads

$$x_{n+1} = \alpha x_n (1 - x_n) . \tag{1.17.3}$$

This equation is plotted in Fig. 1.17.3. In (1.17.3), α serves as a control parameter. If it runs between 0 and 4, any value of $0 \leqq x_n \leqq 1$ is again mapped into (or onto) an interval $0 \leqq x_{n+1} \leqq 1$. Starting from an initial value x_0, it is quite simple to calculate the sequence $x_1, x_2, \ldots$ from (1.17.3), e.g., by means of a pocket computer.

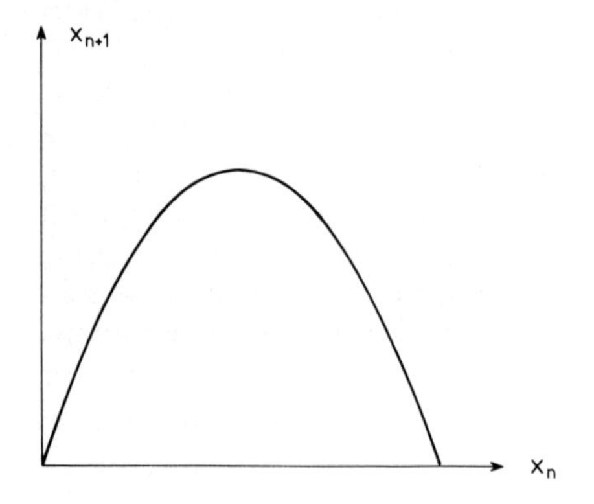

Fig. 1.17.3. A parabolic plot of x_{n+1} versus x_n corresponding to Eq. (1.17.3)

However, a good insight into the behavior of the solutions of (1.17.3) can be gained by a very simple geometric construction explained in Figs. 1.17.4–9. The reader is strongly advised to take a piece of paper and to repeat the individual steps. As it transpires, depending on the numerical value of α, different kinds of behavior of x_n result. For $\alpha < 3$, x_n converges to a fixed point. For $3 < \alpha < \alpha_2$, x_n converges to a periodic hopping motion with period $T = 2$ (Figs. 1.17.10, 11). For $\alpha_2 < \alpha < \alpha_3$, x_n and t_n big enough, x_n comes back to its value for $n+4$, $n+8, \ldots$, i.e., after a period $T = 4$, $T = 8, \ldots$ (Figs. 1.17.12, 13). When we plot the values of x_n reached for $n \to \infty$ for various values of α, Fig. 1.17.14 results.

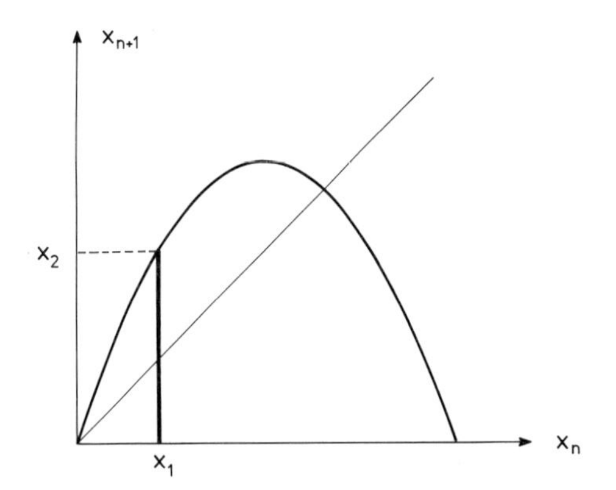

Fig. 1.17.4. The series of Figs. 1.17.4–8 shows how one may derive the sequence x_n, $(n = 2, 3, 4, \ldots)$, once one initial value x_1 is given. Because the parabola represents the mapping function, we have to go from x_1 by a vertical line until we cross the parabola to find x_2

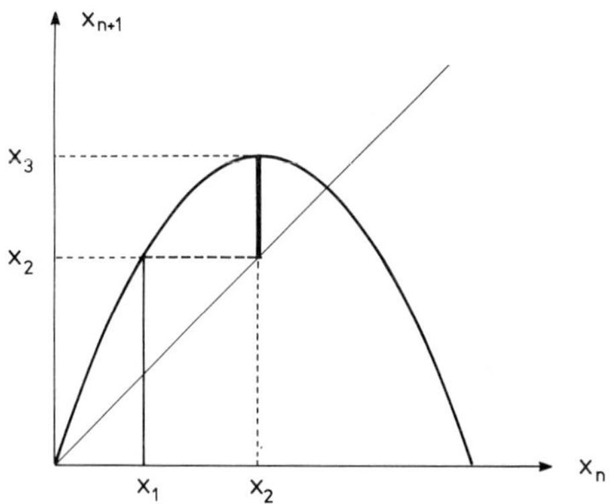

Fig. 1.17.5. Value x_2 found in Fig. 1.17.4 serves now as a new initial value, obtained by projecting x_2 in Fig. 1.17.4 or in the present figure on the diagonal straight line. Going from the corresponding cross section downward, we find x_2 as value on the abscissa. In order to obtain the new value x_3 we go vertically upwards to meet the parabola

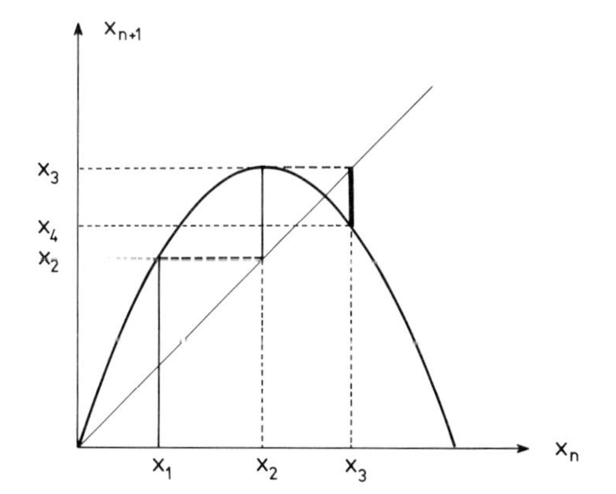

Fig. 1.17.6. To proceed from x_3 to x_4 we repeat the steps done before, i.e., we first go horizontally from x_3 till we hit the diagonal straight line. Going down vertically we may find the value x_3, but on the way we hit the parabola and the cross section gives us x_4

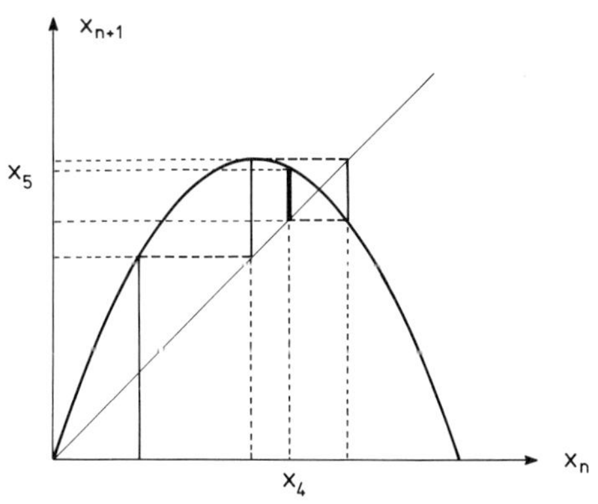

Fig. 1.17.7. The construction is now rather obvious and we merely show how we may proceed from x_4 to x_5

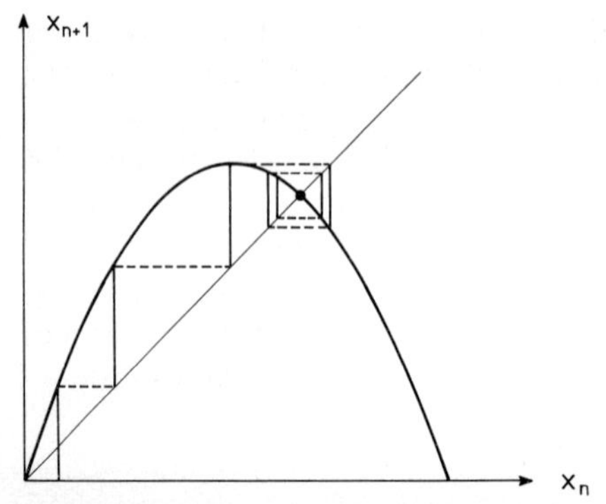

Fig. 1.17.8. When we repeat the above steps many times we approach more and more a limiting point which is just the cross section of the parabola with the diagonal straight line

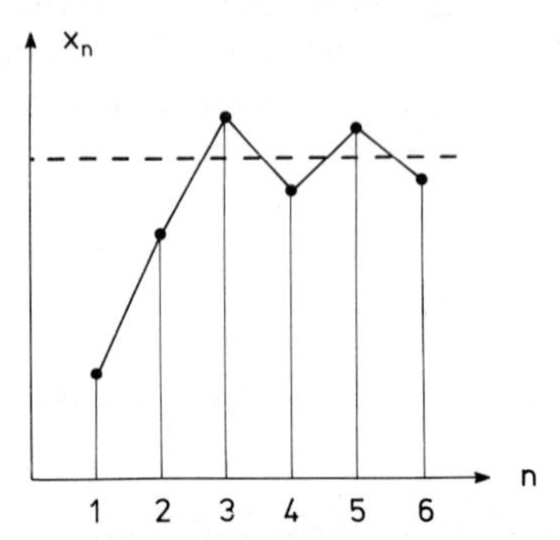

Fig. 1.17.9. The individual values of x_1, x_2, x_3, ... constructed by means of the previous figures. The resulting plot is shown where x_n approaches more and more the dashed line

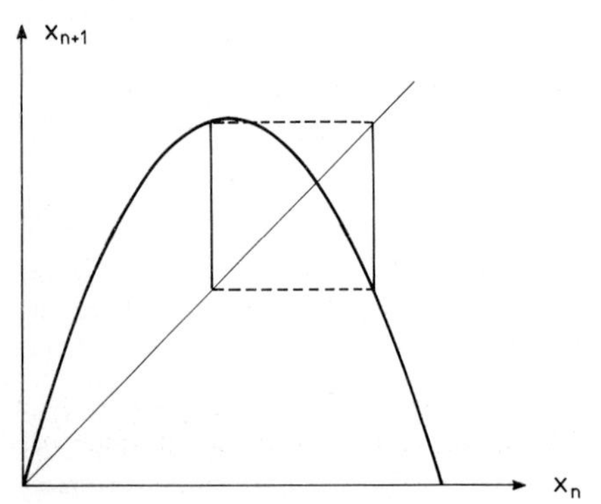

Fig. 1.17.10. When we choose the height of the parabola differently (bigger than that of Fig. 1.17.8), a closed curve arises which indicates a hopping of x_n between two values (see Fig. 1.17.11)

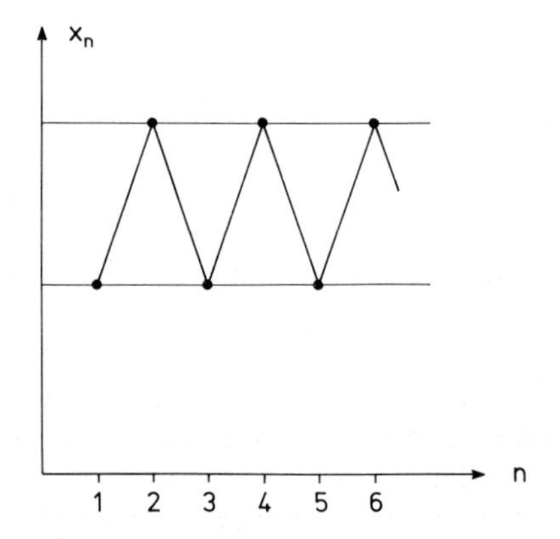

Fig. 1.17.11. A periodic hopping of x_n corresponding to the closed trajectory of Fig. 1.17.10

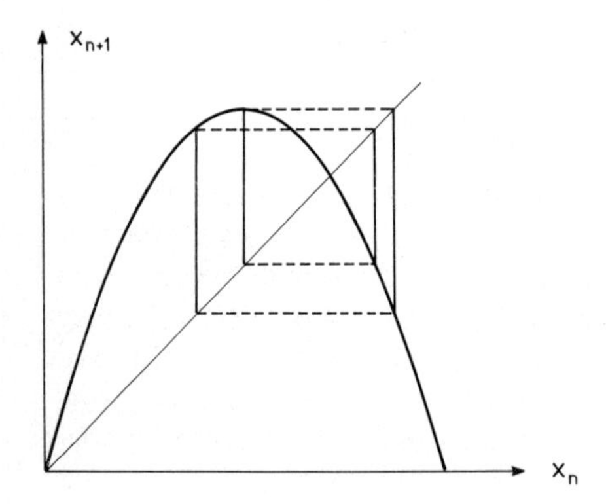

Fig. 1.17.12. The case of period 4

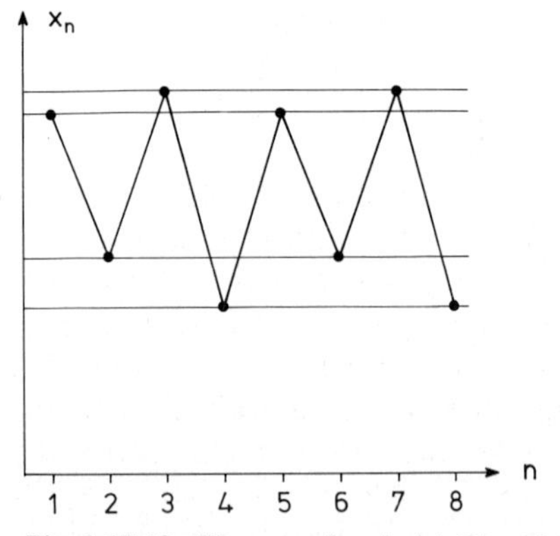

Fig. 1.17.13. The case of period 4. Plot of x_n versus n

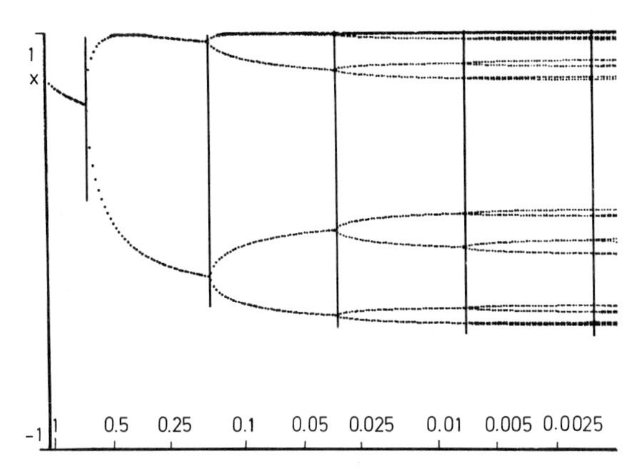

Fig. 1.17.14. The set of possible values of x_n $(n \to \infty)$ (*ordinate*) versus control parameter $\mu_\infty - \mu$ (*abscissa*) on a logarithmic scale. Here the logistic equation is linearly transformed into the equation $x_{n+1} = 1 - \mu x_n^2$. μ_∞ corresponds to the critical value α_∞. [After P. Collet, J. P. Eckmann: In *Progress in Pyhsics*, Vol. 1, ed. by A. Jaffe, D. Ruelle (Birkhäuser, Boston 1980)]

We find a sequence of period doublings which accumulate at $\alpha = \alpha_\infty = 3.569945672\ldots$. The sequence of α_n, at which period doubling occurs, obeys a simple law

$$\lim_{n\to\infty}\left(\frac{\alpha_n - \alpha_{n-1}}{\alpha_{n+1} - \alpha_n}\right) = \delta\,, \quad \text{where} \quad \delta = 4.6692016609\ldots . \tag{1.17.4}$$

The number δ is called the Feigenbaum number and is of a "universal" character, because there is a whole class of maps (1.17.2) which give rise to a sequence of period doublings with that number.

Though experimental results (Sect. 1.2.1) are in qualitative agreement with the theoretically predicted value (1.17.4), one should not be surprised if the agreement is not too good. First of all the Feigenbaum number is derived for $n \to \infty$, while experimental data were obtained for $n = 2, 3, 4, 5$. Furthermore, the control parameter which describes the "real world" and which enters into (1.17.1) need not be proportional to the control parameter α for (1.17.3), but rather the relation can be considerably more complicated.

Beyond the accumulation point α_∞, chaos, i.e., an irregular "motion" of x_n, is observed.

Another important feature of many dynamical systems is revealed by the logistic map: beyond α_∞, with incrasing α, periodic, i.e., regular motion becomes possible in windows of α between chaotic braids. Over the past years, quite a number of regularities often expressed as "scaling properties", have been found, but in the context of this book we shall be concerned rather with the main approaches to cope with discrete maps.

The idea of the Poincaré map can be more generally interpreted. For instance, instead of studying trajectories in the plane as in Fig. 1.17.1, we may also treat trajectories in an n-dimensional space and study the points where the trajectories cut through a hypersurface. This can be visualized in three dimensions according to Fig. 1.17.15. In a number of cases of practical interest it turns out that the cross points can be connected by a smooth curve (Fig. 1.17.16). In such a case one may stretch the curve into a line and use, e.g., the graph of Fig. 1.17.3 again. In general we are led to treat discrete maps of the form

$$\boldsymbol{x}_{n+1} = \boldsymbol{f}(\boldsymbol{x}_n) , \tag{1.17.5}$$

where $\boldsymbol{x}_n$, $n = 1, 2, \ldots$ are vectors in an M-dimensional space. Here $\boldsymbol{f}$ may depend on one or several control parameters α which allows us to study qualitative changes of the "discrete dynamics" of $\boldsymbol{x}_n$ when α is changed. A wide class of such qualitative changes can be put in analogy to nonequilibrium phase transitions, treated by the conventional evolution equations (1.17.1). Thus, we find critical slowing down and symmetry breaking, applicability of the slaving principle (for discrete maps), etc. These analogies become still closer when we treat discrete *noisy* maps (see below). Equations (1.17.5) enable many applications which so far have been studied only partially. For instance, the state vector $\boldsymbol{x}_n$ can symbolize various spatial patterns.

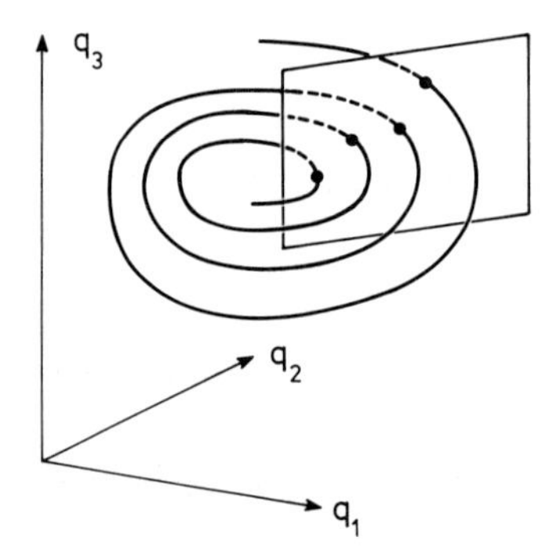

Fig. 1.17.15. Poincaré map belonging to the cross section of a two-dimensional plane with a trajectory in three dimensions

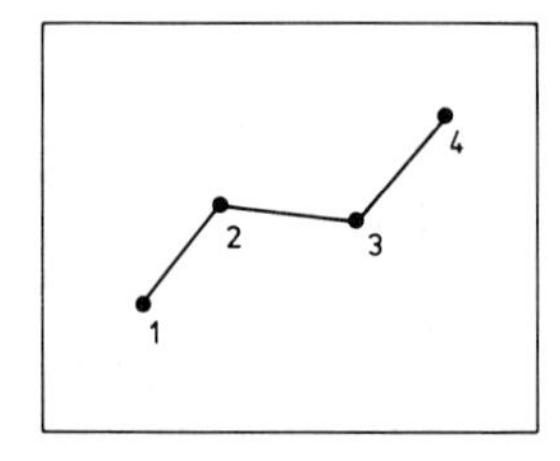

Fig. 1.17.16. The intersection points of Fig. 1.17.15

We finally study how we may formulate *Lyapunov exponents* in the case of a discrete map. To this end we proceed in close analogy to Sect. 1.14.6 where we introduced the concept of Lyapunov exponents for differential equations.

For a discrete map, the "trajectory" consists of the sequence of points $\boldsymbol{x}_n$, $n = 0, 1, \ldots$. We denote the trajectory the neighborhood of which we wish to study, by $\boldsymbol{x}_n^0$. It is assumed to obey (1.17.5). We write the neighboring trajectory as

$$\boldsymbol{x}_n = \boldsymbol{x}_n^0 + \delta\boldsymbol{x}_n \tag{1.17.6}$$

and insert it into (1.17.5). We may expand $\boldsymbol{f}(\boldsymbol{x}_n^0 + \delta\boldsymbol{x}_n)$ into a power series with respect to the components of $\delta\boldsymbol{x}_n$.

Keeping the linear term we obtain

$$\delta \boldsymbol{x}_{n+1} = L(\boldsymbol{x}_n^0)\,\delta \boldsymbol{x}_n \tag{1.17.7}$$

where the matrix L is given by

$$L \equiv (L_{kl}) = (\partial f_k(\boldsymbol{x})/\partial x_l|_{\boldsymbol{x}=\boldsymbol{x}_n^0}) \,. \tag{1.17.8}$$

Equation (1.17.7) can be solved by iteration (provided $\boldsymbol{x}_n^0$ has been determined before). Starting with $\delta \boldsymbol{x}_0$, we obtain

$$\begin{aligned} \delta \boldsymbol{x}_1 &= L(\boldsymbol{x}_0^0)\,\delta \boldsymbol{x}_0\,, \\ \delta \boldsymbol{x}_2 &= L(\boldsymbol{x}_1^0)\,\delta \boldsymbol{x}_1\,, \\ &\vdots \\ \delta \boldsymbol{x}_n &= L(\boldsymbol{x}_{n-1}^0)\,\delta \boldsymbol{x}_{n-1}\,. \end{aligned} \tag{1.17.9}$$

Expressing $\delta \boldsymbol{x}_n$ by $\delta \boldsymbol{x}_{n-1}$, $\delta \boldsymbol{x}_{n-1}$ by $\delta \boldsymbol{x}_{n-2}$ etc., we readily obtain

$$\delta \boldsymbol{x}_n = L(\boldsymbol{x}_{n-1}^0)\,L(\boldsymbol{x}_{n-2}^0)\ldots L(\boldsymbol{x}_0^0)\,\delta \boldsymbol{x}_0\,, \tag{1.17.10}$$

where the L's are multiplied with each other by the rules of matrix multiplication. The Lyapunov exponents are defined by

$$\lambda = \limsup_{n\to\infty} \frac{1}{n} \ln|\delta \boldsymbol{x}_n|. \tag{1.17.11}$$

Depending on different directions of $\delta \boldsymbol{x}_0$, different λ's may result (in multi-dimensional maps).

For a one-dimensional map we may represent λ in a rather explicit way. In this case the L's and δx_0 are mere numbers so that according to (1.17.10)

$$\begin{aligned} |\delta x_n| &= |L(x_{n-1}^0)\ldots L(x_0^0)\,\delta x_0| \\ &= |L(x_{n-1}^0)|\cdot|L(x_{n-2}^0)|\ldots|\delta x_0|. \end{aligned} \tag{1.17.12}$$

Inserting this latter expression into (1.17.11) and using (1.17.8) we obtain

$$\lambda = \limsup_{n\to\infty} \frac{1}{n} \sum_{m=0}^{n-1} \ln|f'(x_m^0)| \tag{1.17.13}$$

where f' is the derivative of $f(x)$ with respect to x.

As a nontrival example we show in Fig. 1.17.17 the Lyapunov exponent of the logistic map (1.17.3) as a function of α. A positive value of the exponent corresponds to chaotic motion whereas negative values indicate the presence of regular (periodic) behavior.

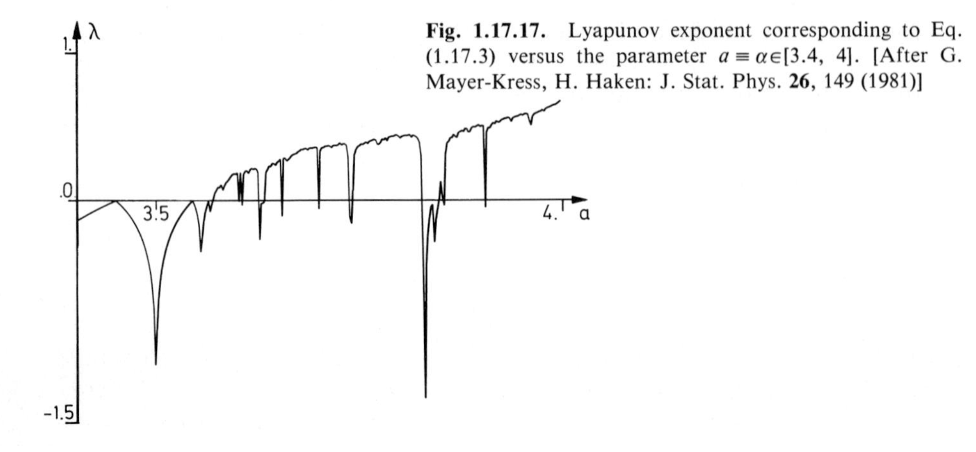

Fig. 1.17.17. Lyapunov exponent corresponding to Eq. (1.17.3) versus the parameter $a \equiv \alpha \in [3.4, 4]$. [After G. Mayer-Kress, H. Haken: J. Stat. Phys. **26**, 149 (1981)]

1.18 Discrete Noisy Maps

The impact of fluctuations on the dynamics of a system can be rather easily modelled within the framework of discrete maps. Each new value x_{n+1} is not only determined by $f(x_n)$, but also by some additional fluctuations η_n, so that

$$x_{n+1} = f(x_n) + \eta_n \, . \tag{1.18.1}$$

Also more complicated maps can be formulated and treated, e.g., maps for vectors $\boldsymbol{x}_n$, and for fluctuations which depend on $\boldsymbol{x}_n$ (Chap. 11).

1.19 Pathways to Self-Organization

In all the cases considered in this book the temporal, spatial or spatio-temporal patterns evolve without being imposed on the system from the outside. We shall call processes, which lead in this way to patterns, "self-organization". Of course, in the evolution or functioning of complex systems, such as biological systems, a whole hierarchy of such self-organizing processes may take place. In a way, in this book we are considering the building blocks of self-organization. But in contrast to other approaches, say by molecular biology which considers individual molecules and their interaction, we are mostly concerned with the interaction of many molecules or many subsystems. The kind of self-organization we are considering here can be caused by different means. We may change the global impact of the surroundings on the system (as expressed by control parameters). Self-organization can be caused also by a mere increase of number of components of a system. Even if we put the same components together, entirely new behavior can arise on the macroscopic level. Finally, self-organization can be caused by a sudden change of control parameters when the system tries to relax to a new state under the new conditions (constraints). This aspect offers us a very broad view on the evolution of structures, including life in the universe. Because the universe is in a transient state, starting from its fireball, and subject to expansion, ordered structures can occur.

In the context of this book we are interested in rigorous mathematical formulations rather than philosophical discussions. Therefore, we briefly indicate here how these three kinds of self-organization can be treated mathematically.

1.19.1 Self-Organization Through Change of Control Parameters

This case has been discussed in extenso above. When we slowly change the impact of the surroundings on a system, at certain critical points the system may acquire new states of higher order or structure. In particular, spatial patterns may evolve although the surroundings act entirely homogeneously on the system.

1.19.2 Self-Organization Through Change of Number of Components

Let us start from two uncoupled systems described by their corresponding state vectors $\boldsymbol{q}^{(1)}$ and $\boldsymbol{q}^{(2)}$ which obey the equations

$$\dot{\boldsymbol{q}}^{(1)} = \boldsymbol{N}^{(1)}(\boldsymbol{q}^{(1)}) , \quad \text{and} \tag{1.19.1}$$

$$\dot{\boldsymbol{q}}^{(2)} = \boldsymbol{N}^{(2)}(\boldsymbol{q}^{(2)}) . \tag{1.19.2}$$

Let us assume that these equations allow stable inactive states described by

$$\boldsymbol{q}_0^{(j)} = 0, \quad j = 1, 2 . \tag{1.19.3}$$

We introduce a coupling between these two systems and describe it by functions $\boldsymbol{K}^{(j)}$ so that the Eqs. (1.19.1, 2) are replaced by

$$\dot{\boldsymbol{q}}^{(1)} = \boldsymbol{N}^{(1)}(\boldsymbol{q}^{(1)}) + \boldsymbol{K}^{(1)}(\boldsymbol{q}^{(1)}, \boldsymbol{q}^{(2)}) \quad \text{and} \tag{1.19.4}$$

$$\dot{\boldsymbol{q}}^{(2)} = \boldsymbol{N}^{(2)}(\boldsymbol{q}^{(2)}) + \boldsymbol{K}^{(2)}(\boldsymbol{q}^{(1)}, \boldsymbol{q}^{(2)}) . \tag{1.19.5}$$

We now have to deal with the total system 1 + 2 described by the equation

$$\dot{\boldsymbol{q}} = \boldsymbol{N}(\boldsymbol{q}, \alpha) , \quad \text{where} \tag{1.19.6}$$

$$\boldsymbol{q} = \begin{pmatrix} \boldsymbol{q}^{(1)} \\ \boldsymbol{q}^{(2)} \end{pmatrix} \quad \text{and} \tag{1.19.7}$$

$$\boldsymbol{N}(\boldsymbol{q}, \alpha) = \begin{pmatrix} \boldsymbol{N}^{(1)} + \alpha \boldsymbol{K}^{(1)} \\ \boldsymbol{N}^{(2)} + \alpha \boldsymbol{K}^{(2)} \end{pmatrix} . \tag{1.19.8}$$

We have introduced a parameter α which ranges from 0 till 1 and which plays the role of a control parameter. Under suitable but realistic conditions, a change of α causes an instability of our original solution (1.19.3) and $q_0 = 0$ is replaced by

$$q \neq 0 , \tag{1.19.9}$$

which indicates some kind of new patterns or active states. This example shows that we may treat this kind of self-organization by the same methods we use when α is treated as a usual control parameter.

Clearly, if already the change of rather unspecific control parameters α can cause patterns, this will be much more the case if control parameters are changed in a specific way. For instance, in a network the couplings between different components may be changed differently. Clearly, this opens new vistas in treating brain functions by means of nonequilibrium phase transitions.

1.19.3 Self-Organization Through Transients

Structures (or patterns) may be formed in a self-organized fashion when the system passes from an initial disordered (or homogeneous) state to another final state which we need not specify or which even does not exist. This is most easily exemplified by a state vector of the form

$$q(x, t) = u(t)\, v(x) , \tag{1.19.10}$$

where $v(x)$ describes some spatial order, and the order parameter equation

$$\dot{u} = \lambda u . \tag{1.19.11}$$

When we change a control parameter α quickly, so that $\lambda < 0$ is quickly replaced by $\lambda > 0$, a transient state vector of the form

$$q(x, t) = e^{\lambda t} v(x) \tag{1.19.12}$$

occurs. It clearly describes some structure, but it does not tend to a new stable state.

This approach hides a deep philosophical problem (which is inherent in all cases of self-organization), because to get the solution (1.19.12) started, some *fluctuations* must be present. Otherwise the solution $u \equiv 0$ and thus $q \equiv 0$ will persist forever.

1.20 How We Shall Proceed

In this book we shall focus our attention on situations where systems undergo qualitative changes. Therefore the main line of our procedure is prescribed by

1) study of the loss of stability
2) derivation of the slaving principle
3) establishment and solution of the order parameter equations.

To this end we first study the *loss of stability*. Therefore Chaps. 2 and 3 deal with linear differential equations. While Chap. 2 contains results well known in mathematics (perhaps represented here, to some extent, in a new fashion), Chap. 3 contains largely new results. This chapter might be somewhat more difficult and can be skipped during a first reading of this book.

Chap. 4 lays the basis for stochastic methods which will be used mainly in Chap. 10. Chaps. 5 and 6 deal with coupled nonlinear oscillators and study quasiperiodic motion. Both Chaps. 5 and 6 serve as a preparation for Chap. 8, especially for its Sects. 8.8 – 11. Chap. 6 presents an important theorem due to Moser. In order not to overload the main text, Moser's proof of his theorem is postponed to the appendix. Chap. 7 then resumes the main line, initiated by Chaps. 2 and 3, and deals with the *slaving principle* (for nonlinear differential

Table 1.20.1. A survey on the relations between Chapters. Note that the catchwords in the boxes don't agree with the chapter headings

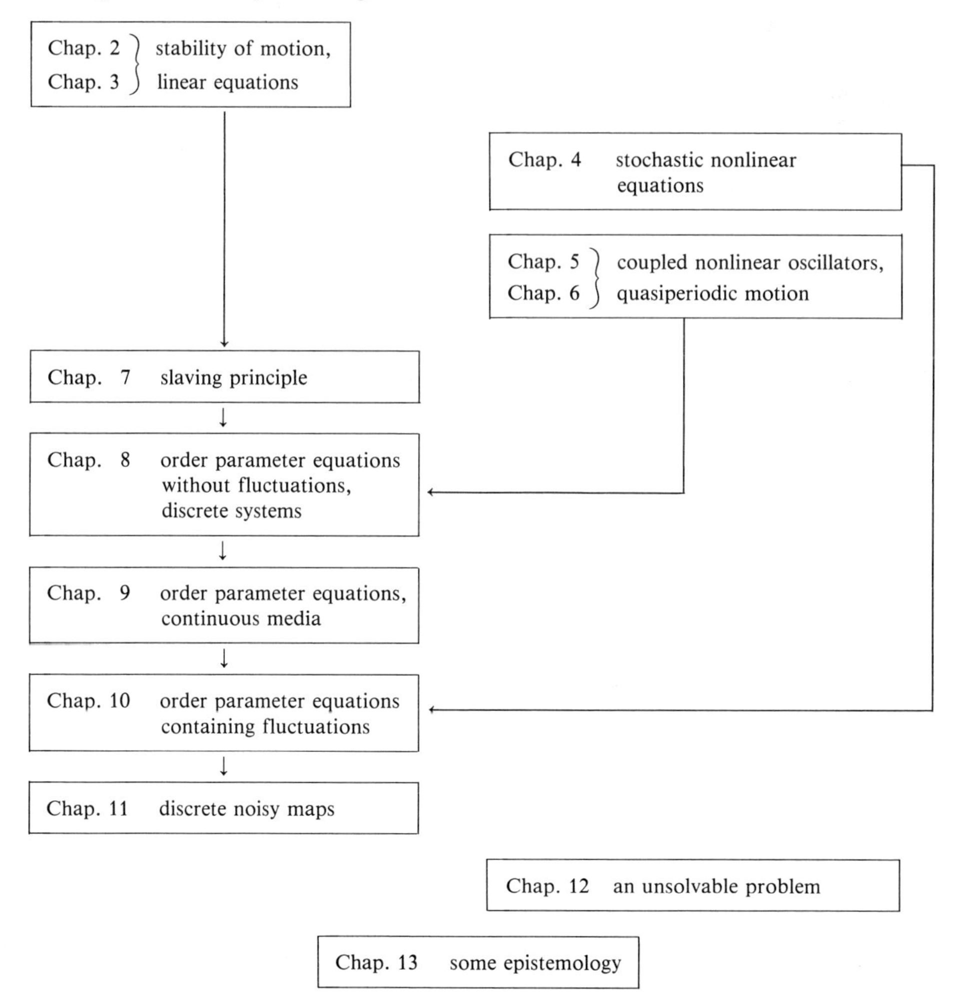

equations) without and with stochastic forces. This chapter, which contains new results, is of crucial importance for the following Chaps. 8 and 9, because it is shown how the number of degrees of freedom can be reduced drastically. Chaps. 8 and 9 then deal with the main problem: the establishment and solution of the *order parameter equations*. Chap. 8 deals with discrete systems while Chap. 9 is devoted to continuously extended media. Again, a number of results are presented here for the first time. Chap. 10 treats the impact of *fluctuations* on systems at instability points. The subsequent chapter is devoted to some general approaches to cope with *discrete noisy maps* and is still in the main line of this book.

Chap. 12 is out of that line – it illustrates that even seemingly simple questions put in dynamic systems theory cannot be answered in principle. Finally Chap. 13 resumes the theme of the introductory chapter – what has synergetics to do with other disciplines. In a way, this short chapter is a little excursion into epistemology. In conclusion it is worth mentioning that Chaps. 2–6 are of use also in problems when systems are away from their instability points.

2. Linear Ordinary Differential Equations

In this chapter we shall present a systematic study of the solutions of linear ordinary differential equations. Such equations continue to play an important role in many branches of the natural sciences and other fields, e.g., economics, so that they may be treated here in their own right. On the other hand, we should not forget that our main objective is to study nonlinear equations, and in the construction of their solutions the solutions of linear equations come in at several instances.

This chapter is organized as follows. Section 2.1 will be devoted to detailed discussions of the solutions of different kinds of homogeneous differential equations. These equations will be distinguished by the time dependence of their coefficients, i.e., constant, periodic, quasiperiodic, or more arbitrary. In Sect. 2.2 we shall point out how to apply the concept of invariance against group operations to the first two types of equations. Section 2.3 deals with inhomogeneous differential equations. Some general theorems from algebra and from the theory of linear ordinary differential equations (coupled systems) will be presented in Sect. 2.4. In Sect. 2.5 dual solution spaces are introduced. The general form of solution for the cases of constant and periodic coefficient matrices will be treated in Sects. 2.6 and 2.7, 8, respectively. Section 2.8 and the beginning of Sect. 2.7 will deal with aspects of group theory, Sect. 2.8 also including representation theory. In Sect. 2.9, a perturbation theory will be developed to get explicit solutions for the case of periodic coefficient matrices.

2.1 Examples of Linear Differential Equations: The Case of a Single Variable

We consider a variable q which depends on a variable t, i.e., $q(t)$, and assume, that q is continuously differentiable. We shall interpret t as time, though in certain applications other interpretations are also possible, for instance as a space coordinate. If not otherwise stated it is taken that $-\infty < t < +\infty$. We consider various typical cases of homogeneous first-order differential equations.

2.1.1 Linear Differential Equation with Constant Coefficient

We consider

$$\dot{q} = aq\,, \tag{2.1.1}$$

where a is a constant. The solution of this differential equation reads

$$q(t) = C\mathrm{e}^{\lambda t}, \tag{2.1.2}$$

where $\lambda = a$ as can be immediately checked by inserting (2.1.2) into (2.1.1). The constant C can be fixed by means of an initial condition, e. g., by the requirement that for $t = 0$

$$q(0) = q_0, \tag{2.1.3}$$

where q_0 is prescribed. Thus

$$C = q_0 \equiv q(0). \tag{2.1.4}$$

Then (2.1.2) can be written as

$$q(t) = q(0)\,\mathrm{e}^{\lambda t}, \quad (\lambda = a). \tag{2.1.5}$$

As shown in Sects. 1.14, 16, linear differential equations are important to determine the stability of solutions of nonlinear equations. Therefore, we shall discuss here and in the following the time dependence of the solutions (2.1.5) for large times t. Obviously for $t > 0$ the asymptotic behavior of (2.1.5) is governed by the sign of $\mathrm{Re}\{\lambda\}$. For $\mathrm{Re}\{\lambda\} > 0$, $|q|$ grows exponentially, for $\mathrm{Re}\{\lambda\} = 0$, $q(t)$ is a constant, and for $\mathrm{Re}\{\lambda\} < 0$, $|q|$ is exponentially damped. Here λ itself is called a *characteristic exponent*.

2.1.2 Linear Differential Equation with Periodic Coefficient

As a further example we consider

$$\dot{q} = a(t)q, \tag{2.1.6}$$

where $a(t)$ is assumed continuous. The solution of (2.1.6) reads

$$q = q(0)\exp\left[\int_0^t a(\tau)\,d\tau\right]. \tag{2.1.7}$$

We now have to distinguish between different kinds of time dependence of $a(t)$. If $a(t)$ is periodic and, for instance, continuously differentiable, we may expand it into a Fourier series of the form

$$a(t) = c_0 + \sum_{\substack{n=-\infty \\ n \neq 0}}^{\infty} c_n \mathrm{e}^{in\omega t}. \tag{2.1.8}$$

To study the asymptotic behavior of (2.1.7), we insert (2.1.8) into the integral occurring in (2.1.7) and obtain

$$\int_0^t a(\tau)d\tau = tc_0 + \sum_{n \neq 0} \frac{c_n}{in\omega}(e^{in\omega t} - 1)\,. \tag{2.1.9}$$

With $n \neq 0$, the sum in (2.1.9) converges at least as well as the sum in (2.1.8). Therefore the sum in (2.1.9) again represents a periodic function. Thus the asymptotic behavior of (2.1.7) is governed by the coefficient c_0 in (2.1.8). Depending on whether its real part is positive, zero, or negative, we find exponential growth, a neutral solution or exponential damping, respectively.

Combining our results (2.1.7 – 9), we may write the solution of (2.1.6) in the form

$$q(t) = e^{\lambda t} u(t) \tag{2.1.10}$$

with the characteristic exponent $\lambda = c_0$ and

$$u(t) = q(0) \exp\left[\sum_{n \neq 0} \frac{c_n}{in\omega}(e^{in\omega t} - 1)\right]. \tag{2.1.11}$$

Because the exponential function of a periodic function is again periodic, $u(t)$ is a periodic function. We therefore have found the result that the solutions of differential equations with periodic coefficients $a(t)$ have the form (2.1.10), where $u(t)$ is again a periodic function. Since $u(t)$ is bounded, the asymptotic behavior of $|q(t)|$ is determined by the exponent $\mathrm{Re}\{\lambda t\}$, as stated above.

2.1.3 Linear Differential Equation with Quasiperiodic Coefficient

The reader may be inclined to believe that our list of examples is about to become boring but the next more complicated example will confront us with an intrinsic difficulty. Assume that in the differential equation

$$\dot{q} = a(t)q \tag{2.1.12}$$

$a(t)$ is quasiperiodic, i.e., we assume that $a(t)$ can be expanded into a multiple Fourier series

$$a(t) = c_0 + \sum_{m \neq 0} c_m \exp(i\boldsymbol{m} \cdot \boldsymbol{\omega} t)\,, \tag{2.1.13}$$

where $\boldsymbol{m}$ is an n-dimensional vector whose components are integers. Further, $\boldsymbol{\omega}$ is a vector of the same dimension with components $\omega_1, \ldots, \omega_n$, so that

$$\boldsymbol{m} \cdot \boldsymbol{\omega} = (\boldsymbol{m} \cdot \boldsymbol{\omega}) = m_1\omega_1 + m_2\omega_2 + \ldots + m_n\omega_n\,. \tag{2.1.14}$$

We shall assume that (2.1.14) vanishes only if $|\boldsymbol{m}| = 0$, because otherwise we can express one or several ω's by the remaining ω's and the actual number of "independent" ω's will be smaller [may be even one, so that (2.1.13) is a periodic function]. Or, in other words, we exclude ω_j's which are rational with respect to each other. The formal solution of (2.1.12) again has the form (2.1.7); it is now appropriate to evaluate the integral in the exponent of (2.1.7). If the series in (2.1.13) converges absolutely we may integrate term by term and thus obtain

$$\int_0^t a(\tau)d\tau = c_0 t + \sum_{m \neq 0} \frac{c_m}{\mathrm{i}\boldsymbol{m}\cdot\boldsymbol{\omega}}[\exp(\mathrm{i}\boldsymbol{m}\cdot\boldsymbol{\omega}t)-1] \,. \tag{2.1.15}$$

In order that the sum in (2.1.15) has the same form as (2.1.13), i.e., that it is a quasiperiodic function, it ought to be possible to split the sum in (2.1.15) into

$$\sum_{m \neq 0} \frac{c_m}{\mathrm{i}\boldsymbol{m}\cdot\boldsymbol{\omega}} \exp(\mathrm{i}\boldsymbol{m}\cdot\boldsymbol{\omega}t) + \text{const} \,. \tag{2.1.16}$$

However, the time-independent terms of (2.1.15) which are formally equal to

$$\text{const} = -\sum_{m \neq 0} \frac{c_m}{\mathrm{i}\boldsymbol{m}\cdot\boldsymbol{\omega}} \tag{2.1.17}$$

and similarly the first sum in (2.1.16) need not converge. Why is this so? The reason lies in the fact that because the m's can take negative values there may be combinations of m's such that

$$\boldsymbol{m}\cdot\boldsymbol{\omega} \to 0 \tag{2.1.18}$$

for

$$|\boldsymbol{m}| \to \infty \,. \tag{2.1.19}$$

[Condition (2.1.18) may be fulfilled even for finite m's if the ratios of the ω's are rational numbers, in which case $\boldsymbol{m}\cdot\boldsymbol{\omega} = 0$ for some $\boldsymbol{m} = \boldsymbol{m}_0$. But this case has been excluded above by our assumptions on the ω_j's.] One might think that one can avoid (2.1.18) if the ω's are irrational with respect to each other. However, even in such a case it can be shown mathematically that (2.1.18, 19) can be fulfilled. Because $\boldsymbol{m}\cdot\boldsymbol{\omega}$ occurs in the denominator, the series (2.1.18) need not converge, even if $\sum_m |c_m|$ converges. Thus the question arises under which conditions the series (2.1.17) still converges. Because c_m and $\boldsymbol{m}\cdot\boldsymbol{\omega}$ occur jointly, this condition concerns both c_m and the ω_j's. Loosely speaking, we are looking for such c_m which converge sufficiently quickly to zero for $|\boldsymbol{m}| \to \infty$ and such ω_j's for which $|\boldsymbol{m}\cdot\boldsymbol{\omega}|$ goes sufficiently slowly to zero for $|\boldsymbol{m}| \to \infty$ so that (2.1.17) converges. Both from the mathematical point of view and that of practical applications, the condition on the ω_j's is more interesting so that we start with this condition.

Loosely speaking, we are looking for such ω_j's which are "sufficiently irrational" with respect to each other. This condition can be cast into various mathematical forms. A form often used is

$$|(\boldsymbol{m} \cdot \omega)| \geqslant K \|\boldsymbol{m}\|^{-(n+1)}, \quad \text{where} \tag{2.1.20}$$

$$\|\boldsymbol{m}\| = |m_1| + |m_2| + \ldots + |m_n|. \tag{2.1.21}$$

Here K is a constant. Though for $\|\boldsymbol{m}\| \to \infty$, (2.1.20) again goes to 0, it may tend to 0 sufficiently slowly. We shall call (2.1.20, 21) the *Kolmogorov-Arnold-Moser condition* or, in short, the KAM condition. When a real system is given, the question arises whether the frequencies occurring meet the KAM condition (2.1.20, 21). While in mathematics this seems to be a reasonable question, it is problematic to decide upon this question in practice. Furthermore, since systems are subject to fluctuations, it is even highly doubtful whether a system will retain its frequencies such that a KAM condition is fulfilled for all times. Rather, it is meaningful to ask how probable it is that the ω's fulfill that condition. This is answered by the following mathematical theorem (which we do not prove here): In the space $\omega = (\omega_1, \omega_2, \ldots, \omega_n)$ the relative measure of the set of those ω's which do not satisfy the condition (2.1.20) tends to 0 together with K. Thus for sufficiently small K most of the ω's satisfy (2.1.20).

We now discuss the second problem, namely the convergence rate of the coefficients c_m in (2.1.13). Because simple Fourier series can be handled more easily than multiple Fourier series of the form (2.1.13), we try to relate (2.1.13) with simple Fourier series. This is achieved by the following trick. We introduce auxiliary variables $\Phi_1, \ldots, \Phi_n$ and replace $a(t)$ in (2.1.13) by

$$a(t, \Phi_1, \Phi_2, \ldots, \Phi_n) = c_0 + \sum_{\boldsymbol{m} \neq 0} c_m \exp(\mathrm{i} m_1 \omega_1 \Phi_1 + \ldots + \mathrm{i} m_n \omega_n \Phi_n + \mathrm{i} \boldsymbol{m} \cdot \omega t), \tag{2.1.22}$$

i.e., by a function of $\Phi_1, \ldots, \Phi_n$ and t. We can take care of t by introducing $\tilde{\Phi}_j = \Phi_j + t$. Now we may keep all $\tilde{\Phi}$'s but one, say $\tilde{\Phi}_j$, fixed. This allows us to apply theorems on simple Fourier series to (2.1.22). We shall use the following (where we make $\omega_j \tilde{\Phi}_j = x$). Let

$$f(x) = \sum_{m=-\infty}^{\infty} \alpha_m \mathrm{e}^{\mathrm{i} m x}, \tag{2.1.23}$$

whose derivatives up to order $(h-1)$ are continuous, and whose h'th derivative is piecewise continuous, then

$$|\alpha_m| \leqslant \frac{C}{|m|^h}. \tag{2.1.24}$$

Consider (2.1.22), and assume that its derivatives fulfill the just-mentioned conditions for all $\tilde{\Phi}_j$'s. Then

$$|c_m| \leqslant \frac{C}{|m_1|^h \ldots |m_n|^h}. \tag{2.1.25}$$

After these preparations we are able to demonstrate how the KAM condition works. To this end we study the convergence of (2.1.17). We start with (Σ' denotes $\boldsymbol{m} \neq 0$)

$$\left| \sum_m{}' \frac{c_m}{\mathrm{i}\boldsymbol{m} \cdot \boldsymbol{\omega}} \right| \leqslant \sum_m{}' \frac{|c_m|}{|\boldsymbol{m} \cdot \boldsymbol{\omega}|}, \tag{2.1.26}$$

or, after use of (2.1.21, 25), we obtain for (2.1.17)

$$\left| \sum_m{}' \frac{c_m}{\mathrm{i}\boldsymbol{m} \cdot \boldsymbol{\omega}} \right| \leqslant \frac{C}{K} \sum_m{}' \frac{(|m_1| + \ldots + |m_n|)^{n+1}}{|m_1|^h \ldots |m_n|^h}. \tag{2.1.27}$$

In order to obtain a sufficient condition for the convergence of (2.1.27), we replace the $|m_j|$'s in the numerator each time by their biggest value $m_{\max}$, so that

$$\left| \sum_m{}' \frac{c_m}{\mathrm{i}\boldsymbol{m} \cdot \boldsymbol{\omega}} \right| \leqslant \frac{C}{K} n^{n+1} \sum_m{}' \frac{m_{\max}^{n+1}}{|m_1|^h \ldots |m_n|^h}. \tag{2.1.28}$$

According to elementary criteria on convergence, (2.1.28) converges, provided that $h \geqslant n+3$ (h being an integer). This exercise illustrates how the KAM condition and the convergence of (2.1.17) are linked.

Now let us assume that the c_m's converge so rapidly that if the KAM condition (2.1.20) is fulfilled the first sum in (2.1.16) converges absolutely. Then the solution of (2.1.12) with the quasiperiodic coefficients (2.1.13) can be written in the form

$$q(t) = \mathrm{e}^{\lambda t} u(t), \tag{2.1.29}$$

where the characteristic exponent is given by

$$\lambda = c_0, \tag{2.1.30}$$

and $u(t)$ is again a quasiperiodic function

$$u(t) = \tilde{q}(0) \exp\left[\sum_{\boldsymbol{m} \neq 0} \frac{c_m}{\mathrm{i}\boldsymbol{m} \cdot \boldsymbol{\omega}} \exp(\mathrm{i}\boldsymbol{m} \cdot \boldsymbol{\omega} t) \right], \tag{2.1.31}$$

where

$$\tilde{q}(0) = q(0) \exp\left(- \sum_{\boldsymbol{m} \neq 0} \frac{c_m}{\mathrm{i}\boldsymbol{m} \cdot \boldsymbol{\omega}} \right). \tag{2.1.32}$$

Because the series in (2.1.31) converges absolutely, $u(t)$ is bounded. Therefore, the asymptotic behavior of (2.1.29) is determined by the exponential function $\exp(\lambda t)$. Let us now finally turn to the general case.

2.1.4 Linear Differential Equation with Real Bounded Coefficient

In the differential equation (2.1.6), $a(t)$ is an arbitrary function of time but continuous and bounded for

$$0 \leqslant t < \infty, \tag{2.1.33}$$

so that

$$|a(t)| \leqslant B. \tag{2.1.34}$$

We assume $a(t)$ real. The general solution reads

$$q(t) = q(0) \exp\left[\int_0^t a(\tau)\, d\tau\right]. \tag{2.1.35}$$

We wish to study how solution (2.1.35) behaves when t goes to infinity. In particular, we shall study whether (2.1.35) grows or decays exponentially. As a first step we form

$$\ln|q(t)| = \ln|q(0)| + \int_0^t a(\tau)\, d\tau, \tag{2.1.36}$$

so that we have to discuss the behavior of the integral in (2.1.36). To this end we first explain the notion of the supremum which is also called the "least upper bound". If A is a set of real numbers $\{a\}$, the supremum of A is the smallest real number b such that $a < b$ for all a. We write "$\sup\{A\}$" to denote the supremum. A similar definition can be given for the infimum or, in other words, for the greatest lower bound ($\inf\{A\}$). If A is an infinite set of real numbers then the symbol "$\limsup\{A\}$" denotes the infimum of all numbers b with the property that only a finite set of numbers in A exceed b. In particular if A is a sequence $\{a_n\}$ then $\limsup\{A\}$ is usually denoted by $\limsup_{n\to\infty}\{a_n\}$.

These definitions are illustrated in Fig. 2.1.1. Starting from (2.1.36) we now form

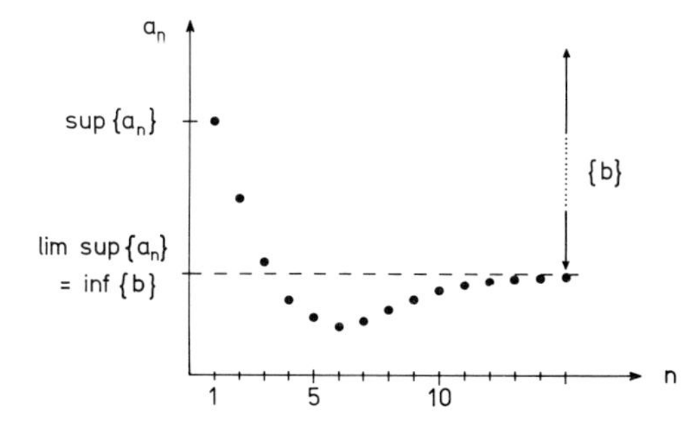

Fig. 2.1.1. Simple behavior of a sequence $\{a_1, a_2, \ldots\}$ is plotted, where $\{b\}$ is the set of all real numbers with the property that only a finite set of numbers $a_1, a_2, \ldots$ exceed b

$$\limsup_{t\to\infty}\left\{\frac{1}{t}\int_0^t a(\tau)\,d\tau\right\}. \tag{2.1.37}$$

Clearly, we get for the first term of the rhs of (2.1.36)

$$\frac{1}{t}\ln|q(0)|\to 0 \quad \text{for} \quad t\to\infty\,. \tag{2.1.38}$$

From (2.1.34) we can construct two bounds, namely

$$\int_0^t a(\tau)\,d\tau \leqslant \int_0^t B\,d\tau = Bt \tag{2.1.39}$$

and

$$\int_0^t a(\tau)\,d\tau \geqslant -Bt\,. \tag{2.1.40}$$

From (2.1.39, 40) it follows that

$$\limsup_{t\to\infty}\left\{\frac{1}{t}\int_0^t a(\tau)\,d\tau\right\} = \lambda \tag{2.1.41}$$

exists with

$$|\lambda| < \infty\,. \tag{2.1.42}$$

Here λ is called *generalized characteristic exponent.* It gives us information about the asymptotic behavior of solution (2.1.35) for large positive times. Its significance can be visualized if we use the fact that the logarithmic function is monotonous and that both (2.1.41, 36) imply that the solution has an upper bound at each time t given by

$$C\,\mathrm{e}^{\lambda t + f(t)}, \quad \text{where} \tag{2.1.43}$$

$$\limsup_{t\to\infty}\left\{\frac{1}{t}f(t)\right\}\to 0\,. \tag{2.1.44}$$

Thus the generalized characteristic exponents have the same significance as the real part of the (characteristic) exponent λ in the simple case of a differential equation with constant coefficients.

2.2 Groups and Invariance

The very simple example of the differential equation (2.1.6) allows us to explain some fundamental ideas about groups and invariance. Let us start with the differential equation for the variable $q(t)$

$$\dot{q}(t) = a(t)\,q(t)\,. \tag{2.2.1}$$

The properties of $a(t)$ (i.e., being constant, periodic, or quasiperiodic) can be characterized by an invariance property also. Let us assume that $a(t)$ remains unaltered if we replace its argument

$$t \to t + t_0\,, \qquad \text{i.e.,} \tag{2.2.2}$$

$$a(t+t_0) = a(t)\,. \tag{2.2.3}$$

If t_0 can be chosen arbitrarily, then (2.2.3) implies that $a(t)$ must be time independent. If (2.2.3) is fulfilled for a certain t_0 (and for integer multiples of it), $a(t)$ is periodic. As shown in Chap. 3 our present considerations can also be extended to quasiperiodic functions $a(t)$, but this requires more mathematics. Now let us subject (2.2.1) to the transformation (2.2.2), which yields

$$\dot{q}(t+t_0) = a(t)\,q(t+t_0)\,, \tag{2.2.4}$$

so that (2.2.4) has exactly the same appearance as (2.2.1). It is the same differential equation for a variable which we may denote by

$$q_{\text{new}}(t) = q_{\text{old}}(t+t_0)\,. \tag{2.2.5}$$

However, it is known from the theory of linear differential equations that the solution of (2.2.1) is unique except for a constant factor. Denoting this constant factor by α, the relation

$$\left.\begin{aligned} &q_{\text{new}}(t) = \alpha q_{\text{old}}(t)\,, \qquad \text{i.e.,}\\ &q_{\text{old}}(t+t_0) = \alpha q_{\text{old}}(t) \end{aligned}\right\} \tag{2.2.6}$$

must hold. We want to show that by means of relation (2.2.6) we may construct the solution of (2.2.1) directly. To this end let us introduce the translation operator T, defined as follows. If T is applied to a function f of time, we replace t by $t + t_0$ in the argument of f

$$Tf(t) = f(t+t_0)\,. \tag{2.2.7}$$

Due to the invariance property (2.2.3) we readily find

$$\begin{aligned} T a(t) q(t) &= a(t+t_0) q(t+t_0) \\ &= a(t) T q(t) . \end{aligned} \tag{2.2.8}$$

Since this relation holds for an arbitrary $q(t)$, (2.2.8) can be written in the form

$$T a(t) = a(t) T \tag{2.2.9}$$

or, in other words, T commutes with $a(t)$. Equation (2.2.6) can now be expressed in the form

$$Tq = \alpha q . \tag{2.2.10}$$

This relation is described by saying that q is an eigenfunction to the (translation) operator T with eigenvalue α and it is a consequence of the invariance of (2.2.1) against T. The operator T allows us to explain what a "group" is.

So far we have attached an operator T to the translation $t \rightarrow t + t_0$ so that $T a(t) = a(t+t_0)$. When we perform the translation n times, we have to replace t by $t + n t_0$ so that $a(t) \rightarrow a(t+n t_0)$. This replacement can be achieved by n times applying the operator T on $a(t)$. The operation can be written as T^n, so that $T^n a(t) = a(t+n t_0)$. Of course, we may also make these substitutions in the reverse direction, $t \rightarrow t - t_0$. This can be expressed by the "inverse" operator T^{-1} which has the property $T^{-1} a(t) = a(t-t_0)$. Finally we note that we may leave t unchanged. This can be expressed in a trivial way by T^0: $T^0 a(t) = a(t)$. We shall call $T^0 \equiv E$ the unity operator. Summing up the various operations we find the following scheme ($n > 0$, integer)

Displacement	Operator	Effect
$t \rightarrow t + t_0$	T	$T a(t) = a(t+t_0)$
$t \rightarrow t + n t_0$	T^n	$T^n a(t) = a(t+n t_0)$
$t \rightarrow t - t_0$	T^{-1}	$T^{-1} a(t) = a(t-t_0)$
$t \rightarrow t - n t_0$	T^{-n}	$T^{-n} a(t) = a(t-n t_0)$
$t \rightarrow t$	$T^0 = E$	$T^0 a(t) = a(t)$

The operations T^n, $n \gtreqless 0$, form a (multiplicative) group, because they fulfill the following axioms.

Individually, T^n, $n \gtreqless 0$ are elements of the group.

1) If we *multiply* (or, more generally speaking, combine) *two elements* with one another, we obtain a *new element* of the group. Indeed $T^n \cdot T^m = T^{n+m}$ is a new element. This relation holds because a displacement by $n t_0$ and one by $m t_0$ yields a new displacement by $(n+m) t_0$.
2) A (right-sided) *unity operator* E exists so that $T^n \cdot E = T^n$.

3) There exists for each element (T^n) a (left-sided) inverse T^{-n} so that $T^{-n}T^n = E$.
4) The associative law holds: $(T^n T^m)T^l = T^n(T^m T^l)$.

The relations (1) – (4) are quite obvious because we can all of them verify by means of the properties of the translations $t \to t + nt_0$. But once we can verify that operators form a group, a number of most valuable theorems may be applied and we shall come across some of them later. Because it does not matter whether we first perform n translations $t \to t + nt_0$ and then m, or vice versa, we immediately verify $T^n T^m = T^m T^n$. In other words, any two elements of the group commute with one another. A group composed of such elements only is called an Abelian group. Because all elements of the present group are generated as powers (positive and negative) of T, we call T the generator of the group.

Construction of the Solution of (2.2.10)

Now we want to show how to construct the solution from the relation (2.2.10). Applying T to (2.2.10) n times and using (2.2.10) repeatedly we readily find

$$T^n q(t) = \alpha^n q(t) . \tag{2.2.11}$$

Since the application of T^n means replacing t by $t + t_0$ n times we obtain instead of (2.2.11)

$$q(t+nt_0) = \alpha^n q(t) . \tag{2.2.12}$$

This is an equation for $q(t)$. To solve it we put

$$q(t) = \mathrm{e}^{\lambda t} u(t) , \tag{2.2.13}$$

where u is still arbitrary. Inserting (2.2.13) into (2.2.12) we obtain

$$\mathrm{e}^{n\lambda t_0} u(t+nt_0) = \alpha^n u(t) . \tag{2.2.14}$$

Putting $\alpha = \exp(\lambda t_0)$ we arrive at the following equation

$$u(t+nt_0) = u(t) . \tag{2.2.15}$$

According to this relation, $u(t)$ is a periodic function of t. Thus we have rederived the general form of the solution of (2.1.6) with periodic coefficients without any need to solve that differential equation directly. If $a(t)$ is invariant against T for any t_0 we can convince ourselves immediately that $u(t)$ must be a constant. The present example teaches us some important concepts which we shall use later in more complicated cases. The first is the concept of invariance. It is expressed by relation (2.2.9), which stems from the fact that $a(t)$ is unchanged under the transformation T. It was possible to derive the general form of the solution (2.2.13) by means of two facts.

i) The solution of (2.2.1) is unique except for a constant factor.
ii) The differential equation is invariant against a certain operation expressed by the operator T.

We then showed that q is an eigenfunction to the operator T. The operators T^n form a group. Furthermore, we have seen by means of (2.2.11) that the number α and its powers can be attached to the operators of the left-hand side of (2.2.11). In particular we have the following correspondence

$$\begin{aligned} &T \to \alpha\,, \quad T^n \to \alpha^n \\ &T^{-1} \to \alpha^{-1}, \quad T^{-n} \to \alpha^{-n} \\ &T^0 \equiv E \to \alpha^0 = 1\,. \end{aligned} \tag{2.2.16}$$

The relations (2.2.16) are probably the simplest example of a group representation. The basis of that representation is formed by q and a certain number, α^n, is attached to each operation T^n so that the numbers α^n obey the same rules as the operators T^n as described by the above axioms (1) – (4), i.e.,

$$\begin{aligned} &T^n T^m = T^{n+m} && \to \alpha^n \alpha^m = \alpha^{n+m} \\ &T^n E = T^n && \to \alpha^n \cdot 1 = \alpha^n \\ &T^{-n} T^n = E && \to \alpha^{-n} \alpha^n = 1 \\ &(T^n T^m) T^l = T^n (T^m T^l) && \to (\alpha^n \alpha^m) \alpha^l = \alpha^n (\alpha^m \alpha^l)\,. \end{aligned} \tag{2.2.17}$$

We shall see later that these concepts can be considerably generalized and will have important applications.

2.3 Driven Systems

We now deal with an inhomogeneous first-order differential equation

$$\dot{q}(t) = a(t)\, q(t) + b(t)\,. \tag{2.3.1}$$

Again it is our aim to get acquainted with general properties of such an equation. First let us treat the case in which $a \equiv 0$ so that (2.3.1) reduces to

$$\dot{q} = b \tag{2.3.2}$$

with the solution

$$q(t) = \int_{t_0}^{t} b(\tau)\, d\tau + c\,, \tag{2.3.3}$$

where c is a constant which can be determined by the initial condition $q(t_0) = q_i$ so that

$$c = q(t_0)\,. \tag{2.3.4}$$

To the integral in (2.3.3) we may apply the same considerations applied in Sects. 2.1.1 – 3 to the integral as it occurs, for instance, in (2.1.7). If $b(t)$ is constant or periodic,

$$q(t) = b_0 t + v(t)\,, \tag{2.3.5}$$

where $v(t)$ is a constant or periodic function, respectively. If $b(t)$ is quasiperiodic, we may prove that $v(t)$ is quasiperiodic also if a KAM condition is fulfilled and if the Fourier coefficients of $v(t)$ (Sect. 2.1.3) converge sufficiently rapidly.

Let us turn to the general case (2.3.1) where $a(t)$ does not vanish everywhere. To solve (2.3.1) we make the hypothesis

$$q(t) = q_0(t)\,c(t)\,, \tag{2.3.6}$$

where $q_0(t)$ is assumed to be a solution of the homogeneous equation

$$\dot{q}_0(t) = a(t)\,q_0(t)\,, \tag{2.3.7}$$

and $c(t)$ is a still unknown function, which would be a constant if we were to solve the homogeneous equation (2.3.7) only.

Inserting (2.3.6) into (2.3.1) we readily obtain

$$\dot{q}_0 c + q_0 \dot{c} = a q_0 c + b \tag{2.3.8}$$

or, using (2.3.7)

$$\dot{c} = q_0^{-1}(t)\,b(t)\,. \tag{2.3.9}$$

Equation (2.3.9) can be immediately integrated to give

$$c(t) = \int_{t_0}^{t} q_0^{-1}(\tau)\,b(\tau)\,d\tau + \alpha\,. \tag{2.3.10}$$

Again, it is our main goal to study what types of solutions from (2.3.1) result if certain types of time dependence of $a(t)$ and $b(t)$ are given. Inserting (2.3.10) into (2.3.6) we obtain the general form of the solution of (2.3.1), namely

$$q(t) = q_0(t) \int_{t_0}^{t} q_0^{-1}(\tau)\,b(\tau)\,d\tau + \alpha q_0(t)\,. \tag{2.3.11}$$

Let us first consider an explicit example, namely

$$\left.\begin{array}{l} a \equiv \lambda = \text{const} \neq 0 \\ b = \text{const} \end{array}\right\}. \tag{2.3.12}$$

In this case we obtain

$$q(t) = \int_{t_0}^{t} e^{\lambda(t-\tau)} b \, d\tau + \alpha e^{\lambda t}, \tag{2.3.13}$$

or after integration

$$q(t) = -\frac{b}{\lambda} + \beta e^{\lambda t} \tag{2.3.14}$$

with

$$\beta = \alpha + \frac{b}{\lambda} e^{-\lambda t_0}. \tag{2.3.15}$$

For $\lambda > 0$, the long-time behavior ($t \to \infty$) of (2.3.14) is dominated by the exponential function and the solution diverges provided β does not vanish, which could occur with a specific initial condition. For $\lambda < 0$ the long-time behavior is determined by the first term in (2.3.14), i.e., $q(t)$ tends to the constant $-b/\lambda$. We can find this solution more directly by putting $\dot{q} = 0$ in (2.3.1) which yields

$$q = -b/\lambda. \tag{2.3.16}$$

The fully time-dependent solution (2.3.14) is called a transient towards the stationary solution (2.3.16). The case $\lambda = 0$ brings us back to the solution (2.3.5) discussed above.

Quite generally we may state that the solution of the inhomogeneous equation (2.3.1) is made up of a particular solution of the inhomogeneous equation, in our case (2.3.16), and a general solution of the homogeneous equation, in our case $\beta \exp(\lambda t)$, where β is a constant which can be chosen arbitrarily. It can be fixed, however, for instance if the value of $q(t)$ at an initial time is given.

Let us now consider the general case in which $a(t)$ and $b(t)$ are constant, periodic or quasiperiodic. In order to obtain results of practical use we assume that in the quasiperiodic case a KAM condition is fulfilled and the Fourier series of $a(t)$ converges sufficiently rapidly, so that the solution of the homogeneous equation (2.3.7) has the form

$$q_0(t) = e^{\lambda t} u(t), \tag{2.3.17}$$

with $u(t)$ constant, periodic, or quasiperiodic. From the explicit form of $u(t)$ and from the fact that the quasiperiodic function in the exponential function of (2.3.17) is bounded, it follows that $|u|$ is bounded from below, so that its inverse exists.

We first consider a particular solution of the inhomogeneous equation (2.3.1) which by use of (2.3.11) and (2.3.17) can be written in the form

$$q(t) = e^{\lambda t} u(t) \int_{t_0}^{t} e^{-\lambda \tau} u^{-1}(\tau) b(\tau) \, d\tau. \tag{2.3.18}$$

In the following we treat the case

$$\mathrm{Re}\{\lambda\} < 0 \tag{2.3.19}$$

which allows us to take

$$t_0 \to -\infty \,. \tag{2.3.20}$$

Making further a change of the integration variable

$$\tau = \tau' + t \tag{2.3.21}$$

we transform (2.3.18) into

$$q(t) = u(t) \int_{-\infty}^{0} \mathrm{e}^{-\lambda\tau'} u^{-1}(\tau' + t)\, b(\tau' + t)\, d\tau' \,. \tag{2.3.22}$$

We are now interested in what kind of temporal behavior the particular solution of the inhomogeneous equation (2.3.22) represents. Let $u(t)$ be quasiperiodic with basic frequencies $\omega_1, \ldots, \omega_n$. Since u^{-1} exists (as stated above) and is quasiperiodic, we may write it in the form

$$u^{-1}(\tau) = \sum_{m} d_m \mathrm{e}^{\mathrm{i}\boldsymbol{m}\cdot\boldsymbol{\omega}\tau} \,. \tag{2.3.23}$$

Similarly, we expand $b(\tau)$ into a multiple Fourier series

$$b(\tau) = \sum_{n} f_n \mathrm{e}^{\mathrm{i}\boldsymbol{n}\cdot\boldsymbol{\Omega}\tau} \,. \tag{2.3.24}$$

Inserting (2.3.23, 24) into (2.3.22) we readily obtain

$$q(\tau) = \sum_{m,n} d_m f_n \underbrace{[-\lambda + \mathrm{i}\,[(\boldsymbol{m}\cdot\boldsymbol{\omega} + \boldsymbol{n}\cdot\boldsymbol{\Omega})]}_{D}{}^{-1} \exp\,[(\mathrm{i}\boldsymbol{m}\cdot\boldsymbol{\omega} + \boldsymbol{n}\cdot\boldsymbol{\Omega})\,t] \,. \tag{2.3.25}$$

Because evidently

$$|D| \geqslant |\mathrm{Re}\{\lambda\}| > 0 \,, \tag{2.3.26}$$

(2.3.25) represents a converging multiple Fourier series provided (2.3.23, 24) were absolutely converging. (In this notion "quasiperiodic" implies that the series converges absolutely.) In such a case (2.3.25) converges absolutely also. From this and the general form of (2.3.25) it follows that $q(t)$ is a quasiperiodic function with basic frequencies $\omega_1, \ldots, \omega_M, \Omega_1, \ldots, \Omega_N$. Since for $\mathrm{Re}\{\lambda\} < 0$ the solution of the homogeneous equation vanishes when $t \to \infty$, the temporal behavior of the solution $q(t)$ of the inhomogeneous equation (2.3.1) for large t is determined by the particular solution we have just discussed, i.e.,

$$q(t) = \text{quasiperiodic with basic frequencies } \omega_1, \ldots, \omega_M, \Omega_1, \ldots, \Omega_N . \tag{2.3.27}$$

2.4 General Theorems on Algebraic and Differential Equations

2.4.1 The Form of the Equations

In this section some general theorems of algebra and the theory of linear differential equations are recapitulated. The purpose will be the following. In the subsequent section we shall study sets of coupled ordinary differential equations which are of the general form

$$\begin{aligned}
\dot{q}_1(t) &= a_{11}q_1(t) + a_{12}q_2(t) + \ldots + a_{1n}q_n(t) \\
\dot{q}_2(t) &= a_{21}q_1(t) + a_{22}q_2(t) + \ldots + a_{2n}q_n(t) \\
\vdots \quad & \quad \vdots \qquad\qquad \vdots \qquad\qquad\qquad \vdots \\
\dot{q}_n(t) &= a_{n1}q_1(t) + a_{n2}q_2(t) + \ldots + a_{nn}q_n(t) .
\end{aligned} \tag{2.4.1}$$

The unknown variables are the q's. The coefficients a_{ik} are assumed to be given. They may be either constants, or time-dependent functions of various types discussed below. The set of Eqs. (2.4.1) can most concisely be written in matrix notation

$$\dot{q}(t) = L\, q(t) , \tag{2.4.2}$$

where the vector $\boldsymbol{q}$ is defined by

$$q(t) = \begin{pmatrix} q_1 \\ \vdots \\ q_n \end{pmatrix} \tag{2.4.3}$$

and the matrix L is given by

$$L = \begin{pmatrix} a_{11} \ldots a_{1n} \\ \vdots \qquad \vdots \\ a_{n1} \ldots a_{nn} \end{pmatrix} . \tag{2.4.4}$$

2.4.2 Jordan's Normal Form

For the moment we are not concerned with the time dependence of the coefficients of L, i.e., we assume that they are constant or are taken at a fixed value of time t. Let us consider a quadratic matrix L of the form (2.4.4).

The maximal numbers of linear independent rows or equivalently columns of a matrix L are called the rank of that matrix or, symbolically, $\mathrm{Rk}\{L\}$. A quadratic matrix with n rows is called regular if $\mathrm{Rk}\{L\} = n$.

Two quadratic matrices L and $\tilde{L}$ each with n rows are called similar if there exists a regular matrix S with $\tilde{L} = S^{-1}LS$. After this reminder of some simple definitions we now formulate the following important theorem concerning Jordan's normal form.

Each quadratic matrix L whose elements are complex numbers is similar to a normal matrix of the following form

$$\tilde{L} = \begin{pmatrix} \boxed{A_1} & & & 0 \\ & \boxed{A_2} & & \\ & & \ddots & \\ 0 & & & \boxed{A_r} \end{pmatrix} . \tag{2.4.5}$$

In this normal matrix the square submatrices are uniquely determined by L except for their sequence. The submatrices A_ρ possess the form

$$\begin{pmatrix} \lambda_\rho & 1 & & 0 \\ & \lambda_\rho & 1 & \\ & & \ddots & \ddots \\ & 0 & & \vdots & 1 \\ & & & & \lambda_\rho \end{pmatrix} \tag{2.4.6}$$

where λ_ρ is an eigenvalue of L.

This theorem can be stated using the above definitions also as follows. For each quadratic matrix L of complex numbers we may find a regular matrix S such that

$$\tilde{L} = S^{-1}LS \tag{2.4.7}$$

holds and $\tilde{L}$ is of the form (2.4.5). Because S is regular, its determinant does not vanish so that S^{-1} exists.

2.4.3 Some General Theorems on Linear Differential Equations

We consider the differential equation (2.4.2) (also called "differential system") where

$$L(t) = (a_{ij}(t)) \tag{2.4.8}$$

is a $n \times n$ complex-valued matrix, whose elements a_{ij} are continuous in the real variable t on the interval I defined by

$$-\infty < \alpha \leqslant t \leqslant \beta < \infty . \tag{2.4.9}$$

A solution of (2.4.2) is a complex column vector $\boldsymbol{q}(t) \in E^n$ (E^n: n-dimensional complex vector space), differentiable and satisfying the differential system for each $t \in I$. The theory of linear differential equations assures us that for each $t_0 \in I$ and each column vector $\boldsymbol{q}_0 \in E^n$ a unique solution $\boldsymbol{q}(t)$ exists such that $\boldsymbol{q}(t_0) = \boldsymbol{q}_0$.

The solutions of $d\boldsymbol{q}/dt = L(t)\boldsymbol{q}$ form an n-dimensional complex linear vector space (compare Fig. 2.4.1). In other words, n linearly independent solutions of (2.4.2) exist which we may label $\boldsymbol{q}^{(1)}(t), \ldots, \boldsymbol{q}^{(n)}(t)$. Any solution $\boldsymbol{q}(t)$ of (2.4.2) can be written as a linear combination

$$\boldsymbol{q}(t) = \sum_{j=1}^{n} c_j \boldsymbol{q}^{(j)}(t), \tag{2.4.10}$$

where the coefficients c_j are time independent. The $\boldsymbol{q}^{(j)}$'s are not determined uniquely because by a transformation (2.4.10) we may go from one basis $\{\boldsymbol{q}^{(j)}\}$ to another one $\{\tilde{\boldsymbol{q}}^{(j)}\}$.

It is often convenient to lump the individual solution vectors $\boldsymbol{q}$ together to a "solution matrix" Q

$$Q = (\boldsymbol{q}^{(1)}, \boldsymbol{q}^{(2)}, \ldots, \boldsymbol{q}^{(n)}). \tag{2.4.11}$$

Writing the matrix Q in the form

$$Q = (Q_{ij}), \tag{2.4.12}$$

we find by comparison

$$Q_{ij} = q_i^{(j)}. \tag{2.4.13}$$

The solution matrix then obeys the equation

$$\dot{Q}(t) = LQ(t). \tag{2.4.14}$$

Our above statement on the transformations from one basis $\{q^{(j)}\}$ to another can now be given a more precise formulation.

Theorem 2.4.1. Let $Q(t)$ be a solution matrix of $dQ/dt = L(t)Q$ which is nonsingular. The set of all nonsingular matrix solutions is formed by precisely the

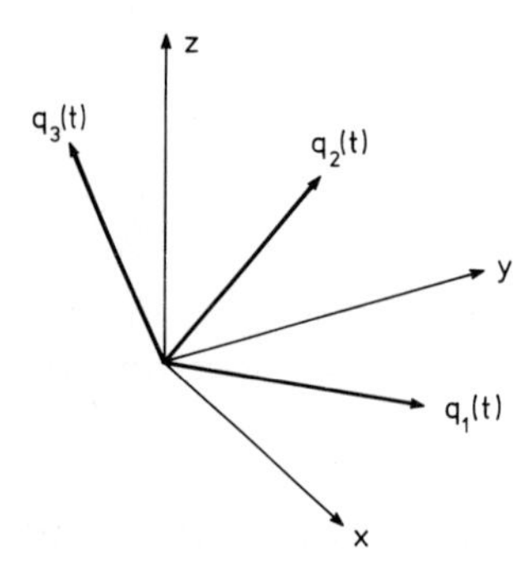

Fig. 2.4.1. Illustration of a three-dimensional real linear vector space

matrices $Q(t)C$ where C is any $n \times n$ constant, nonsingular matrix. For each $t_0 \in I$ and each complex constant matrix Q_0 a unique solution matrix $Q(t)$ exists such that $Q(t_0) = Q_0$. A set of solution vectors $\boldsymbol{q}^{(1)}(t), \ldots, \boldsymbol{q}^{n)}(t)$ of $d\boldsymbol{q}/dt = L(t)\boldsymbol{q}$ form a basis for the solution space if and only if they form the columns of a solution matrix $Q(t)$ of $dQ/dt = L(t)Q$ which corresponds to a *nonsingular* initial matrix Q_0.

The differential system $d\boldsymbol{q}/dt = L(t)\boldsymbol{q}$, where $\alpha < t < \beta = \infty$, is called stable if every solution remains bounded as $t \to \infty$, i.e.,

$$\limsup_{t \to \infty} \{|\boldsymbol{q}(t)|\} < \infty . \tag{2.4.15}$$

The following theorem is sometimes useful. Let $Q(t)$ be a solution matrix of $dQ/dt = L(t)Q$. Then $(d/dt) \det \{Q(t)\} = \operatorname{tr} \{L(t)\} \det \{Q(t)\}$ at each point $t \in I$ where $Q(t)$ is nonsingular, and $\operatorname{tr} \{L(t)\}$ is the sum of the entries on the principal diagonal. A solution matrix $Q(t)$ is nonsingular everywhere on I if it is nonsingular at one point on I.

In Sect. 2.1 we studied the asymptotic behavior of the solution of the differential equation $\dot{q} = a(t)q$. This led us to the concept of characteristic exponents and generalized characteristic exponents λ. A similar statement can be made in the general case of the system (2.4.2).

2.4.4 Generalized Characteristic Exponents and Lyapunov Exponents

Suppose $L(t)$ is continuous and

$$\sup \{|a_{ij}(t)|\} < B \quad \text{for all} \quad \alpha < t < \infty \tag{2.4.16}$$

and some constant B. Then for every solution vector $q^{(i)}(t)$ of $d\boldsymbol{q}/dt = L(t)\boldsymbol{q}$ it holds that

$$\limsup_{t \to \infty} \{t^{-1} \ln |\boldsymbol{q}^{(i)}(t)|\} = \lambda_i , \quad |\lambda_i| < Bn . \tag{2.4.17}$$

The real numbers λ_i which arise in this way are called the generalized characteristic exponents. There are at most n distinct generalized characteristic exponents. The differential system is stable whenever all the λ's are negative.

A special case of the "generalized characteristic exponents" is that of *Lyapunov exponents.*

To this end let us consider *nonlinear* equations of the form

$$\dot{\boldsymbol{q}}(t) = \boldsymbol{N}(\boldsymbol{q}(t)) , \tag{2.4.18}$$

where $\boldsymbol{N}$ is a nonlinear vector function of $\boldsymbol{q}$. Let $\boldsymbol{q}_0$ be a solution of (2.4.18) and consider the linearization of (2.4.18) around $\boldsymbol{q}_0$, i.e., we put $\boldsymbol{q} = \boldsymbol{q}_0 + \delta\boldsymbol{q}$ and retain only linear terms

$$\delta\dot{\boldsymbol{q}} = L(\boldsymbol{q}_0(t))\,\delta\boldsymbol{q} , \tag{2.4.19}$$

where the matrix $L(\boldsymbol{q}_0(t))$ is defined by $L = (L_{kl})$,

$$L_{kl} = \frac{\partial N_k(\boldsymbol{q}_0(t))}{\partial q_{0,l}(t)}. \qquad (2.4.20)$$

In such a case the generalized characteristic exponents of $\delta \boldsymbol{q}$ are called *Lyapunov exponents* of $\boldsymbol{q}_0(t)$.

In conclusion of this section we present the following theorem: At least one Lyapunov exponent vanishes if the trajectory $\boldsymbol{q}(t)$ of an autonomous system remains in a bounded region for $t \to \infty$ and does not contain a fixed point.

In order to prove this theorem we make the following additional assumptions on $N(\boldsymbol{q}(t))$ in (2.4.18). $N(\boldsymbol{q}(t))$ is a continuous function of $\boldsymbol{q}(t)$ and it possesses only finitely many zeros (corresponding to fixed points).

The proof of this theorem is very simple. For brevity we put $\delta \boldsymbol{q} \equiv \boldsymbol{u}$ and drop the index 0 of $\boldsymbol{q}_0$ so that according to (2.4.19) $\boldsymbol{u}$ obeys the equation

$$\dot{\boldsymbol{u}} = L(\boldsymbol{q}(t))\boldsymbol{u}. \qquad (2.4.21)$$

The Lyapunov exponent is then defined by

$$\lambda = \limsup_{t \to \infty} \frac{1}{t} \ln |\boldsymbol{u}(t)|. \qquad (2.4.22)$$

We now construct a solution of (2.4.21) which possesses a vanishing Lyapunov exponent. We can immediately verify by differentiating both sides of (2.4.18) with respect to time that

$$\boldsymbol{u} \equiv \dot{\boldsymbol{q}} \qquad (2.4.23)$$

is a solution to (2.4.21). Because N is continuous and $\boldsymbol{q}(t)$ is bounded for $t \to \infty$ we have

$$|N(\boldsymbol{q}(t))| < D, \quad D > 0. \qquad (2.4.24)$$

Using (2.4.18) and (2.4.23) it follows that

$$|\boldsymbol{u}| < D. \qquad (2.4.25)$$

As a result we obtain

$$\lambda = \limsup_{t \to \infty} \frac{1}{t} \ln |\boldsymbol{u}(t)| \leqslant \limsup_{t \to \infty} \frac{1}{t} \ln D, \qquad (2.4.26)$$

from which we conclude

$$\lambda \leqslant 0. \qquad (2.4.27)$$

Now let us assume

$$\lambda < 0 \,. \tag{2.4.28}$$

According to the definition of lim sup, for any $\varepsilon > 0$ there exists a t_0 so that for any $t > t_0$ the inequality

$$\frac{1}{t} \ln |\dot{q}| < \lambda + \varepsilon \tag{2.4.29}$$

can be satisfied. We shall choose ε so small that

$$\lambda' \equiv \lambda + \varepsilon < 0 \tag{2.4.30}$$

holds. We now choose a function

$$v = v_0 \mathrm{e}^{-|\lambda'|t} \tag{2.4.31}$$

which majorizes $\dot{q}$, so that

$$|\dot{q}| \leqslant |v_0| \mathrm{e}^{-|\lambda'|t} \quad \text{for all} \quad t > t_0 \,. \tag{2.4.32}$$

Therefore

$$|\dot{q}| \to 0 \quad \text{for} \quad t \to \infty \tag{2.4.33}$$

which according to (2.4.18) implies

$$N(q) \to 0 \,. \tag{2.4.34}$$

Because N is continuous, vanishes only at a finite number of points q and approaches $0, q(t)$ must approach one of the singular points

$$q_j = \text{const.}\,, \tag{2.4.35}$$

i.e., the trajectory terminates at a fixed point. Therefore, if the trajectory $q(t)$ does not terminate at a fixed point, the only remaining possibility for the Lyapunov exponent under study is $\lambda = 0$, so that our theorem is proved.

2.5 Forward and Backward Equations: Dual Solution Spaces

For later purposes we slightly change the notation, writing $w^{(k)}$ instead of $q^{(j)}$. Equations (2.4.14, 11) read

$$\dot{w}^{(k)}(t) = L(t)\, w^{(k)}(t) \,, \qquad \text{where} \tag{2.5.1}$$

$$w^{(k)} = \begin{pmatrix} w_1^{(k)} \\ \vdots \\ w_n^{(k)} \end{pmatrix} . \tag{2.5.2}$$

We distinguish the different solution vectors by an upper index k and assume that these solution vectors form a complete set, i.e., that they span the solution space. The matrix L

$$L = (L_{jj'}) \tag{2.5.3}$$

may have any time dependence. Of course, we may represent (2.5.1) in the somewhat more suggestive form

$$\underset{\dot{w}}{\Big(\quad\Big)} = \underset{L}{\Big(\qquad\Big)}\underset{w}{\Big(\quad\Big)} . \tag{2.5.4}$$

As is known from algebra we may construct a dual space to $w^{(k)}(0)$ spanned by vectors $\bar{w}^{(k')}(0)$ in such a way that

$$\langle \bar{w}^{(k')}(0)\, w^{(k)}(0) \rangle = \delta_{kk'} \text{ for all } k, k' , \qquad \text{where} \tag{2.5.5}$$

$$\delta_{kk'} = \begin{cases} 1 \text{ for } k = k' \\ 0 \text{ for } k \neq k' . \end{cases} \tag{2.5.6}$$

The scalar product $\langle \ldots \rangle$ between $\bar{w}$ and w is defined by

$$\langle \bar{w}^{(k')} w^{(k)} \rangle = \sum_j \bar{w}_j^{(k')} w_j^{(k)} . \tag{2.5.7}$$

All our considerations can be easily extended to partial differential equations in which L is an operator acting on spatial coordinates. In such a case the sum on the rhs of (2.5.7) must be replaced by a sum over j and in addition by an integral over space

$$\sum_j \to \sum_j \int d^3x . \tag{2.5.8}$$

We now ask whether we may find an equation for the basis vectors, cf. (2.5.5), of the dual space which guarantees that its solutions are for all times $t \gtrless 0$ orthogonal to the basic solutions (2.5.2) of the original equations (2.5.1). To this end we define $\bar{w}^{(k')}$ as the vector

$$\bar{w}^{(k')} = (w_1^{(k')}, \ldots, w_n^{(k')}) , \tag{2.5.9}$$

which obeys the equation

$$\dot{\bar{w}}^{(k')} = \bar{w}^{(k')} \bar{L} . \tag{2.5.10}$$

We require that $\bar{L}$ has matrix elements connected with those of L by

$$\bar{L}_{jj'} = -L_{jj'}\,, \tag{2.5.11}$$

or, in short, we require

$$\bar{L} = -L\,. \tag{2.5.12}$$

In analogy to (2.5.4) we may write (2.5.10) in the form

$$\underset{\dot{\bar{w}}}{\frown} = \underset{\bar{w}}{\frown} \underset{\bar{L}}{\left(\quad\right)}. \tag{2.5.13}$$

We want to show that (2.5.5) is fulfilled for all times provided $\bar{w}$ obeys the equations (2.5.10). To this end we differentiate the scalar product (2.5.7), where we use $\bar{w}^{(k')}$ and $w^{(k)}$ at time t. We obtain

$$\begin{aligned}\frac{d}{dt}&\langle \bar{w}^{(k')}(t)\, w^{(k)}(t)\rangle\\ &= \langle \dot{\bar{w}}^{(k')}(t)\, w^{(k)}(t)\rangle + \langle \bar{w}^{(k')}(t)\, \dot{w}^{(k)}(t)\rangle\end{aligned} \tag{2.5.14}$$

which by use of (2.5.10 and 1)is transformed into

$$\begin{aligned}\frac{d}{dt}&\langle \bar{w}^{(k)}(t)\, w^{(k)}(t)\rangle\\ &= \langle \bar{w}^{(k')}(t)\bar{L}(t)\, w^{(k)}(t)\rangle + \langle \bar{w}^{(k')}(t)\, L(t) w^{(k)}(t)\rangle\end{aligned} \tag{2.5.15}$$

On account of (2.5.12), the rhs of (2.5.15) vanishes. This tells us that (2.5.5) is fulfilled for all later times (or previous times when going in backward direction). Thus we may formulate our final result by

$$\langle \bar{w}^{(k')}(t)\, w^{(k)}(t)\rangle = \delta_{kk'}\,. \tag{2.5.16}$$

In later sections we shall show that for certain classes of $L(t)$ we may decompose $w^{(k)}$ into

$$w^{(k)} = e^{\lambda_k t} v^{(k)}(t)\,, \tag{2.5.17}$$

where $v^{(k)}$ has certain properties. In this section we use the decomposition (2.5.17), but in an entirely arbitrary fashion so that $v^{(k)}$ need not be specified in any way. All we want to show is that whenever we make a decomposition of the form (2.5.17) and a corresponding one for $\bar{w}$,

$$\bar{w}^{(k')} = e^{-\lambda_{k'} t} \bar{v}^{(k)}(t)\,, \tag{2.5.18}$$

the relations (2.5.16) are also fulfilled by the v's. The λ_k's are quite arbitrary. Inserting (2.5.17 and 18) in (2.5.16) we immediately find

$$\exp[(\lambda_k - \lambda_{k'})t]\langle \bar{v}^{(k')} v^{(k)} \rangle = \delta_{kk'} , \tag{2.5.19}$$

which due to the property of the Kronecker symbol can also be written in the form

$$\langle \bar{v}^{(k')} v^{(k)} \rangle = \delta_{kk'} . \tag{2.5.20}$$

The orthogonality relations (2.5.16, 20) will be used in later chapters.

2.6 Linear Differential Equations with Constant Coefficients

Now we wish to derive the form of the solutions of (2.5.1) if L is a constant matrix. If $\boldsymbol{q}$ is prescribed at time $t = 0$, the formal solution of (2.5.1) with a constant matrix L can be written in the form

$$\boldsymbol{q}(t) = \mathrm{e}^{Lt} \boldsymbol{q}(0) , \tag{2.6.1}$$

which is in formal analogy to solution (2.1.5). Note, however, that L is a matrix so that the formal solution (2.6.1) needs further explanation. It is provided by the definition of the exponential function of an operator by means of the power series expansion

$$\mathrm{e}^{Lt} = \sum_{\nu=0}^{\infty} \frac{1}{\nu!} (Lt)^{\nu} . \tag{2.6.2}$$

Since each power of a matrix is defined, (2.6.2) is defined too, and one can show especially that the series (2.6.2) converges in the sense that each matrix element of $\exp(Lt)$ is finite for any finite time t. By inserting (2.6.2) into (2.6.1) we may readily convince ourselves that (2.6.1) fulfills (2.5.1) because

$$\dot{\boldsymbol{q}} = \mathrm{L} \mathrm{e}^{Lt} \boldsymbol{q}(0) . \tag{2.6.3}$$

While this kind of solution may be useful for some purposes, we wish to derive a more explicit form of $\boldsymbol{q}$. Depending on different initial vectors $\boldsymbol{q}^{(j)}(0)$ different solutions $\boldsymbol{q}^{(j)}(t)$ evolve. Provided the $\boldsymbol{q}^{(j)}(0)$'s are linearly independent, then so are the $\boldsymbol{q}^{(j)}$ at all other times.

In the following we shall prove the following theorem:
We can choose the solutions $\boldsymbol{q}^{(j)}(t)$ in such a way that they have the form

$$\boldsymbol{q}^{(j)}(t) = \mathrm{e}^{\lambda_j t} \boldsymbol{v}^{(j)}(t) . \tag{2.6.4}$$

The exponents λ_j, often called "characteristic exponents", are the eigenvalues of the matrix L. The vectors $v^{(j)}(t)$ have the form

$$v^{(j)}(t) = v_0^{(j)} + v_1^{(j)} t + \ldots + v_{m_j}^{(j)} t^{m_j} , \tag{2.6.5}$$

i.e., they are polynomials in t where the highest power m_j is smaller or equal to the degeneracy of λ_j. If all λ_j's are different from each other, $v^{(j)}(t)$ must then be a constant. If several λ_j's coincide it might but need not happen that powers of t occur in v^j. We shall see later how to decide by construction up to which power t occurs in (2.6.5). Let us start to prove these statements. We start from the formal solution of $\dot{Q} = LQ$, namely

$$Q(t) = \mathrm{e}^{Lt} Q(0) , \tag{2.6.6}$$

and leave the choice of $Q(0)$ open. We now choose a regular matrix S so that L is brought to Jordan's normal form $\tilde{L}$,

$$S^{-1} L S = \tilde{L} . \tag{2.6.7}$$

Multiplying this equation by S from the left and S^{-1} from the right we obtain

$$L = S \tilde{L} S^{-1} . \tag{2.6.8}$$

Thus (2.6.6) acquires the form

$$Q(t) = \mathrm{e}^{S \tilde{L} S^{-1} t} Q(0) . \tag{2.6.9}$$

Making use of the expansion of the exponential function (2.6.2), one may readily convince oneself that (2.6.9) can be replaced by

$$Q(t) = S \, \mathrm{e}^{\tilde{L} t} S^{-1} Q(0) . \tag{2.6.10}$$

By the special choice of the initial solution matrix $Q(0)$,

$$Q(0) = S , \tag{2.6.11}$$

a particularly simple form for (2.6.10) is found.

To discuss the resulting form of the solution matrix

$$Q(t) = S \, \mathrm{e}^{\tilde{L} t} \tag{2.6.12}$$

we use the explicit form of Jordan's normal form of $\tilde{L}$, namely

$$\tilde{L} = \begin{pmatrix} \boxed{1} & & 0 & \\ & \boxed{2} & & \\ & & \boxed{3} & \\ & 0 & & \ddots \end{pmatrix} . \tag{2.6.13}$$

In it each box may be either a single matrix element of the form

$$\boxed{j} = \lambda_j\,, \tag{2.6.14}$$

or may represent a $m_j \times m_j$ matrix of the form

$$\boxed{j} = \begin{pmatrix} \lambda_j & 1 & & & 0 \\ & \lambda_j & 1 & & \\ & & \lambda_j & 1 & \\ & & & \ddots & \ddots \\ & 0 & & & 1 \\ & & & & \lambda_j \end{pmatrix}. \tag{2.6.15}$$

In it, in the main diagonal all λ_j's are equal. Above this diagonal we have another diagonal with 1's. All other matrix elements are equal to zero.

We first show that $\exp(\tilde{L}t)$ has the same shape as (2.6.13). Using the rules of matrix multiplication it is demonstrated that

$$\tilde{L}^2 = \begin{pmatrix} \boxed{1}^2 & & & \\ & \boxed{2}^2 & & 0 \\ & & \boxed{3}^2 & \\ & 0 & & \ddots \end{pmatrix} \tag{2.6.16}$$

and quite generally for an arbitrary power m

$$\tilde{L}^m = \begin{pmatrix} \boxed{1}^m & & & \\ & \boxed{2}^m & & 0 \\ & & \boxed{3}^m & \\ & 0 & & \ddots \end{pmatrix}. \tag{2.6.17}$$

When we multiply both sides of (2.6.17) by $t^m/m!$ and sum up over m we obtain $e^{\tilde{L}t}$ which yields the same structure as (2.6.13), namely

$$e^{\tilde{L}t} = \begin{pmatrix} \boxed{H_1} & & 0 \\ & \boxed{H_2} & \\ 0 & & \ddots \end{pmatrix} \equiv H\,. \tag{2.6.18}$$

Therefore, for our study of the form of the solution it will be sufficient to focus our attention on each of the matrices H_j which are of the form

$$H_j = e^{M_j t}\,, \tag{2.6.19}$$

where we have denoted the box j by M_j. For what follows we shall put

$$e^{\tilde{L}t} = \tilde{Q}(t)\,, \tag{2.6.20}$$

where we wish to use the normal form (2.6.13). Dropping the index j, we write the matrix (2.6.15) in the form

$$M = \lambda \cdot 1 + K\,. \tag{2.6.21}$$

In it 1 is the $m_j \times m_j$ unity matrix

$$1 = \begin{bmatrix} 1 & & & 0 \\ & 1 & & \\ & & 1 & \\ 0 & & & \ddots \end{bmatrix}, \tag{2.6.22}$$

whereas K is defined by

$$K = \begin{bmatrix} 0 & 1 & & & 0 \\ & 0 & 1 & & \\ & & \ddots & \ddots & \\ & 0 & & \ddots & 1 \\ & & & & 0 \end{bmatrix}, \tag{2.6.23}$$

with m_j rows and columns. Because the matrix 1 commutes with all other matrices we may split the exponential function according to

$$e^{Mt} = e^{\lambda t} e^{Kt} . \tag{2.6.24}$$

Let us consider the exponential function of Kt by expanding it into a series

$$e^{Kt} = 1 + Kt + \frac{1}{2!} K^2 t^2 + \dots . \tag{2.6.25}$$

What happens can be seen immediately by means of examples. If K is one-dimensional, i.e.,

$$K = 0 , \tag{2.6.26}$$

(2.6.25) reduces to a constant. If K is two-dimensional,

$$K = \begin{pmatrix} 0 & 1 \\ 0 & 0 \end{pmatrix}, \tag{2.6.27}$$

we readily obtain by matrix multiplication

$$K^2 = 0 , \tag{2.6.28}$$

i.e., (2.6.25) contains a constant and a term linear in t, i.e.,

$$e^{Kt} = 1 + t \begin{pmatrix} 0 & 1 \\ 0 & 0 \end{pmatrix} = \begin{pmatrix} 1 & t \\ 0 & 1 \end{pmatrix}. \tag{2.6.29}$$

In the case of a 3 by 3 matrix

$$K = \begin{pmatrix} 0 & 1 & 0 \\ 0 & 0 & 1 \\ 0 & 0 & 0 \end{pmatrix} \tag{2.6.30}$$

we readily obtain

$$K^2 = \begin{pmatrix} 0 & 0 & 1 \\ 0 & 0 & 0 \\ 0 & 0 & 0 \end{pmatrix}, \tag{2.6.31}$$

$$K^3 = (0) \tag{2.6.32}$$

so that $\exp(Kt)$ acquires the form

$$\mathrm{e}^{Kt} = 1 + t \begin{pmatrix} 0 & 1 & 0 \\ 0 & 0 & 1 \\ 0 & 0 & 0 \end{pmatrix} + \frac{t^2}{2} \begin{pmatrix} 0 & 0 & 1 \\ 0 & 0 & 0 \\ 0 & 0 & 0 \end{pmatrix} = \begin{pmatrix} 1 & t & t^2/2 \\ 0 & 1 & t \\ 0 & 0 & 1 \end{pmatrix}. \tag{2.6.33}$$

After these preparations we may find $\tilde{Q}$ (2.6.20). To illustrate what happens, consider an example where λ_1 belongs to a one-by-one box, λ_2 to a two-by-two box, etc. Again the general structure is obvious. In such a case we have

$$\tilde{Q}(t) \equiv \mathrm{e}^{\tilde{L}t} \equiv H = \begin{pmatrix} \mathrm{e}^{\lambda_1 t} & 0 & 0 & \\ 0 & \mathrm{e}^{\lambda_2 t} & t\mathrm{e}^{\lambda_2 t} & \\ 0 & 0 & \mathrm{e}^{\lambda_2 t} & \\ & & & \ddots \end{pmatrix}. \tag{2.6.34}$$

In order to obtain $\tilde{q}^{(j)}$ we decompose (2.6.34) into its column vectors $\tilde{q}^{(j)}$ which yields

$$\tilde{q}^{(1)} = \mathrm{e}^{\lambda_1 t} \begin{pmatrix} 1 \\ 0 \\ 0 \\ \vdots \end{pmatrix}, \quad \tilde{q}^{(2)} = \mathrm{e}^{\lambda_2 t} \begin{pmatrix} 0 \\ 1 \\ 0 \\ \vdots \end{pmatrix}, \quad q^{(3)} = \mathrm{e}^{\lambda_2 t} \begin{pmatrix} 0 \\ t \\ 1 \\ \vdots \end{pmatrix}. \tag{2.6.35}$$

From these considerations we see how Q can be constructed. Now we must remember that we still have to form (2.6.12) to obtain our solution matrix $Q(t)$. In order to explore its shape we have to perform the product

$$S\tilde{Q} = \begin{pmatrix} S_{11} & S_{12} \dots S_{1n} \\ S_{21} & S_{22} & \\ \vdots & \ddots & \\ S_{n1} & & S_{nn} \end{pmatrix} \begin{pmatrix} \mathrm{e}^{\lambda_1 t} & 0 & 0 & \\ 0 & e^{\lambda_2 t} & t\mathrm{e}^{\lambda_2 t} & \\ 0 & 0 & \mathrm{e}^{\lambda_2 t} & \\ & & & \ddots \end{pmatrix}. \tag{2.6.36}$$

Taking the same example as before we immediately find

$$q^{(1)} = \mathrm{e}^{\lambda_1 t} \begin{pmatrix} S_{11} \\ S_{21} \\ S_{31} \\ \vdots \end{pmatrix}, \quad q^{(2)} = \mathrm{e}^{\lambda_2 t} \begin{pmatrix} S_{12} \\ S_{22} \\ S_{32} \\ \vdots \end{pmatrix},$$

$$q^{(3)} = \mathrm{e}^{\lambda_2 t} \left[\begin{pmatrix} S_{13} \\ S_{23} \\ S_{33} \\ \vdots \end{pmatrix} + t \begin{pmatrix} S_{12} \\ S_{22} \\ S_{32} \\ \vdots \end{pmatrix} \right]. \tag{2.6.37}$$

Evidently all these q's have the form (2.6.4) with (2.6.5), as the theorem stated at the beginning of this section. The way in which this result was derived provides an explicit construction method of these solutions. This procedure can be generalized to the case of a time-periodic matrix L, to which case we now turn.

2.7 Linear Differential Equations with Periodic Coefficients

We wish to derive the general form of the solutions of the equation

$$\dot{q} = L(t)q\,, \tag{2.7.1}$$

where the coefficient matrix L is periodic. In other words, L is invariant against the replacement

$$T\colon \quad t \to t + t_0\,, \tag{2.7.2}$$

i.e., it commutes with the operator defined by (2.7.2)

$$TL = LT\,. \tag{2.7.3}$$

Multiplying

$$\dot{Q} = LQ \tag{2.7.4}$$

on both sides by T and using (2.7.3) we immediately obtain

$$(TQ)^{\boldsymbol{\cdot}} = L(TQ)\,. \tag{2.7.5}$$

This equation tells us that TQ is again a solution of (2.7.4). According to Theorem 2.4.1 (Sect. 2.4.3) we know that this solution matrix can be expressed by the old solution matrix $Q(t)$ by means of a constant transformation matrix C

$$TQ \equiv Q(t + t_0) = Q(t)C\,. \tag{2.7.6}$$

Let us assume that this transformation matrix C is known[1]. Instead of solving (2.7.4) we explore the solution of (2.7.6). First of all we note that according to Theorem 2.4.1 C is a regular matrix. As is shown in mathematics and as shown below quite explicitly we can always find a matrix Λ so that

$$e^{\Lambda t_0} = C . \tag{2.7.8}$$

In order to solve (2.7.6) we make the hypothesis

$$Q(t) = U(t)\, e^{\Lambda t} . \tag{2.7.9}$$

Inserting this hypothesis into (2.7.6) we immediately find

$$U(t+t_0) \exp [\Lambda(t+t_0)] = U(t)\, e^{\Lambda t} C \tag{2.7.10}$$

or due to (2.7.8)

$$U(t+t_0) = U(t) . \tag{2.7.11}$$

Equation (2.7.11) tells us that the solution matrix U is periodic.

We now introduce a matrix S which brings Λ into Jordan's normal form

$$\Lambda = S \tilde{\Lambda} S^{-1} . \tag{2.7.12}$$

This allows us to perform steps which are quite similar to those of the preceding section. Due to (2.7.12) we have

$$e^{\Lambda t} = S e^{\tilde{\Lambda} t} S^{-1} . \tag{2.7.13}$$

Inserting this into (2.7.9) we obtain for the solution matrix

$$Q(t) = \underbrace{U(t) S e^{\tilde{\Lambda} t}}_{\tilde{Q}(t)} S^{-1} , \tag{2.7.14}$$

which we may write in the form

$$Q(t) = \tilde{Q}(t)\, S^{-1} . \tag{2.7.15}$$

---

[1] Since in most cases (2.7.4) cannot be solved analytically one may resort to computer calculations. One takes $Q(0)$ as the unit matrix and lets the computer calculate $Q(t)$ by standard iteration methods until the time $t = t_0$ is reached. Specializing (2.7.6) to this case we obtain

$$Q(t_0) = C , \tag{2.7.7}$$

which gives us directly the transformation matrix C.

According to Theorem 2.4.1 (where $S^{-1} \equiv C$), $\tilde{Q}$ is a solution matrix if $Q(t)$ is such a matrix. Because it will turn out that $\tilde{Q}$ has a simpler form than $Q(t)$ we now treat this latter matrix.

With the abbreviation

$$\tilde{U}(t) = U(t)S \tag{2.7.16}$$

$\tilde{Q}$ acquires the form

$$\tilde{Q}(t) = \tilde{U}(t)\,\mathrm{e}^{\tilde{\Lambda}t}. \tag{2.7.17}$$

The form of this solution exhibits a strong resemblance to the form (2.6.12) of the solution of a differential equation with constant coefficients. The only difference consists in the fact that the matrix S of (2.6.12) is now replaced by a matrix $\tilde{U}$, whose coefficients are periodic in time. This allows us to repeat all former steps of Sect. 2.6 and to derive a standard form of the solution vectors. They read

$$\boldsymbol{q}^{(j)}(t) = \mathrm{e}^{\lambda_j t}\boldsymbol{v}^{(j)}(t)\,, \tag{2.7.18}$$

where

$$\boldsymbol{v}^{(j)}(t) = \boldsymbol{v}_0^{(j)}(t) + \boldsymbol{v}_1^{(j)}(t)\,t + \ldots + \boldsymbol{v}_{m_j}^{(j)}(t)\,t^{m_j}. \tag{2.7.19}$$

The characteristic exponents λ_j in (2.7.18) are called *Floquet exponents*; they are the eigenvalues of Λ, cf. (2.7.12).

The coefficients $\boldsymbol{v}_l^{(j)}(t)$ are periodic functions of time with period t_0. For m_j we have the rule

$$m_j \leqslant \text{degree of degeneracy of } \lambda_j\,. \tag{2.7.20}$$

This is the final result of this section. For readers who are interested in all details we now turn to the question how to determine Λ in (2.7.8) if C is given. To this end we introduce a matrix V which brings C into Jordan's normal form. From (2.7.8) we find

$$V^{-1}\mathrm{e}^{\Lambda t_0}V = V^{-1}CV \equiv \tilde{C}\,. \tag{2.7.21}$$

Introducing the abbreviation

$$V^{-1}\Lambda V = \hat{\Lambda} \tag{2.7.22}$$

we obtain for the lhs of (2.7.21)

$$\mathrm{e}^{\hat{\Lambda}t_0} = \tilde{C}\,. \tag{2.7.23}$$

In order to fulfill (2.7.23) where $\tilde{C}$ has the shape

$$\begin{pmatrix} \boxed{\tilde{C}_1} & & \\ & \boxed{\tilde{C}_2} & 0 \\ 0 & & \ddots \end{pmatrix} \tag{2.7.24}$$

it is sufficient to assume that the still unknown $\hat{\Lambda}$ (which stems from the still unknown Λ) has the same decomposition corresponding to

$$\hat{\Lambda} = \begin{pmatrix} \boxed{\hat{\Lambda}_1} & & \\ & \boxed{\hat{\Lambda}_2} & 0 \\ 0 & & \ddots \end{pmatrix} \tag{2.7.25}$$

[compare steps (2.6.16 to 18)]. Therefore our problem reduces to one which refers to any of the submatrices of (2.7.24 or 25) or, in other words, we have to solve an equation of the form

$$\exp(\square t_0) = \begin{pmatrix} \mu & 1 & & & \\ & \mu & 1 & 0 & \\ & & \ddots & \ddots & \\ & 0 & & \ddots & 1 \\ & & & & \mu \end{pmatrix}. \tag{2.7.26}$$

where the box on the left-hand side represents a matrix still to be determined.

As C is a regular matrix,

$$\mu \neq 0 \tag{2.7.27}$$

in (2.7.26) holds. We put

$$e^{\lambda t_0} = \mu \tag{2.7.28}$$

and introduce the decomposition

$$\square = \lambda 1 + \square'/t_0\,. \tag{2.7.29}$$

This allows us to write (2.7.26) as

$$\mu \exp(\square') = \begin{pmatrix} \mu & 1 & & & \\ & \mu & 1 & 0 & \\ & & \ddots & \ddots & \\ & 0 & & \ddots & 1 \\ & & & & \mu \end{pmatrix} \tag{2.7.30}$$

or

$$\exp(\square') = 1 + \mu^{-1} \begin{pmatrix} 0 & 1 & & 0 & \\ & 0 & 1 & & \\ & & \ddots & \ddots & \\ & 0 & & \ddots & 1 \\ & & & & 0 \end{pmatrix}. \tag{2.7.31}$$

Though we are dealing here with matrices, we have already seen that we may also use functions of matrices in analogy to usual numbers by using power series expansions. Because the logarithm can be defined by a power series expansion, we take the logarithm of both sides of (2.7.31) which yields

$$\square' = \ln(1 + \mu^{-1}K) \tag{2.7.32}$$

with K as defined in (2.6.23). Expanding the logarithm of the right-hand side we immediately obtain

$$\square' = \mu^{-1}K - \tfrac{1}{2}\mu^{-2}K^2 + \tfrac{1}{3}\mu^{-3}K^3 - \dots \tag{2.7.33}$$

Fortunately enough we need not worry about the convergence of the power series on the rhs of (2.6.33) because it follows from formulas (2.6.28, 32) and their generalization that powers higher than a fixed number vanish. Equation (2.7.33) gives us an explicit solution of our problem to determine the square in (2.7.29). We have thus shown how we can explicitly calculate Λ when C is given.

2.8 Group Theoretical Interpretation

Results of Sects. 2.2, 7 may serve as an illustration of basic concepts of the theory of group representations. In complete analogy to our results of Sect. 2.2, the operator T generates an Abelian group with elements T^n, $n \gtrless 0$, n integer. But what happens to the correspondence $T \to \alpha$ we established in (2.2.16)? There our starting point was the relation (2.2.10), i.e.,

$$Tq(t) = \alpha q(t)\,. \tag{2.8.1}$$

In the present case, the analog to this relation is provided by

$$TQ(t) = Q(t)C\,, \tag{2.8.2}$$

i.e., (2.7.6) where C is a *matrix*. By applying T on both sides of (2.8.2) and using (2.8.2) again we obtain

$$T^2Q(t) = TQ(t)C = (Q(t)C)C = Q(t)C^2\,, \tag{2.8.3}$$

and similarly

$$T^nQ(t) = Q(t)C^n\,. \tag{2.8.4}$$

Clearly

$$T^0Q(t) = Q(t)C^0 \tag{2.8.5}$$

if we define C^0 as unit matrix. Multipyling (2.8.2) on both sides by T^{-1} yields

$$Q(t) = T^{-1}Q(t)C. \tag{2.8.6}$$

Following from Theorem 2.4.1 (Sect. 2.4.3), C is a regular matrix. Thus we may form the inverse C^{-1} and multiply both sides of (2.8.6) from the left with that matrix. We then obtain

$$T^{-1}Q(t) = Q(t)C^{-1}. \tag{2.8.7}$$

Similarly, we obtain from (2.8.4)

$$T^{-n}Q(t) = Q(t)C^{-n}. \tag{2.8.8}$$

It is now quite obvious what the analog of relation (2.2.16) looks like:

$$\begin{aligned} T &\to C \\ T^n &\to C^n \\ T^0 &\to I \\ T^{-n} &\to C^{-n} \end{aligned} \tag{2.8.9}$$

and relation (2.2.17) now is generalized to

$$\begin{aligned} T^nT^m &= T^{n+m} \to C^nC^m = C^{n+m} \\ T^nE &= T^n \quad\;\; \to C^nI \;\;= C^n \\ T^{-n}T^n &= E \quad\;\;\;\, \to C^{-n}C^n = I \\ (T^nT^m)T^l &= T^n(T^mT^l) \to (C^nC^m)C^l = C^n(C^mC^l). \end{aligned} \tag{2.8.10}$$

But the fundamental difference between (2.2.16, 17) on the one hand and (2.8.9, 10) on the other rests on the fact that α was a number, whereas C is a matrix. Therefore, the abstract transformations T^n are now represented by the matrices C^n, and the multiplication of elements of the T-group is represented by multiplication of the matrices C^n. Since in mathematics one knows quite well how to deal with matrices, one can use them to study the properties of abstract groups (in our case the group generated by T). This is one of the basic ideas of group representation theory.

In the foregoing section we have seen that by the transformation (2.7.14), i.e.,

$$Q(t) = \tilde{Q}(t)S^{-1}, \tag{2.8.11}$$

we can reduce L into Jordan's normal form

$$S^{-1}LS = \tilde{L}, \tag{2.8.12}$$

i.e., $\tilde{L}$ is decomposed into individual boxes along its diagonal [cf. (2.6.13)]. Now we insert (2.8.11) into (2.8.2) and multiply both sides from the right by S,

$$T\tilde{Q}(t) = \tilde{Q}(t)\underbrace{S^{-1}CS}_{\tilde{C}}. \tag{2.8.13}$$

This means that by a change of the basis of the solution matrix $Q \to \tilde{Q}$, C in (2.8.2) is transformed into

$$\tilde{C} = S^{-1}CS. \tag{2.8.14}$$

But according to (2.7.8)

$$C = \mathrm{e}^{\Lambda t_0} \tag{2.8.15}$$

holds so that

$$\tilde{C} = S^{-1}\mathrm{e}^{\Lambda t_0}S = \mathrm{e}^{S^{-1}\Lambda S t_0} = \mathrm{e}^{\tilde{\Lambda} t_0}. \tag{2.8.16}$$

But because $\tilde{\Lambda}$ is in Jordan's normal form, so is $\tilde{C}$, on account of our considerations in Sect. 2.6. We then see that by the proper choice of $\tilde{Q}$ the matrix $\tilde{C}$ of the group representation can be reduced to the simple form

$$C = \begin{pmatrix} \boxed{1} & & 0 \\ & \boxed{2} & \\ 0 & & \ddots \end{pmatrix}. \tag{2.8.17}$$

Since the individual boxes (i.e., matrices) cannot be reduced further (due to algebra), they are called an irreducible representation. Because the boxes are multiplied individually if we form $\tilde{C}^n$, we obtain

$$\tilde{C}^n = \begin{pmatrix} \boxed{1}^n & & 0 \\ & \boxed{2}^n & \\ 0 & & \ddots \end{pmatrix}. \tag{2.8.18}$$

In this way we may attach an individual box k, or an *irreducible representation*, to T:

$$T \to \boxed{k} \tag{2.8.19}$$

and

$$T^n \to \boxed{k}^n. \tag{2.8.20}$$

Generally speaking, it is one of the main goals of group representation theory to establish the irreducible representations of an abstract group (here T^n).

2.9 Perturbation Approach*

Since in general it is not possible to solve a set of differential equations with periodic coefficients explicitly, in some cases a perturbation method may be useful. To this end we decompose the matrix L of the differential equation

$$\dot{Q} = LQ \tag{2.9.1}$$

into

$$L = L_0 + L_1 , \tag{2.9.2}$$

where L_0 is a constant matrix whereas L_1 contains no constant terms. This can be most easily achieved by expanding the matrix elements of L into a Fourier series in which the constant term is clearly exhibited.

We now insert the ansatz

$$Q = S\tilde{Q} , \tag{2.9.3}$$

where S is a constant regular matrix, into (2.9.1), which yields after multiplying both sides by S^{-1}

$$\dot{\tilde{Q}} = S^{-1}LS\tilde{Q} . \tag{2.9.4}$$

We choose S in such a way that

$$S^{-1}L_0S = J \tag{2.9.5}$$

assumes Jordan's normal form. To simplify subsequent discussion, we shall assume that J contains only diagonal elements. We further put

$$S^{-1}L_1S = \tilde{M} \tag{2.9.6}$$

so that

$$\dot{\tilde{Q}} = (J + \tilde{M})\tilde{Q} . \tag{2.9.7}$$

Inserting the hypothesis

$$\tilde{Q} = \mathrm{e}^{tJ}\hat{Q} \tag{2.9.8}$$

into (2.9.7) we readily obtain after performing the differentiation with respect to t

$$\dot{\hat{Q}} = \underbrace{e^{-tJ}\tilde{M}e^{tJ}}_{M}\hat{Q}. \tag{2.9.9}$$

It may be shown that the matrix elements M_{kl} of M have the form

$$M_{kl} = \exp[-(J_k - J_l)t]\tilde{M}_{kl}, \tag{2.9.10}$$

where J_l are the diagonal elements of the diagonal matrix J. So far all transformations were exact. Now to start with perturbation theory we assume that the periodic part of L, i.e., L_1 or eventually M in (2.9.9), is a small quantity. In order to exhibit this explicitly we introduce a small parameter ε. Furthermore, we decompose the differential equation (2.9.9) for the solution matrix into equations for the individual solution vectors. Therefore we write

$$\dot{\hat{q}} = \varepsilon M\hat{q}, \tag{2.9.11}$$

where the matrix M is of the form

$$\begin{pmatrix} M_{11} & M_{12} & M_{13} & \dots \\ M_{21} & M_{22} & M_{23} & \dots \\ \vdots & & & \\ M_{n1} & \dots & \dots & M_{nn} \end{pmatrix}, \tag{2.9.12}$$

and in particular

$$\bar{M}_{jj} = 0, \quad \bar{M}_{jj} = \frac{1}{2\pi}\int_0^{2\pi} M_{jj}\,d\varphi, \quad \varphi = \omega t. \tag{2.9.13}$$

The nondiagonal elements may be rewritten as

$$M_{ij} = e^{\Delta_{ij}t}P_{ij}^{(0)}(t), \quad \Delta_{ij} = -(J_i - J_j), \tag{2.9.14}$$

where $P_{ij}^{(0)}$ denote periodic functions of time.

Furthermore, we shall assume

$$\Delta_{ij} + \mathrm{i}m\omega \neq 0 \quad \text{for} \quad m = 0, \pm 1, \pm 2, \dots. \tag{2.9.15}$$

To solve (2.9.11) we make the following hypothesis

$$\hat{q} = \exp[t(a_2\varepsilon^2 + a_3\varepsilon^3 + \dots)]\left[\begin{pmatrix} 1 \\ 0 \\ \vdots \\ 0 \end{pmatrix} + \varepsilon\begin{pmatrix} A_1^{(1)} \\ A_2^{(1)} \\ \vdots \\ A_n^{(1)} \end{pmatrix} + \varepsilon^2\begin{pmatrix} A_1^{(2)} \\ A_2^{(2)} \\ \vdots \\ A_n^{(2)} \end{pmatrix} + \dots\right] \tag{2.9.16}$$

with time-dependent vectors $A^{(\varkappa)}$. For $\varepsilon = 0$ (2.9.16) reduces to the special vector

$$\begin{pmatrix} 1 \\ 0 \\ \vdots \\ 0 \end{pmatrix} . \tag{2.9.17}$$

Since we may numerate the components of $\hat{q}$ arbitrarily, our treatment is, however, quite general. Inserting (2.9.16) into (2.9.11), performing the differentiation and dividing both sides of the resulting equation by $\exp[t(a_2\varepsilon^2 + \dots)]$ we obtain

$$\dot{\hat{q}} \equiv (a_2\varepsilon^2 + a_3\varepsilon^3 + a_4\varepsilon^4 + \dots) \left[\begin{pmatrix} 1 \\ 0 \\ \vdots \\ 0 \end{pmatrix} + \varepsilon \begin{pmatrix} A_1^{(1)} \\ A_2^{(1)} \\ \vdots \\ A_n^{(1)} \end{pmatrix} + \dots \right]$$

$$+ \varepsilon \begin{pmatrix} \dot{A}_1^{(1)} \\ \dot{A}_2^{(1)} \\ \vdots \\ \dot{A}_n^{(1)} \end{pmatrix} + \varepsilon^2 \begin{pmatrix} \dot{A}_1^{(2)} \\ \dot{A}_2^{(2)} \\ \vdots \\ \dot{A}_n^{(2)} \end{pmatrix} + \dots$$

$$= \varepsilon \begin{pmatrix} M_{11} \\ M_{21} \\ \vdots \\ M_{n1} \end{pmatrix} + \varepsilon^2 \begin{pmatrix} \sum_{l=1}^{n} M_{1l} A_l^{(1)} \\ \sum_{l=1}^{n} M_{2l} A_l^{(1)} \\ \vdots \\ \sum_{l=1} M_{nl} A_l^{(1)} \end{pmatrix} + \dots + \varepsilon^{k+1} \begin{pmatrix} \sum_{l=1}^{n} M_{1l} A_l^{(k)} \\ \vdots \\ \vdots \\ \sum_{l=1} M_{nl} A_l^{(k)} \end{pmatrix} . \tag{2.9.18}$$

We now compare the coefficients of the same powers of ε on both sides of (2.9.18). This yields the set of differential equations discussed below.

In lowest order ε, the first row of the resulting matrix equation reads

$$\varepsilon: \quad \dot{A}_1^{(1)} = M_{11} \equiv \sum_{m \neq 0} c_m^{(11)} \mathrm{e}^{\mathrm{i}m\omega t} . \tag{2.9.19}$$

Equation (2.9.19) can be fulfilled by

$$A_1^{(1)} = \sum_{m \neq 0} \frac{c_m^{(11)}}{\mathrm{i}m\omega} \mathrm{e}^{\mathrm{i}m\omega t} . \tag{2.9.20}$$

It is not necessary to choose[1] $\bar{A}_1^{(1)} \neq 0$ because this would alter only a normalization. In the same order ε but for the other rows with $l \neq 1$ we obtain the relations

[1] The bar above $A_1^{(1)}$ denotes the averaging defined by (2.9.13).

$$\dot{A}_l^{(1)} = M_{l1}\,, \quad l = 2, 3, \dots\,. \tag{2.9.21}$$

We can solve for $A_l^{(1)}$ by integrating (2.9.21). Now it is quite important to choose the integration constant properly, because we intend to develop a perturbation theory which yields solutions with the form

$$\boldsymbol{q}(t) = \mathrm{e}^{\lambda t}\boldsymbol{v}(t)\,, \tag{2.9.22}$$

$\boldsymbol{v}(t)$ periodic. Making a specific choice of the integration constant means that we choose a specific initial condition. Because M_{l1} has the form

$$M_{l1} = \mathrm{e}^{\Delta_{l1}t} \sum_{m \neq 0} c_m^{(l1)} \mathrm{e}^{\mathrm{i}m\omega t}\,, \tag{2.9.23}$$

we may choose $A_l^{(1)}$ in the form

$$A_l^{(1)}(t) = \mathrm{e}^{\Delta_{l1}t} \sum_{m \neq 0} \frac{c_m^{(l1)}}{\Delta_{l1} + \mathrm{i}m\omega} \mathrm{e}^{\mathrm{i}m\omega t}\,, \tag{2.9.24}$$

i.e., the exponential function $\exp(\Delta_{l1}t)$ is multiplied by a periodic function. As it will transpire below, such a form will secure that $\boldsymbol{q}(t)$ acquires the form (2.9.22).

Let us now consider the next order in ε^2. For the first row of the resulting matrix equation we obtain the relation

$$\varepsilon^2\colon \qquad a_2 + \dot{A}_1^{(2)} = \sum_{l=1}^{n} M_{1l} A_l^{(1)}\,. \tag{2.9.25}$$

To study the structure of the rhs of (2.9.25) more closely, we make use of the explicit form of M_{ij} (2.9.14) and of $A_l^{(1)}$ (2.9.20, 24).

If we decompose the periodic functions contained in M and $A_l^{(1)}$ into their Fourier series, we encounter products of the form

$$\exp(\mathrm{i}\omega m t + \mathrm{i}\omega m' t)\,. \tag{2.9.26}$$

For

$$m + m' \neq 0 \tag{2.9.27}$$

the result is again a periodic function which truly depends on time t, whereas for

$$m + m' = 0 \tag{2.9.28}$$

a constant results. Consequently, the rhs of (2.9.25) can be written in the form

$$\exp[(\Delta_{1l} + \Delta_{l1})t](P_1^{(2)} + C_1^{(2)})\,. \tag{2.9.29}$$

Because the exponential functions cancel, (2.9.29) consists of a sum of a constant $C_1^{(2)}$ and a periodic function $P_1^{(2)}$ containing no constant terms. We choose

$$a_2 = C_1^{(2)} = \sum_{l=1}^{n} \sum_{m \neq 0} \frac{c_m^{(1l)} c_m^{(l1)}}{\Delta_{l1} + \mathrm{i} m \omega} \tag{2.9.30}$$

so that the constant terms on both sides of (2.9.25) cancel.

The remaining equation for $A_1^{(2)}$ can be solved in a way similar to (2.9.20) so that $A_1^{(2)}$ can be chosen as a periodic function with no constant term. In the same order ε^2 we obtain for the other rows, i. e., $k \neq 1$, the following equations

$$\dot{A}_k^{(2)} = \sum_{l=1}^{n} M_{kl} A_l^{(1)}, \tag{2.9.31}$$

where the rhs can be cast into the form

$$\exp\left[(\Delta_{kl} + \Delta_{l1})t\right](P_k^{(2)} + C_k^{(2)}) \tag{2.9.32}$$

or

$$\mathrm{e}^{\Delta_{k1}t}(P_k^{(2)} + C_k^{(2)}). \tag{2.9.33}$$

Equation (2.9.31) has a solution of the form

$$A_k^{(2)} = \mathrm{e}^{\Delta_{k1}t}(\hat{P}_k^{(2)} + \hat{C}_k^{(2)}), \tag{2.9.34}$$

where $\hat{P}_k^{(2)}$ is periodic without a constant term and $\hat{C}_k^{(2)}$ is constant. Now the structure of the evolving equations is clear enough to treat the general case with powers $\nu \geqq 3$ of ε. For the first row we obtain for $\nu \geqslant 3$ ($A^{(\varkappa)} \equiv 0$ for $\varkappa \leqslant 0$, $\alpha_\varkappa = 0$ for $\varkappa \leqslant 1$)

$$\begin{aligned} \varepsilon^\nu: \quad & a_\nu + a_2 A_1^{(\nu-2)} + a_3 A_1^{(\nu-3)} + \ldots + a_{\nu-2} A_1^{(2)} \\ & + a_{\nu-1} A_1^{(1)} + \dot{A}_1^{(\nu)} = \sum_{l=1}^{n} M_{1l} A_l^{(\nu-1)}. \end{aligned} \tag{2.9.35}$$

Here the constant a_ν and the function $A_1^{(\nu)}$ are still unknown. All other functions, including $A_l^{(\varkappa)}$, $1 \leqq \varkappa \leqq \nu - 2$, have been determined in the previous steps, assuming that it was shown that these A's are periodic functions without constant terms. Substituting these previously determined A's into the rhs of (2.9.35) we readily find that the rhs has the form

$$\exp\left[(\Delta_{1l} + \Delta_{l1})t\right] P_{1l}^{(0)}(P_l^{(\nu-1)} + C_l^{(\nu-1)}) = P_1^{(\nu)} + C_1^{(\nu)}, \tag{2.9.36}$$

i. e., it consists of a constant and of a purely periodic function without a constant term. We may now choose a_ν equal to that constant term, whereas $A_1^{(\nu)}$ can be chosen as a purely periodic function without constant term. Clearly $A_1^{(\nu)}$ can be determined explicitly by mere integration of (2.9.35).

Let us now consider in the same order ν the other rows of (2.9.18). Then we obtain

$$a_2 A_k^{(\nu-2)} + a_3 A_k^{(\nu-3)} + \ldots + a_{\nu-1} A_k^{(1)} + \dot{A}_k^{(\nu)} = \sum_{l=1}^{n} M_{kl} A_l^{(\nu-1)} . \tag{2.9.37}$$

In it the function $A_k^{(\nu)}$ is unknown. All others are determined from previous steps and have the general form

$$A_l \propto \mathrm{e}^{\Delta_{l1} t} \times \text{periodic function} . \tag{2.9.38}$$

From this it follows that the rhs of (2.9.37) can be written in the form

$$\exp\left[(\Delta_{kl} + \Delta_{l1}) t\right] P_{kl}^{(0)} (P_l^{(\nu-1)} + C_l^{(\nu-1)}) , \tag{2.9.39}$$

or more concisely

$$\mathrm{e}^{\Delta_{k1} t} (P_k^{(\nu)} + C_k^{(\nu)}) . \tag{2.9.40}$$

The solution of (2.9.37) can therefore be written as

$$A_k^{(\nu)} = \mathrm{e}^{\Delta_{k1} t} (\hat{P}_k^{(\nu)} + \hat{C}_k^{(\nu)}) , \tag{2.9.41}$$

where $\hat{P}_k^{(\nu)}$ is a periodic function without a constant and $\hat{C}_k^{(\nu)}$ is a constant.

These considerations tell us several things. Firstly, we see that we can determine the subsequent contributions to the shift of the exponent λ, i.e., the terms $a_2, a_3, \ldots,$ explicitly by an iteration procedure. Furthermore, the expressions in square brackets in (2.9.16) can be constructed explicitly yielding each time a periodic function. Putting these ingredients together we find the general structure of the solution $\hat{\boldsymbol{q}}$

$$\hat{\boldsymbol{q}} = \mathrm{e}^{\gamma t} \begin{pmatrix} P_1 + C_1 \\ \mathrm{e}^{\Delta_{21} t} (P_2 + C_2) \\ \vdots \\ \mathrm{e}^{\Delta_{n1} t} (P_n + C_n) \end{pmatrix} \tag{2.9.42}$$

where all terms can be constructed explicitly by an iteration procedure. From this solution vector of (2.9.11) we can go back via (2.9.8) to the solution of (2.9.4). We then find that the first column of the solution matrix has the form

$$\tilde{\boldsymbol{q}}^{(1)} = \mathrm{e}^{J_1 t} \mathrm{e}^{\gamma t} \begin{pmatrix} P_1 + C_1 \\ P_2 + C_2 \\ \vdots \\ P_n + C_n \end{pmatrix} . \tag{2.9.43}$$

The C_j's are constants, while the P_j's are periodic functions (without additive constants).

Similarly, we may determine the other column vectors by a mere interchange of appropriate indices. In a last step we may return to Q by means of (2.9.3). As we have seen in Sects. 2.6 (cf. (2.6.36)) and 2.7, this transformation leaves the *structure* of the solution vector (2.9.22) unaltered. Our present procedure can be generalized to the case of degeneracy of the λ's in which case we find terms in (2.9.43) which contain not only periodic functions but also periodic functions multiplied by finite powers of t. For practical purposes and for the lowest few orders this procedure is quite useful. On the other hand, the convergence of this procedure is difficult to judge because in each subsequent iteration step the number of terms increases. In Sect. 3.9 another procedure will be introduced which may be somewhat more involved but which converges rapidly (and even works in the quasiperiodic case).

3. Linear Ordinary Differential Equations with Quasiperiodic Coefficients*

3.1 Formulation of the Problem and of Theorem 3.1.1

In this section we wish to study the general form of the solution matrix $Q(t)$ of the differential equation

$$\dot{Q}(t) = M(t)Q(t)\,, \tag{3.1.1}$$

where M is a complex-valued $m \times m$ matrix which can be expressed as a Fourier series of the form

$$M(t) = \sum_{n_1, n_2, \dots, n_N} M_{n_1, n_2, \dots, n_N} \exp(\mathrm{i}\,\omega_1 n_1 t + \mathrm{i}\,\omega_2 n_2 t + \dots + \mathrm{i}\,\omega_N n_N t)\,. \tag{3.1.2}$$

In it each Fourier coefficient $M_{n_1, n_2, \dots}$ is an $m \times m$ matrix. In view of the preceding sections the question arises whether the solution vectors $\boldsymbol{q}(t)$ can be brought into the form

$$\boldsymbol{q}(t) = \mathrm{e}^{\lambda t} \boldsymbol{v}(t)\,, \tag{3.1.3}$$

where $\boldsymbol{v}(t)$ is quasiperiodic or a polynomial in t with quasiperiodic coefficients. Though in the literature a number of efforts have been devoted to this problem, it has not been solved entirely. In fact, other forms of the solution vector may also be expected. Thus in this and the following sections we are led to the forefront of mathematical research. The solution matrices $Q(t)$ will be classified according to their transformation properties. Among these classes we shall find a class which implies the form (3.1.3). In order to study the transformation properties we wish to introduce a translation operator T.

As one may readily convince oneself it is not possible to extend the translation operator of Chap. 2 to the present case in a straightforward way. However, the following procedure has proven to be successful. We consider (3.1.1) as a special case of a larger set of equations in which instead of (3.1.2) we use the matrix

$$M(t, \varphi) = \sum_{n_1, n_2, \dots, n_N} M_{n_1, n_2, \dots, n_N} \exp[\mathrm{i}\,\omega_1 n_1(-\varphi_1 + t) + \mathrm{i}\,\omega_2 n_2(-\varphi_2 + t) + \dots] \tag{3.1.4}$$

which contains the phase angles $\varphi_1, \ldots, \varphi_N$. In other words, we are embedding the problem of solving (3.1.1) in the problem of solving

$$\dot{Q}(t, \varphi) = M(t, \varphi) Q(t, \varphi) . \tag{3.1.5}$$

We introduce the translation operator

$$T_\tau : \begin{cases} t \to t + \tau \\ \boldsymbol{\varphi} \to \boldsymbol{\varphi} + \boldsymbol{\tau}, \boldsymbol{\tau} = \tau \boldsymbol{e} , \end{cases} \tag{3.1.6}$$

where τ is an arbitrary shift and $\boldsymbol{e}$ is the vector $(1, 1, \ldots, 1)$ in $\boldsymbol{\varphi}$ space. One sees immediately that M is invariant against (3.1.6) or, in other words, that M commutes with T_τ,

$$T_\tau M = M T_\tau . \tag{3.1.7}$$

As a consequence of (3.1.7)

$$T_\tau Q(t, \varphi) \tag{3.1.8}$$

is again a solution of (3.1.5). Therefore the relation

$$T_\tau Q(t, \varphi) = Q(t, \varphi) C(\tau, \varphi) \tag{3.1.9}$$

must hold (compare Theorem 2.4.1), where $C(\tau, \varphi)$ is a matrix independent of t. The difficulty rests on the fact that the matrix C still depends on τ *and* φ. Using the lhs of (3.1.9) in a more explicit form we find

$$Q(t + \tau, \boldsymbol{\varphi} + \boldsymbol{\tau}) = Q(t, \boldsymbol{\varphi}) C(\tau, \boldsymbol{\varphi}) . \tag{3.1.10}$$

After these preparatory steps we are able to formulate Theorem 3.1.1 which we shall prove in the following.

Theorem 3.1.1. Let us make the following assumptions on (3.1.4 and 5):

1) The frequencies $\omega_1, \ldots, \omega_N$ are irrational with respect to each other (otherwise we could choose a smaller basic set of ω's).

2) $M(t, \varphi) = M(0, \varphi - t)$ is T_j-periodic in φ_j, $T_j = 2\pi/\omega_j$, and C^k with respect to φ ($k \geqslant 0$) ("C^k" means as usual "k times differentiable with continuous derivatives").

3) For some $\varphi = \varphi_0$ the generalized characteristic exponents λ_j of $\boldsymbol{q}^{(j)}(t, \varphi_0)$, $j = 1, \ldots, m$, are different from each other. We choose $\lambda_1 > \lambda_2 > \ldots$.

4) We use the decomposition

$$\boldsymbol{q}^{(j)}(t, \varphi_0) = \exp [z_j(t)] \boldsymbol{u}^{(j)}(t) , \tag{3.1.11}$$

where $|\boldsymbol{u}^{(j)}(t)| = 1$, z_j real, $j = 1, \ldots, m$, and, clearly, z_j is connected with λ_j by

$$\limsup_{t\to\infty}\left\{\frac{1}{t}z_j(t)\right\} = \lambda_j . \tag{3.1.12}$$

We then require

$$|\det \boldsymbol{u}^{(j)}(t)| > d_0 > 0 \quad \text{for all times } t . \tag{3.1.13}$$

This implies that the unit vectors $\boldsymbol{u}(t)$ keep minimum angles with each other, or, in other words, that the $\boldsymbol{u}$'s never become collinear, i.e., they are linearly independent.

5) $z_j(t)$ possesses the following properties. A sequence t_n, $t_n \to \infty$ exists such that

$$\left|\frac{1}{t_n+\tau} z_j(t_n+\tau) - \lambda_j\right| < \delta_1 \quad \text{and} \tag{3.1.14}$$

$$\frac{1}{t_n+\tau}[z_j(t_n+\tau) - z_j(\tau)] > \frac{1}{t_n+\tau}[z_{j+1}(t_n+\tau) - z_{j+1}(\tau)] + \delta_2 , \tag{3.1.15}$$

where

$$\delta_{1,2} > 0 , \quad j = 1, \ldots, m-1 , \quad \text{and} \quad \tau \geqslant 0 . \tag{3.1.16}$$

(In particular, the conditions (3.1.14) and (3.1.15) are fulfilled for any sequence t_n if

$$z_j(t) = \lambda_j t + w_j(t) , \tag{3.1.17}$$

where $w_j(t)$ is bounded.)

6) The ω's fulfill a KAM condition [cf. (2.1.20 and 21)] jointly with a suitably chosen k in assumption (2).

Then the following assertions hold:

a) Under conditions (1 – 4), $k \geqq 0$, $Q(t, \varphi)$ can be chosen in such a way that in

$$T_\tau Q(t, \varphi) = Q(t, \varphi)C(\tau, \varphi) \tag{3.1.18}$$

the matrix C can be made triangular,

$$C^T = \begin{pmatrix} \diagdown\!\!\!\!\!\!\!\!\!/\!/\!/ \\ 0 \end{pmatrix} , \tag{3.1.19}$$

with coefficients quasiperiodic in τ.

b) Under conditions (1 – 5), the coefficients of C are C^k with respect to φ and T_j-periodic in φ_j, $T_j = 2\pi/\omega_j$.

c) Under conditions (1 – 6), C can be made diagonal, and the solutions $q^{(j)}$ can be chosen such that

$$q^{(j)} = e^{\hat{\lambda}_j t} v^{(j)}(t, \varphi), \quad \mathrm{Re}\{\hat{\lambda}_j\} = \lambda_j, \tag{3.1.20}$$

where $v^{(j)}$ is quasiperiodic in t and T_j-periodic in φ_j, and C^k with respect to φ.

In particular

$$T_\tau v^{(j)}(t, \varphi) = v^{(j)}(t, \varphi), \tag{3.1.21}$$

i.e., $v^{(j)}$ is invariant against T_τ.

In Sect. 3.8 a generalization of this theorem will be presented for the case where some of the λ's coincide.

The proof of Theorem 3.1.1 will be given in Sects. 3.2 – 5, 3.7 in several steps (Sect. 3.6 is devoted to approximation methods – linked with the ideas of the proof – for constructing the solutions $q^{(j)}$). After stating some auxiliary theorems ("lemmas") in Sect. 3.2, we first show how C can be brought to triangular form (exemplified in Sect. 3.3 by the case of a 2×2 matrix, and in Sect. 3.5 by the case of an $m \times m$ matrix). This implies that we can choose $q^{(j)}(t, \varphi)$ such that $\lambda_1 > \lambda_2 > \ldots > \lambda_m$ for all φ. Then we show that the elements of the triangular matrix can be chosen according to statements (a) and (b) of Theorem 3.1.1 (in Sect. 3.4 by the case of a 2×2 matrix, and in Sect. 3.5 by the case of a $m \times m$ matrix). Finally in Sect. 3.7 we prove assertion (c).

3.2 Auxiliary Theorems (Lemmas)

Lemma 3.2.1. If M (3.1.4) is C^k, $k \geqslant 0$ with respect to φ, then M is bounded for all t with $-\infty < t < +\infty$ and all φ. Therefore the conclusions of Sect. 2.4.4 apply and we can define generalized characteristic exponents.

Proof: Since M is periodic in each φ_j or, equivalently, in each $\varphi_j' = \varphi_j + t$, it is continuous in the *closed* intervals $0 \leqslant \varphi_j' (\mathrm{mod}\, 2\pi/\omega_j) \leqslant 2\pi/\omega_j$ and therefore bounded for all t and φ.

Lemma 3.2.2. If M (3.1.4) and the initial solution matrix $Q(0, \varphi)$ are C^k with respect to φ, then the solution matrix $Q(t, \varphi)$ is also C^k with respect to φ for $-\infty < t < +\infty$.

Proof: We start from (3.1.5) where the matrix M is given by (3.1.4). The solution of (3.1.5) can be expressed in the formal way

$$Q(t, \varphi) = \hat{T} \exp\left[\int_0^t M(s, \varphi)\, ds\right] Q(0, \varphi), \tag{3.2.1}$$

where $\hat{T}$ is the time-ordering operator. The exponential function is defined by

$$\hat{T}\exp\left[\int_0^t M(s,\varphi)\,ds\right] = 1 + \sum_{n=1}^{\infty}\frac{1}{n!}\hat{T}\left[\int_0^t M(s,\varphi)\,ds\right]^n . \tag{3.2.2}$$

According to the time-ordering operator we have to arrange a product of matrices in such a way that later times stand to the left of earlier times. This means especially

$$I^{(n)} \equiv \frac{1}{n!}\hat{T}\left[\int_0^t M(s,\varphi)\,ds\right]^n = \int_0^t ds_n M(s_n,\varphi)\int_0^{s_n} ds_{n-1}M(s_{n-1},\varphi)\dots \int_0^{s_2} ds_1 M(s_1,\varphi) . \tag{3.2.3}$$

Because M is a matrix with matrix elements M_{jk} we can also write (3.2.3) in the form

$$I_{jk}^{(n)} \equiv \frac{1}{n!}\hat{T}\left\{\left[\int_0^t M(s,\varphi)\,ds\right]^n\right\}_{jk} = \sum_{l_1,\dots,l_{n-1}}\int_0^t ds_n M_{j,l_{n-1}}(s_n,\varphi) \cdot \int_0^{s_n} ds_{n-1}M_{l_{n-1},l_{n-2}}(s_{n-1},\varphi)\dots\int_0^{s_2} ds_1 M_{l_1,k}(s_1,\varphi) . \tag{3.2.4}$$

According to Lemma 3.2.1 M is bounded

$$|M_{jk}| \leqslant \bar{M} . \tag{3.2.5}$$

Therefore the estimate

$$|I_{jk}^{(n)}| \leqslant \underbrace{\sum_{l_1,\dots,l_{n-1}}}_{m^{n-1}} \bar{M}^n \underbrace{\int_0^t ds_n \int_0^{s_n} ds_{n-1}\dots\int_0^{s_2} ds_1}_{t^n/n!} \tag{3.2.6}$$

holds. Because according to (3.2.1) we must eventually multiply (3.2.2) by an initial solution vector q_k at time $t=0$, we need to know the norm of

$$I^{(n)} q_k(0,\varphi) . \tag{3.2.7}$$

Assume that each component q_k is bounded,

$$|q_k(0,\varphi)| \leqslant \bar{q} , \quad k=1,\dots,m . \tag{3.2.8}$$

Then we readily obtain

$$|I^{(n)} q_k(0,\varphi)| \leqslant m^{n-1}\frac{t^n}{n!}\bar{M}^n m\bar{q} \tag{3.2.9}$$

from which

$$|q_k(t,\varphi)| \leqslant \sum_{n=0}^{\infty} \frac{t^n}{n!} \bar{M}^n m^n \bar{q} = \mathrm{e}^{t\bar{M}m} \bar{q}, \quad k = 1, \ldots, m. \tag{3.2.10}$$

Evidently the series converges for all times $-\infty < t < +\infty$.

These considerations can be easily extended to prove the differentiability of (3.2.1) with respect to φ. Denoting the derivative with respect to a specific φ_j by a prime

$$Q'_{\varphi_j}(t,\varphi) \equiv \frac{\partial}{\partial \varphi_j} Q(t,\varphi) \tag{3.2.11}$$

and dropping the index φ_j, we obtain the following expressions for the derivative of each matrix element jk of a member of the series (3.2.2)

$$\left\{ \frac{\partial}{\partial \varphi_j} I^n \right\}_{jk} \equiv \left\{ \frac{\partial}{\partial \varphi_j} \frac{1}{n!} \hat{T} \left[\int_0^t M(s,\varphi)\, ds \right]^n \right\}_{jk}$$

$$= \sum_{l_1,\ldots,l_{n-1}} \left\{ \int_0^t ds_n M'_{j,l_{n-1}}(s_n,\varphi) \int_0^{s_n} \ldots + \int_0^t ds_n M_{j,l_{n-1}} \int_0^{s_n} ds_{n-1} M'_{l_{n-1},l_{n-2}} \ldots + \ldots \right\}_{jk}. \tag{3.2.12}$$

Let

$$|M'| \leqslant K \quad \text{then} \tag{3.2.13}$$

$$\left| \left\{ \frac{\partial}{\partial \varphi_j} I^{(n)} \right\}_{jk} \right| \leqslant n m^{n-1} \bar{M}^{n-1} K. \tag{3.2.14}$$

Let us consider (3.2.11) in more detail. Denoting derivatives with respect to φ_j by a prime,

$$\left. \frac{\partial}{\partial \varphi_j} Q(t,\varphi) \right|_{ik} = \left. \left[\hat{T} \exp \int_0^t \ldots \right]' Q(0,\varphi) \right|_{ik} + \left. \underbrace{\left[\hat{T} \exp \int_0^t \ldots \right]}_{\text{exists for all } t} Q'(0,\varphi) \right|_{ik}, \tag{3.2.15}$$

we may use the estimate (3.2.14) and obtain

$$\left| \frac{\partial}{\partial \varphi_j} Q(t,\varphi) \right|_{ik} \leqslant K \exp(m\bar{M}t) m\bar{q} + \left| \left[\hat{T} \exp \int_0^t \ldots \right] \frac{\partial}{\partial \varphi_j} Q(0,\varphi) \right|_{ik}, \tag{3.2.16}$$

where the second term on the rhs can be estimated in a way analogous to (3.2.9). Thus if M is C^1 with respect to φ, cf. (3.2.13), then (3.2.11) exists for all times $-\infty < t < \infty$. Similarly, one may prove the convergence (and existence) of the k'th derivative of $Q(t, \varphi)$ provided M is correspondingly often continuously differentiable with respect to φ.

Lemma 3.2.3. If the initial matrix $Q(0, \varphi)$ is periodic in φ with periods $T_j = 2\pi/\omega_j$, then $Q(t, \varphi)$ has the same property.

Proof: In (3.2.1, 2) each term has this periodicity property.

Lemma 3.2.4. This lemma deals with the choice of the initial matrix as unity matrix. We denote the (nonsingular) initial matrix of the solution matrix with $\varphi = \varphi_0$ by $\hat{Q}$,

$$Q(0, \varphi_0) = \hat{Q}(0, \varphi_0) . \tag{3.2.17}$$

For all other φ's we assume the same initial condition for the solution matrix for $t = 0$

$$Q(0, \varphi) = \hat{Q}(0, \varphi_0) \equiv Q_0 . \tag{3.2.18}$$

For what follows it will be convenient to transform the solutions to such a form that the initial solution matrix (3.2.17) is transformed into a unity matrix. To this end we write the solution matrix in the form (omitting the argument φ)

$$Q(t) = U(t) Q_0 \quad \text{with} \quad U(0) = 1 . \tag{3.2.19}$$

From (3.2.19) follows

$$\dot{Q} = \dot{U}(t) Q_0 \quad \text{with} \quad \dot{U} = MU . \tag{3.2.20}$$

Inserting (3.2.19) into (3.2.20) and multiplying both sides from the left by Q_0^{-1} we obtain

$$Q_0^{-1} \frac{d}{dt} U Q_0 = \underbrace{Q_0^{-1} M Q_0}_{\tilde{M}} \, \underbrace{Q_0^{-1} U Q_0}_{Q} . \tag{3.2.21}$$

Introducing

$$Q_0^{-1} Q = \tilde{Q} \tag{3.2.22}$$

as a new solution matrix we may write (3.2.21) in the form [compare (3.1.1)]

$$\dot{\tilde{Q}}(t) = \tilde{M} \tilde{Q} , \tag{3.2.23}$$

where $\tilde{Q}$ obeys the initial condition

$$\tilde{Q}(0) = 1 . \tag{3.2.24}$$

Because each of the transformations can be made in a backwards direction it suffices for our discussion to consider the new problem (3.2.23, 24). Entirely analogous to the procedure in Sect. 2.7, we can show that the new solution vectors $\tilde{q}^{(j)}$ are again of the form (3.1.11) with the same generalized characteristic exponents as before.

From now on we shall use the transformed system (3.2.23, 24), but drop the tilde.

Lemma 3.2.5. The transformation matrix C (cf. 2.7.7) for the system (3.2.23, 24) is given by

$$C(\tau, \varphi) = Q(\tau, \varphi + \tau) . \tag{3.2.25}$$

Proof: Using the definition of T_τ (3.1.6) we may write (3.1.9) in the form

$$Q(t+\tau, \varphi+\tau) = Q(t, \varphi) C(\tau, \varphi) . \tag{3.2.26}$$

When we choose $t = 0$ and apply (3.2.24), we obtain (3.2.25). Because $Q(t, \varphi)$ is a nonsingular matrix (cf. Theorem 2.4.1),

$$C^{-1}(\tau, \varphi) = Q^{-1}(\tau, \varphi + \tau) \tag{3.2.27}$$

exists for all τ and φ.

By means of (3.2.25) we cast (3.2.26) into the form

$$Q(t+\tau, \varphi+\tau) = Q(t, \varphi) Q(\tau, \varphi+\tau) . \tag{3.2.28}$$

Replacing everywhere φ by $\varphi - \tau$ we find after a slight rearrangement

$$Q(t, \varphi - \tau) = Q(t+\tau, \varphi) Q^{-1}(\tau, \varphi) . \tag{3.2.29}$$

In the special case $\varphi = \varphi_0$ we shall put

$$C(\tau, \varphi_0) = C(\tau) . \tag{3.2.30}$$

Lemma 3.2.6. Let us introduce the transposed matrix belonging to $Q(\tau, \varphi)$

$$\Gamma = (Q^{-1})^T \tag{3.2.31}$$

so that we arrive at the following form of (3.2.29) with $\varphi = \varphi_0$

$$q^{(j)}(t, \varphi_0 - \tau) = \sum_{k=1}^{m} \Gamma_{jk}(\tau) q^{(k)}(t+\tau, \varphi_0) , \quad j = 1, \ldots, m . \tag{3.2.32}$$

We put

$$\Gamma_{jk}(\tau) \equiv \Gamma_{jk}(\tau, \varphi_0) = \mathrm{e}^{-z_k(\tau)} D_{jk}(\tau) . \tag{3.2.33}$$

Then the assertions hold:

1) $|D_{jk}(\tau)| < d_1$, d_1 independent of τ;
2) $D_{jk}(\tau)$ is C^1 with respect to τ.

Proof: Assertion (1) follows immediately from the definition of the inverse of a matrix from (3.2.30, 25) and from assumption (3) of Theorem 3.1.1. The proof of assertion (2) follows from the decomposition of $\boldsymbol{q}^{(j)}$ (which is at least C^1 with respect to τ) in a real factor $\exp(z_j t)$ and a vector $\boldsymbol{u}_j(t)$ of unit length.

3.3 Proof of Assertion (a) of Theorem 3.1.1: Construction of a Triangular Matrix: Example of a 2 × 2 Matrix

Since all the basic ideas and essential steps can be seen by the simple case in which M and thus Q are 2×2 matrices, we start with this example and put

$$Q(t, \varphi_0) = \{e^{z_1(t)}\boldsymbol{u}_1(t), e^{z_2(t)}\boldsymbol{u}_2(t)\}. \tag{3.3.1}$$

We may write (3.2.32) in the form

$$\boldsymbol{q}^{(1)}(t, \varphi_0 - \tau) = \Gamma_{11}(\tau)\boldsymbol{q}^{(1)}(t+\tau, \varphi_0) + \Gamma_{12}(\tau)\boldsymbol{q}^{(2)}(t+\tau, \varphi_0) \tag{3.3.2}$$

$$\boldsymbol{q}^{(2)}(t, \varphi_0 - \tau) = \Gamma_{21}(\tau)\boldsymbol{q}^{(1)}(t+\tau, \varphi_0) + \Gamma_{22}(\tau)\boldsymbol{q}^{(2)}(t+\tau, \varphi_0). \tag{3.3.3}$$

Let us consider (3.3.2, 3) in more detail. On the lhs, $\boldsymbol{q}^{(j)}(t, \varphi_0 - \tau)$ is a subset of the functions $\boldsymbol{q}^{(j)}(t, \varphi)$ which are periodic in φ_j and C^k with respect to φ. In particular, the functions on the lhs of (3.3.2, 3) are quasiperiodic in τ. Because the ω's are assumed irrational with respect to each other, $\varphi_0 - \tau$ lies dense in φ or, more precisely,

$$(\varphi_{0,j} - \tau) \bmod \frac{2\pi}{\omega_j} \quad \text{lies dense in } \varphi_j, \quad 0 \leqq \varphi_j \leqq \frac{2\pi}{\omega_j}. \tag{3.3.4}$$

As a consequence of this and the C^k property of $\boldsymbol{q}^{(j)}(t, \varphi)$, $\boldsymbol{q}^{(j)}(t, \varphi_0 - \tau)$ lies dense in the space $\boldsymbol{q}^{(j)}(t, \varphi)$. On the rhs the asymptotic behavior of $\boldsymbol{q}^{(j)}$ for $t \to \infty$ is known. According to our assumption, $\boldsymbol{q}^{(j)}$, $j = 1, 2$ possess different generalized characteristic exponents λ_j. We wish to form new solutions of (3.1.5), $\hat{\boldsymbol{q}}^{(1)}$ and $\hat{\boldsymbol{q}}^{(2)}$, which combine both these features, namely asymptotic behavior ($\hat{\boldsymbol{q}}^{(j)}$ shall possess the generalized characteristic exponent λ_j) and quasiperiodicity in the argument $\varphi_0 - \tau$. Take $\lambda_1 > \lambda_2$. In order to construct $\hat{\boldsymbol{q}}^{(2)}$ we multiply (3.3.2) by $\alpha(\tau)$ and (3.3.3) by $\beta(\tau)$, forming

$$\hat{\boldsymbol{q}}^{(2)}(t, \varphi_0 - \tau, \tau) = \alpha(\tau)\boldsymbol{q}^{(1)}(t, \varphi_0 - \tau) + \beta(\tau)\boldsymbol{q}^{(2)}(t, \varphi_0 - \tau). \tag{3.3.5}$$

In order that $\hat{\boldsymbol{q}}^{(2)}$ no longer contains the generalized characteristic exponent λ_1, we require

$$\alpha(\tau)\Gamma_{11}(\tau)\boldsymbol{q}^{(1)}(t+\tau,\varphi_0)+\beta(\tau)\Gamma_{21}(\tau)\boldsymbol{q}^{(1)}(t+\tau,\varphi_0)=0\,, \tag{3.3.6}$$

which can obviously be fulfilled because the vector $\boldsymbol{q}^{(2)}$ drops out of this equation.

By means of (3.2.33) (Lemma 3.2.6) we may transform (3.3.6) into

$$\alpha(\tau)\boldsymbol{u}_1(t+\tau)D_{11}(\tau)+\beta(\tau)\boldsymbol{u}_1(t+\tau)D_{21}(\tau)=0\,, \tag{3.3.7}$$

where the D's are bounded.

The solution of (3.3.7) can be written in the form

$$\alpha(\tau)=-D_{21}(\tau)[|D_{11}(\tau)|^2+|D_{21}(\tau)|^2]^{-1/2} \tag{3.3.8}$$

$$\beta(\tau)=\quad D_{11}(\tau)[|D_{11}(\tau)|^2+|D_{21}(\tau)|^2]^{-1/2}\,, \tag{3.3.9}$$

where due to the arbitrariness of the solution of the homogeneous equations we may but need not include the denominators. Because of the linear independence of the solutions the denominator does not vanish.

Using the same α and β we may construct $\hat{q}_1$ by means of

$$\hat{\boldsymbol{q}}^{(1)}(t,\varphi_0-\tau,\tau)=\beta^*(\tau)\boldsymbol{q}^{(1)}(t,\varphi_0-\tau)-\alpha^*(\tau)\boldsymbol{q}^{(2)}(t,\varphi_0-\tau)\,. \tag{3.3.10}$$

Namely, using (3.3.8, 9) and (3.3.2, 3) with (3.2.33), $\hat{\boldsymbol{q}}^{(1)}$ reads explicitly

$$\begin{aligned}\hat{\boldsymbol{q}}^{(1)}(t,\varphi_0-\tau,\tau)=&\,(|D_{11}(\tau)|^2+|D_{21}(\tau)|^2)^{1/2}\mathrm{e}^{-z_1(\tau)}\boldsymbol{q}^{(1)}(t+\tau,\varphi_0)\\&+\text{terms containing }\boldsymbol{q}^{(2)}(t+\tau,\varphi_0)\,.\end{aligned} \tag{3.3.11}$$

(Here again $D_{11}(\tau)$ and $D_{21}(\tau)$ cannot vanish simultaneously, because otherwise the solution vectors $\boldsymbol{q}^{(j)}(t,\varphi_0-\tau)$ would become linearly dependent in contrast to our assumption. Therefore the factor in front of $\boldsymbol{q}^{(1)}(t+\tau,\varphi_0)$ does not vanish.) Therefore the choice (3.3.10) secures that $\hat{q}_1$ possesses the characteristic exponent λ_1. In this way we have constructed two new solutions $\hat{q}_1$, $\hat{q}_2$ which are connected with the generalized characteristic exponents λ_1 and λ_2, respectively.

When we use these new solutions vectors instead of the old ones, the matrix C (3.1.18) appears in triangular form, as can be easily demonstrated and as will be shown explicitly in Sect. 3.7.

3.4 Proof that the Elements of the Triangular Matrix C are Quasiperiodic in τ (and Periodic in φ_j and C^k with Respect to φ): Example of a 2 × 2 Matrix

Firstly we show that (perhaps up to a common factor) $\alpha(\tau)$ and $\beta(\tau)$ can be approximated asymptotically, i.e., to any desired degree of accuracy by functions which are quasiperiodic in τ. To this end we write the coefficients of $\alpha(\tau)$ in (3.3.6) or (3.3.7) by use of (3.3.2) in the form

$$\Gamma_{11}(\tau)\boldsymbol{q}^{(1)}(t+\tau,\varphi_0) = \boldsymbol{q}^{(1)}(t,\varphi_0-\tau) - \Gamma_{12}(\tau)\boldsymbol{q}^{(2)}(t+\tau,\varphi_0)\,. \tag{3.4.1}$$

Similarly, we proceed with the coefficients of $\beta(\tau)$ in (3.3.6). For the following it will be sufficient to demonstrate how to deal with the coefficient of $\alpha(\tau)$ as an example. In order to ensure that the coefficients of $\alpha(\tau)$ and $\beta(\tau)$ remain bounded for all positive times t and τ, instead of (3.4.1) we form

$$\begin{aligned}\mathcal{N}(t,\varphi_0-\tau)\Gamma_{11}(\tau)\boldsymbol{q}^{(1)}(t+\tau,\varphi_0) &= \mathcal{N}(t,\varphi_0-\tau)\boldsymbol{q}^{(1)}(t,\varphi_0-\tau)\\ &\quad - \mathcal{N}(t,\varphi_0-\tau)\Gamma_{12}(\tau)\boldsymbol{q}^{(2)}(t+\tau,\varphi_0)\,,\end{aligned} \tag{3.4.2}$$

where

$$\mathcal{N}(t,\varphi_0-\tau) = [|q^{(1)}(t,\varphi_0-\tau)|^2 + |q^{(2)}(t,\varphi_0-\tau)|^2]^{-1/2}\,. \tag{3.4.3}$$

We first present an argument, rigorously developed later, whereby we assume that for $t\to\infty$ the time dependence of $z_2(t+\tau)$ is given by $z_2(t+\tau)\approx\lambda_2\cdot(t+\tau)$ and that of $\boldsymbol{q}^{(1)}(t,\varphi_0-\tau)$ by $\exp(\lambda_1 t)$. Let us assume that $\lambda_1>\lambda_2$. Then for $t\to\infty$ the second term in (3.4.2) vanishes. That means that we can approximate the coefficient of $\alpha(\tau)$ in (3.3.7) and correspondingly of β in (3.3.7) by expressions which contain $q^{(1)}(t,\varphi_0-\tau)$ and $q^{(2)}(t,\varphi_0-\tau)$ alone. But because $q^{(j)}(t,\varphi_0-\tau)$ is dense in $q^{(j)}(t,\varphi)$ we may express the coefficients of $\alpha(\tau)$, $\beta(\tau)$ to any desired degree of accuracy by functions which are quasiperiodic in τ and even C^k with respect to φ. Therefore we may embed $\alpha(\tau)$ and $\beta(\tau)$ in the set of functions $\alpha(\varphi)$ and $\beta(\varphi)$ which are quasiperiodic ($\varphi=\varphi_0-\tau$) in τ, and T_j-periodic and C^k in φ.

Now let us cast that idea in a more precise form. Using (3.1.11, 2.33) we write (3.3.2 and 3) in the form

$$\begin{aligned}\boldsymbol{q}^{(j)}(t,\varphi_0-\tau) &= \exp[z_1(t+\tau)-z_1(\tau)]\boldsymbol{u}^{(1)}(t+\tau)D_{j1}(\tau)\\ &\quad + \exp[z_2(t+\tau)-z_2(\tau)]\boldsymbol{u}^{(2)}(t+\tau)D_{j2}(\tau)\,,\quad j=1,2\,.\end{aligned} \tag{3.4.4}$$

We first choose τ within

$$\tau_1<\tau<\tau_2\,. \tag{3.4.5}$$

With help of $\mathcal{N}$ (3.4.3) we form

$$\mathcal{N}(t,\varphi_0-\tau)\boldsymbol{q}^{(j)}$$

and the corresponding expressions on the rhs of (3.4.4).

When both sides of the numerator and denominator of this new equation are divided by

$$\exp[z_1(t+\tau)-z_1(\tau)]\,, \tag{3.4.6}$$

we see that the coefficients of $D_{j2}(\tau)$ in (3.4.4) are multiplied by

$$\exp\,[z_2(t+\tau) - z_2(\tau) - z_1(t+\tau) + z_1(\tau)]\,. \tag{3.4.7}$$

We now recall the definition of the generalized characteristic exponents, according to which

$$\limsup_{t\to\infty}\left[\frac{1}{t}z_j(t)\right] = \lambda_j\,. \tag{3.4.8}$$

Accordingly, we may find a sequence of times

$$t_n \to \infty \quad \text{that} \quad \frac{1}{t_n}z_1(t_n) > \lambda_1 - \delta\,, \quad \delta > 0 \tag{3.4.9}$$

and for each τ, $\tau_1 < \tau < \tau_2$, a corresponding sequence so that

$$\frac{1}{t_n}[z_1(t_n+\tau) - z_1(\tau)] > \lambda_1 - \delta'\,, \quad \delta' > 0\,. \tag{3.4.10}$$

For t_n sufficiently large, δ' can be chosen as small as we wish. On the other hand, because of

$$\limsup\left\{\frac{1}{t}z_2(t)\right\} = \lambda_2\,, \tag{3.4.11}$$

we may find some $\delta'' > 0$ so that for the same sequence t_n and $\tau_1 < \tau < \tau_2$

$$\frac{1}{t_n}[z_2(t_n+\tau) - z_2(\tau)] < \lambda_2 + \delta''\,, \tag{3.4.12}$$

and δ'' may be chosen arbitrarily small as t_n is sufficiently large. Taking (3.4.10 and 12) together we see that (3.4.7) goes to zero for $t_n \to \infty$.

Consequently, we may replace everywhere in (3.3.7) for $\tau_1 < \tau < \tau_2$

$$u_1(t+\tau)D_{j1}(\tau) \tag{3.4.13}$$

by

$$\limsup_{t\to\infty}\{\mathcal{N}(t, \varphi_0 - \tau)q^{(j)}(t, \varphi_0 - \tau)\}\,, \tag{3.4.14}$$

which is quasiperiodic in τ. This allows us to construct α and β as quasiperiodic functions by making the corresponding replacement of D_{j1} in (3.3.8, 9) where we may eventually let $\tau_2 \to \infty$.

Provided assumption (5) of Theorem 3.1.1 is fulfilled, we may again replace (3.4.13) by (3.4.14) in addition embedding $\mathcal{N}(t, \varphi_0 - \tau)q_j(t, \varphi_0 - \tau)$ within

$\mathcal{N}(t,\varphi)q^{(j)}(t,\varphi)$. Because this expression is periodic and C^k in φ, so are $\alpha(\tau)\rightarrow\alpha(\varphi)$ and $\beta(\tau)\rightarrow\beta(\varphi)$.

Let us summarize what we have achieved so far. Under assumptions (1 – 4) of Theorem 3.1.1, we have constructed a complete set of solutions to (3.1.5),

$$\tilde{q}^{(j)}(t,\varphi_\tau)\,,\quad \varphi_\tau\equiv\varphi_0-\tau\,, \tag{3.4.15}$$

with the following properties: $\tilde{q}^{(j)}$ is (at least) C^1 with respect to t and τ, and it is quasiperiodic in τ.

Under assumptions (1 – 5) of Theorem 3.1.1 we have constructed a complete set of solutions to (3.1.5),

$$\tilde{q}^{(j)}(t,\varphi) \tag{3.4.16}$$

with the following properties: $\tilde{q}^{(j)}$ is (at least) C^1 with respect to t, and T_j-periodic and C^k with respect to φ. The set $\tilde{q}^{(j)}(t,\varphi_\tau)$ lies densely in the set $\tilde{q}^{(j)}(t,\varphi)$. In both (3.4.15, 16) the generalized characteristic exponents (for $t\rightarrow+\infty$) of $\tilde{q}^{(j)}$ are given by λ_j.

3.5 Construction of the Triangular Matrix C and Proof that Its Elements are Quasiperiodic in τ (and Periodic in φ_j and C^k with Respect to φ): The Case of an $m\times m$ Matrix, all λ's Different

So far we have considered the special case in which M is a 2×2 matrix. Now let us turn to the general case of an $m\times m$ matrix, again assuming that the characteristic exponents are all different from each other

$$\lambda_1>\lambda_2>\ldots>\lambda_m\,. \tag{3.5.1}$$

We wish to construct linear combinations of $q^{(j)}(t,\varphi_0-\tau)$ in the form

$$\hat{q}^{(l)}(t,\varphi_0-\tau,\tau)=\sum_j\alpha_j^{(l)}(\tau)q^{(j)}(t,\varphi_0-\tau)\,,\quad l=1,\ldots,m\,, \tag{3.5.2}$$

such that only one of these linear combinations has the generalized characteristic exponent λ_1 but that all other $m-1$ linear combinations have at maximum a generalized characteristic exponent λ_2.

Furthermore, we want to show that $\alpha_j^{(l)}$ can be chosen as a quasiperiodic function in τ or, more specifically, that we can embed $\alpha_j^{(l)}(\tau)$ in functions $\alpha_j^{(l)}(\varphi)$ which are T_j-periodic and C^k with respect to φ. For simplicity, we shall drop the index l in the following. We first introduce a normalization factor by

$$\mathcal{N}(t,\varphi_0-\tau)=\left[\sum_{j=1}^m|q^{(j)}(t,\varphi_0-\tau)|^2\right]^{-1/2} \tag{3.5.3}$$

and form

$$\chi^{(j)}(t, \varphi_0-\tau) = \mathcal{N}(t, \varphi_0-\tau)\, q^{(j)}(t, \varphi_0-\tau)\,. \tag{3.5.4}$$

According to (3.1.11), (3.2.32, 33) we may replace the rhs of (3.5.4) by

$$\sum_k \mathcal{N}(t, \varphi_0-\tau) D_{jk}(\tau) \exp\,[z_k(t+\tau) - z_k(\tau)]\, u^{(k)}(t+\tau)\,. \tag{3.5.5}$$

Let us use a sequence $t_n \to \infty$ defined by (3.4.10). We shall denote such a sequence by $\underset{t\to\infty}{\text{Lim Sup}}$. We readily obtain

$$\underset{t\to\infty}{\text{Lim Sup}}\,[\chi^{(j)}(t, \varphi_0-\tau)] = \underset{t\to\infty}{\text{Lim Sup}} \left\{ \frac{D_{j1}(\tau) u^{(1)}(t+\tau)}{\left[\sum\limits_{j=1}^{m} |D_{j1}(\tau)|^2\right]^{1/2}} \right\} \tag{3.5.6}$$

which may also be written in the form

$$\xi^{(j)} = \frac{D_{j1}(\tau)}{\left[\sum\limits_{j=1}^{m} |D_{j1}(\tau)|^2\right]^{1/2}} \underset{t\to\infty}{\text{Lim Sup}}\, \{u^{(1)}(t+\tau)\}\,. \tag{3.5.7}$$

When the coefficients $\alpha_j(\tau)$ of (3.5.2) are subjected to the equation

$$\sum_j \alpha_j \xi^{(j)} = 0\,, \tag{3.5.8}$$

where according to (3.5.7) all vectors $\xi^{(j)}$ are parallel, we can determine $m-1$ linearly independent solution vectors

$$\alpha^{(l)} = (\alpha_1^{(l)}, \alpha_2^{(l)}, \dots, \alpha_m^{(l)})\,, \quad l = 2, \dots, m \tag{3.5.9}$$

which fulfill (3.5.8).

The use of these α's in (3.5.2) guarantees that $q^{(l)}$ in (3.5.2) possesses a generalized characteristic exponent λ_2 or smaller. Choosing $\sum_j \alpha_j \xi^{(j)} \neq 0$ for any nonvanishing vector component of $\xi^{(j)}$, we can construct a $\hat{q}^{(1)}$ with the generalized characteristic exponent λ_1.

We now want to show that we may choose $\alpha_j(\tau)$ as quasiperiodic in τ, or that we may even embed $\alpha_j(\tau)$ in $\alpha_j(\varphi)$. We note that due to its construction [cf. (3.5.6)], $\xi^{(j)}$ is quasiperiodic in τ and can be embedded in a set of functions $\xi^{(j)}(\varphi)$ which are T_j-periodic and C^k with respect to φ. Under assumption (5) of Theorem 3.1.1, Lim Sup converges uniformly in τ towards the rhs of (3.5.7). This allows one to exchange the performance of "Lim Sup" and the dense embedding of $\xi_j(\tau)$ in $\xi_j(\varphi)$. Using linear algebra we may construct $\alpha^{(l)}$ in such a way that it possesses the same differentiability and periodicity properties as the coefficients $\xi^{(j)}$.

Since a detailed presentation of this construction will diverge too far from our main line, we merely indicate how this construction can be visualized. Let us consider one vector component of the vector equation (3.5.8) [in which all vectors $\xi^{(j)}$ are parallel, cf. (3.5.7)]. Let us form a vector $\hat{\xi} = (\xi_k^{(1)}, \xi_k^{(2)}, \ldots, \xi_k^{(m)})$, where k is chosen such that $\hat{\xi} \neq 0$ which is always possible. Then the corresponding vector component of (3.5.8) can be written as

$$\alpha \hat{\xi} = 0 , \tag{3.5.10}$$

i.e., as the scalar product between $\hat{\xi}$ and the vector $\alpha (= \alpha^{(l)})$ as introduced by (3.5.9). In other words, we seek the $m-1$ vectors $\alpha^{(l)}$, $l = 2, \ldots, m$, which span a vector space orthogonal to the vector $\hat{\xi}$. Now, when we smoothly change the direction of $\hat{\xi}$, we may smoothly change the vectors of the space orthogonal to $\hat{\xi}$. (More generally, "smoothly" can be replaced by C^k.) A simple example of an algebraic construction will be given below.

For pedagogical reasons it must be added that it seems tempting to divide (3.5.8) by the common "factor" Lim Sup $\{u^{(1)}(t_n + \tau)\}$. Then the difficulty occurs that it is not proven that $D_{j1}(\tau)$ and this factor have the required differentiability properties with respect to τ (or φ) individually. Therefore we must not divide (3.5.8), which contains (3.5.7) via ξ^j, by $\operatorname{Lim\,Sup}_{t_n \to \infty} \{u^{(1)}(t_n + \tau)\}$. If we want to get rid of $u^{(1)}$, we may form the quantities

$$\underbrace{\operatorname*{Lim\,Sup}_{t \to \infty} \{\chi^{(j)}(t, \varphi_0 - \tau)\, \chi^{(k)*}(t, \varphi_0 - \tau)\}}_{C^k \text{ in } \varphi} = \eta_{jk}(\tau) , \tag{3.5.11}$$

where $\eta_{jk}(\tau)$ is C^k with respect to τ, provided M was C^k with respect to φ. Since $\varphi_0 - \tau$ lies dense on the torus we can embed (3.5.11) in functions which are C^k with respect to φ and T_j-periodic. Then (3.5.11) can be evaluated in analogy to (3.5.6) which yields

$$\frac{D_{j1}(\tau)}{\left(\sum_{j=1}^{m} |D_{j1}(\tau)|^2\right)} D_{k1}^*(\tau) \underbrace{|u^{(1)}(t+\tau) u^{(1)*}(t+\tau)|}_{=1} . \tag{3.5.12}$$

Now $u^{(1)}$ drops out. In order to construct new linear combinations of (3.5.2) which have a characteristic exponent λ_2 or smaller we require

$$\sum_{j=1}^{m} \alpha_j(\tau) \eta_{jk}(\tau) = 0 \quad \text{for} \quad k = 1, \ldots, m . \tag{3.5.13}$$

We introduce the vectors

$$\eta_k = (\eta_{1k}, \eta_{2k}, \ldots, \eta_{mk}) \quad \text{for} \quad k = 1, \ldots, m \tag{3.5.14}$$

and readily deduce from (3.5.11) that in the limit $t \to \infty$ all these vectors are parallel.

Furthermore we introduce (3.5.9) so that (3.5.13) can be written as

$$\alpha \eta_k = 0 . \tag{3.5.15}$$

Since all η_k are parallel, the requirement (3.5.15) can be fulfilled by $m-1$ different vectors α. In this way we can construct $m-1$ new solutions (3.5.2) or, more generally, we can embed

$$\tilde{q}^{(l)}(t, \varphi) = \sum \alpha_j^{(l)} q^{(j)}(t, \varphi) . \tag{3.5.16}$$

Though for some or all τ some of the vectors η_k may vanish, on account of our previous assumptions at least one η_k must be unequal to 0.

As mentioned before, the α's of (3.5.15) can always be chosen in such a way that they have the same differentiability properties with respect to parameters as η_k. This statement is illustrated by a simple example in the case of an even m:

$$\alpha_j = \prod_{l \neq j} \eta_{lk} (-1)^j . \tag{3.5.17}$$

If this is inserted into (3.5.15) we readily convince ourselves that (3.5.15) is fulfilled and that α has the same differentiability properties as η_k. If one of the η_k vanishes one can go over to another η_k by multiplying α_j with a well-defined factor.

Summary of Results. We have shown how to construct new solutions $\tilde{q}$ of the original equation (3.1.5) which have the following properties. One of these solutions has the generalized characteristic exponent λ_1, all other $m-1$ solutions have the characteristic exponent λ_2 or still smaller ones. We can now perform precisely the same procedure with a group of the remaining $m-1$ solutions and reduce this group to one which contains only one solution with generalized characteristic exponent λ_2. The continuation of this procedure is then obvious. Clearly, the resulting coefficient matrix C is triangular as is explicitly shown in Sect. 3.7.

Since the original $q^{(j)}$ are C^k and T_j-periodic with respect to φ and the coefficients α can be constructed in such a way that they retain this property also, then the newly derived solutions are C^k and T_j-periodic with respect to φ.

3.6 Approximation Methods. Smoothing

We make two remarks on possible approximative methods.

3.6.1 A Variational Method

Instead of (3.5.15) we may also require

$$\sum_{k=1}^{m} |\boldsymbol{\alpha}\boldsymbol{\eta}_k|^2 = 0 \,. \tag{3.6.1}$$

If in an approximation scheme $\underset{t\to\infty}{\text{Lim Sup}}$ is not performed entirely but only up to $t = t_{n_0}$ then

$$\eta_{jk}(\tau) \to \eta_{jk}(\tau, t_{n_0}) \tag{3.6.2}$$

and we cannot fulfill (3.6.1) exactly because the vectors defined by (3.6.2) are not completely parallel. But now we can require

$$\sum_{k=1}^{m} |\boldsymbol{\alpha}\boldsymbol{\eta}_k|^2 = \text{minimum!}\,, \qquad |\boldsymbol{\alpha}|^2 = \text{fixed}\,. \tag{3.6.3}$$

The variational principle defined by (3.6.3) leads to a set of eigenvalue equations

$$\sum_{k=1}^{m} \eta_{jk}^*(\boldsymbol{\alpha}\boldsymbol{\eta}_k) = \lambda\,\alpha_j\,, \qquad j = 1, \ldots, m \tag{3.6.4}$$

which are linear in λ. Multiplying (3.6.4) by α_j^* and summing over j we obtain

$$\sum_{k=1}^{m} (\boldsymbol{\alpha}^*\boldsymbol{\eta}_k^*)(\boldsymbol{\alpha}\boldsymbol{\eta}_k) = \lambda\,|\boldsymbol{\alpha}|^2\,. \tag{3.6.5}$$

We thus see that the eigenvalues λ will be a measure as to how far we are still away from the exact equation (3.6.1).

3.6.2 Smoothing

Since some ξ_k or, equivalently, some $\boldsymbol{\eta}_k$, k fixed, in (3.5.15) can vanish for some τ, we have to switch to another equation stemming from another $\boldsymbol{\eta}_j$ (or ξ_j). If we have not performed the limit $t_n \to \infty$ in Lim Sup, these two $\boldsymbol{\eta}$'s may not entirely coincide. This in turn would mean that at those switching points $\boldsymbol{\alpha}$ becomes discontinuous and nondifferentiable. We want to show that we can easily smooth the switching. To this end we introduce the new vector

$$\hat{\boldsymbol{\eta}}_k = [1 - \beta(\tau)]\,\boldsymbol{\eta}_k + \beta(\tau)\,\boldsymbol{\eta}_j(\tau)\,, \tag{3.6.6}$$

where β is given qualitatively in Fig. 3.6.1, and replace (3.5.15) by

$$\boldsymbol{\alpha}\,\hat{\boldsymbol{\eta}}_k = 0\,. \tag{3.6.7}$$

Since due to the linear independence of the solutions not all $\boldsymbol{\eta}_j$ can vanish for the same τ, we may choose such an index j that (3.6.6) does not vanish where $\boldsymbol{\eta}_k$ vanishes. We define a variable x by

$$x = \frac{\tau - \tau_0}{\tau_1 - \tau_0} \tag{3.6.8}$$

which secures

$$x = 0\,, \qquad \tau = \tau_0 \tag{3.6.9}$$

and

$$x = 1\,, \qquad \tau = \tau_1\,. \tag{3.6.10}$$

We choose τ_0 and τ_1 such that η_k is replaced by η_j before η_k vanishes. On the other hand, we choose τ_0 and τ_1 such that $|\eta_k|$ is smaller than $|\eta_j|$ so that (3.6.6) cannot vanish due to compensation of the η's. We define β by

$$\begin{aligned} &\beta = 0\,, && -\infty < x \leqslant 0 \\ &\beta = h(x)\,, && 0 \leqslant x \leqslant 1 \\ &\beta = 1\,, && 1 \leqslant x < \infty\,. \end{aligned} \tag{3.6.11}$$

In order that the "spline function" (3.6.6) is sufficiently often continuously differentiable with respect to τ, we require in addition for a given n

$$\begin{aligned} &\beta^{(m)} = 0\,, \quad 1 \leqslant m \leqslant n-1 \\ &\text{for} \quad x = 0 \quad \text{and} \quad x = 1\,. \end{aligned} \tag{3.6.12}$$

Here and in the following the superscript (m) denotes the m'th derivative.

Since the points τ at which we have to apply this smoothing procedure have the same properties of quasiperiodicity as η_k, η_j, the newly constructed β's will also have this property. When we let $t \to \infty$, our above procedure guarantees that during the whole approach β remains C^n and quasiperiodic.

The following short discussion shows how to construct β in (3.6.11, 12) explicitly. This outline is not, however, important for what follows. In order to fulfill (3.6.12) we put

$$h(x) = x^n f(x)\,, \tag{3.6.13}$$

where f is a polynomial of order n

$$f(x) = \sum_{l=0}^{n} a_l x^l \tag{3.6.14}$$

where we require

$$a_0 = 1 \; . \tag{3.6.15}$$

To determine f we introduce the variable

$$x = 1 + \xi \tag{3.6.16}$$

and put

$$f(1+\xi) = g(\xi) \; , \tag{3.6.17}$$

where g is again a polynomial which we may write in the form

$$g(\xi) = \sum_{l=0}^{n} b_l \xi^l \; . \tag{3.6.18}$$

We now choose the coefficients a or b, respectively, in such a way that

$$h(1) = 1 \quad \text{and} \tag{3.6.19}$$

$$h^{(m)}(1) = 0 \; , \quad m = 1, \ldots, n-1 \tag{3.6.20}$$

are fulfilled, where the superscript m in (3.6.20) means derivative with respect to x.

From (3.6.19) we immediately find

$$g(0) = 1 \tag{3.6.21}$$

and thus

$$b_0 = 1 \; . \tag{3.6.22}$$

From (3.6.20) with $m = 1$ we readily obtain

$$n g(0) + g'(0) = 0 \quad \text{and} \tag{3.6.23}$$

$$b_1 = -n \; . \tag{3.6.24}$$

One may determine all other coefficients b_l by use of (3.6.20) consecutively. Thus one may indeed find a polynomial with the required properties. We still have to convince ourselves that the thus constructed solution (3.6.13) has everywhere a positive slope in the interval [0, 1], i.e., that $h(x)$ has no wiggles. Since h is a polynomial of order $2n$, it can have $2n-1$ different extrema at most, which are just the zeros of $h^{(1)}$. According to our construction $n-1$ extrema coincide at $x = 0$ and another $n-1$ extrema coincide at $x = 1$. Therefore only one additional extremum remains at most. One readily shows that

$$h'(0+\varepsilon) > 0 \quad \text{and} \tag{3.6.25}$$

$$h'(1-\varepsilon) > 0 \tag{3.6.26}$$

hold for small positive ε. This implies that a wiggled curve must have at least two more extrema as shown in Fig. 3.6.2, in contrast to our results predicting that there could be at most only one additional extremum. Therefore, the case of Fig. 3.6.2 cannot apply and (3.6.13) can represent only the curve like Fig. 3.6.1. This completes the excursion.

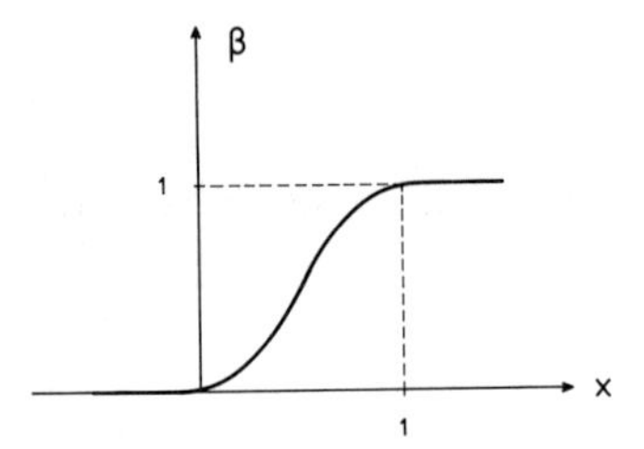

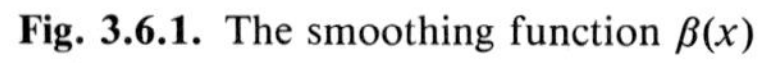
Fig. 3.6.1. The smoothing function $\beta(x)$

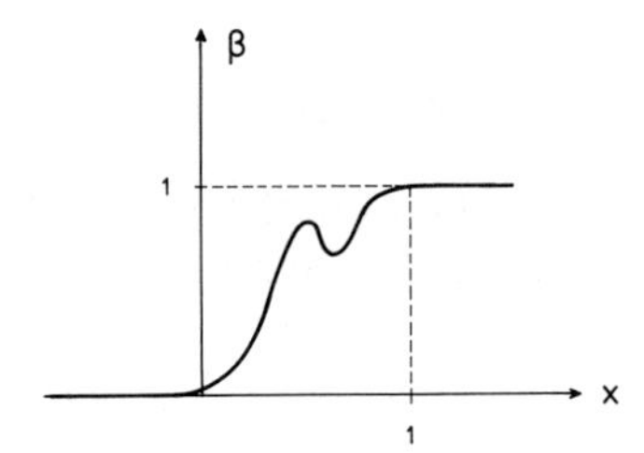

Fig. 3.6.2. This figure shows that β has two extrema in contrast to our assumption made in the text

So far we have discussed how we can bridge the jumps when η_k vanishes as a function of τ. Now we want to show that a similar procedure holds when we treat η_k as a function of φ via embedding. Consider a $\varphi \leftrightarrow \varphi_0 - \tau$ for which $D^*_{k1}(\tau) \neq 0$, k fixed. Because $q^{(j)}(t,\varphi)$ is C^k, $k \geqslant 0$, a whole surrounding $S(\varphi)$ of φ exists where each point in $S(\varphi)$ is approximated by sequences $\varphi - \tau_n$ (mod $\boldsymbol{T}$), $\boldsymbol{T} = (T_1, \ldots, T_n)$, $T_j = 2\pi/\omega_j$, and for which $D^*_{k1}(\tau_n) \neq 0$, k fixed. In this way, the whole torus can be covered by overlapping patches of nonvanishing $D^*_{j1}(\tau)$. On each of these patches at least one nonvanishing vector η_k exists. Now the only thing to do is to smooth the transition when going from one patch to a neighboring one. But the size of these patches, being defined by $D^*_{j1}(\tau) \neq 0$, is independent of t and thus fixed, whereas the misalignment of the vectors η_k tends to zero for $t \to \infty$. Consequently the smoothing functions remain sufficiently often differentiable for $t \to \infty$.

3.7 The Triangular Matrix C and Its Reduction

Since $Q(t,\varphi)$ or $\tilde{Q}(t,\varphi)$ (3.5.16) are solutions to (3.1.5) we may again invoke the general relation (3.1.10). Writing down this relation for the solution vectors explicitly we obtain ($G = C^T$)

$$\tilde{q}^{(j)}(t+\tau, \varphi+\tau) = \sum_k G_{jk}(\tau,\varphi)\tilde{q}^{(k)}(t,\varphi), \quad j = 1, \ldots, m, \tag{3.7.1}$$

where the generalized characteristic exponents λ_j which belong to $\tilde{q}^{(j)}$ are ordered according to

$$\lambda_1 > \lambda_2 > \lambda_3 \ldots > \lambda_m . \tag{3.7.2}$$

Because the solution matrix is nonsingular we may derive from the properties of $\tilde{Q}(t, \varphi)$ in Sect. 3.5 the following properties for the coefficients G_{jk}.

If assumptions (1 – 5) of Theorem 3.1.1 hold, the coefficients $G_{jk}(\tau, \varphi)$ are differentiable with respect to τ and are T_j-periodic and C^k with respect to φ. If the weaker assumptions (1 – 4) of Theorem 3.1.1 hold, the coefficients are of the form $G_{jk}(\tau, \varphi_0 - \tau)$, they are (at least) C^1 with respect to τ and C^0 with respect to the argument $\varphi_\tau (= \varphi_0 - \tau)$ and quasiperiodic in τ. Now the asymptotic behavior of the lhs of (3.7.1) must be the same as that of the rhs of (3.7.1) for $t \to \infty$. This implies that

$$G_{jk} = 0 \quad \text{for} \quad k < j . \tag{3.7.3}$$

The reasoning is similar to that performed above and is based on the study of the asymptotic behavior of $\tilde{q}^{(k)}$ for $t \to \infty$.

In the next step we want to show that it is possible to transform the $\tilde{q}$'s in such a way that eventually the matrix G_{jk} can be brought into diagonal form. To this end we need to know the asymptotic behavior of $\tilde{q}$ for $t \to -\infty$. In order to obtain such information we solve the equations (3.7.1) step by step starting with the equation for $j = m$,

$$\tilde{q}^{(m)}(t+\tau, \varphi+\tau) = G_{mm}(\tau, \varphi)\tilde{q}^{(m)}(t, \varphi) . \tag{3.7.4}$$

We assume that τ becomes an infinitesimal. Because G_{mm} is differentiable with respect to τ and for $\tau = 0$ the q's on both sides of (3.7.4) coincide, we may write

$$G_{mm}(\tau, \varphi) = 1 + a_m(\varphi)\tau . \tag{3.7.5}$$

Replacing t by $t + \tau$ and φ by $\varphi + \tau$ we obtain

$$\tilde{q}^{(m)}(t+2\tau, \varphi+2\tau) = G_{mm}(\tau, \varphi+\tau)\tilde{q}^{(m)}(t+\tau, \varphi+\tau) \tag{3.7.6}$$

instead of (3.7.4). On the rhs of (3.7.6) we may replace $\tilde{q}$ by the rhs of (3.7.4) which yields

$$\tilde{q}^{(m)}(t+2\tau, \varphi+2\tau) = G_{mm}(\tau, \varphi+\tau)G_{mm}(\tau, \varphi)\tilde{q}^{(m)}(t, \varphi) . \tag{3.7.7}$$

Repeating this procedure N times we readily find

$$\tilde{q}^{(m)}(t+\underbrace{N\tau}_{t_0}, \varphi+\underbrace{N\tau}_{t_0}) = \prod_{l=1}^{N} G_{mm}(\tau, \varphi+(l-1)\tau)\tilde{q}^{(m)}(t, \varphi) . \tag{3.7.8}$$

Using (3.7.5) and the fact that τ is infinitesimal we may replace the product over l by the exponential function which yields

$$\tilde{q}^{(m)}(t+t_0, \varphi+t_0) = \exp\left\{\sum_{l=1}^{N} \tau a(\varphi+(l-1)\tau)\right\}\tilde{q}^{(m)}(t, \varphi) . \tag{3.7.9}$$

Eventually taking the limit $\tau \to 0$ we may replace the sum in (3.7.9) by an integral

$$\tilde{q}^{(m)}(t+t_0, \varphi+t_0) = \exp\left[\int_0^{t_0} d\sigma\, a_m(\varphi+\sigma)\right]\tilde{q}^{(m)}(t, \varphi) . \tag{3.7.10}$$

We now make the replacement

$$t=0\,, \quad t_0 \to t\,, \quad \varphi \to \varphi - t\,, \tag{3.7.11}$$

which after a change of variables of the integral brings us from (3.7.10) to

$$\tilde{q}^{(m)}(t, \varphi) = \exp\left[\int_0^{t} d\sigma\, a_m(\varphi-\sigma)\right]\tilde{q}^{(m)}(0, \varphi-t) . \tag{3.7.12}$$

This equation can be considered as a functional equation for $\tilde{q}^{(m)}$. To solve this equation we make the hypothesis

$$\tilde{q}^{(m)}(t, \varphi) = \exp\left[\int_0^{t} d\sigma\, a_m(\varphi-\sigma)\right] w^{(m)}(t, \varphi) . \tag{3.7.13}$$

Inserting this into (3.7.12) we readily obtain the relation

$$w^{(m)}(t, \varphi) = w^{(m)}(0, \varphi-t) . \tag{3.7.14}$$

Since (3.7.13) was C^k with respect to φ and T_j-periodic in φ, we obtain the following result:

$$w^{(m)}(t, \varphi) \text{ is a quasiperiodic function in } t . \tag{3.7.15}$$

Thus the explicit solution of (3.7.12) reads

$$\tilde{q}^{(m)}(t, \varphi) = \exp\left[\int_0^{t} d\sigma\, a_m(\varphi-\sigma)\right] w^{(m)}(0, \varphi-t) . \tag{3.7.16}$$

We note that in it a_m is a quasiperiodic function of σ.

Now let us turn to the solution of the next equation for $q^{(m-1)}$. To elucidate the general idea we consider the example $m=2$. Then (3.7.1) acquires the form

$$\tilde{q}^{(1)}(t+\tau, \varphi+\tau) = G_{11}(\tau, \varphi)\tilde{q}^{(1)}(t, \varphi) + G_{12}(\tau, \varphi)\tilde{q}^{(2)}(t, \varphi) \tag{3.7.17}$$

$$\tilde{q}^{(2)}(t+\tau, \varphi+\tau) = G_{22}(\tau, \varphi)\tilde{q}^{(2)}(t, \varphi) . \tag{3.7.18}$$

We want to show that we can construct a new solution to our original equation (3.1.5) with the solution vector $\hat{q}^{(1)}$ so that the Eqs. (3.7.17 and 18) acquire a diagonal form. We note that any linear combination of $\tilde{q}^{(1)}$ and $\tilde{q}^{(2)}$ is again a solution of (3.1.5) provided the coefficients are time independent. We therefore make the hypothesis

$$\tilde{q}^{(1)}(t,\varphi) = \hat{q}^{(1)}(t,\varphi) + h(\varphi)\tilde{q}^{(2)}(t,\varphi) . \tag{3.7.19}$$

Inserting it into (3.7.17) we immediately obtain

$$\underline{\hat{q}^{(1)}(t+\tau,\varphi_0+\tau)} + h(\varphi+\tau)\tilde{q}^{(2)}(t+\tau,\varphi+\tau)$$
$$= \underline{G_{11}(\tau,\varphi)\hat{q}^{(1)}(t,\varphi)} + G_{11}(\tau,\varphi)h(\varphi)\tilde{q}^{(2)}(t,\varphi) + G_{12}(\tau,\varphi)\tilde{q}^{(2)}(t,\varphi) . \tag{3.7.20}$$

Since the underlined expressions are those which we want to keep finally, we require that the rest vanishes. Making use of (3.7.18) we are then led to

$$h(\varphi+\tau)G_{22}(\tau,\varphi) = G_{11}(\tau,\varphi)h(\varphi) + G_{12}(\tau,\varphi) . \tag{3.7.21}$$

We want to show that this equation can indeed be fulfilled, which is by no means obvious because G_{jk} are functions of both τ and φ whereas h is assumed to be a function of φ alone. In the following we shall show that the G's have such a structure that h can indeed be chosen as a function of φ alone.

With this task in mind, we turn to a corresponding transformation of the Eqs. (3.7.1), starting with the second last

$$\tilde{q}^{(m-1)}(t+\tau,\varphi+\tau) = G_{m-1,m-1}(\tau,\varphi)\tilde{q}^{(m-1)}(t,\varphi)$$
$$+ G_{m-1,m}(\tau,\varphi)\tilde{q}^{(m)}(t,\varphi) . \tag{3.7.22}$$

This equation is considered in the limit $\tau \to 0$. For $\tau = 0$, $G_{m-1,m-1} = 1$. Because $G_{m-1,m-1}$ is differentiable with respect to τ, for small τ

$$G_{m-1,m-1}(\tau,\varphi) = 1 + a_{m-1}(\varphi)\tau . \tag{3.7.23}$$

In precisely the same way as in the case of the last equation ($j = m$) of (3.7.1), we readily obtain for finite τ

$$G_{m-1,m-1}(\tau,\varphi) = \exp\left[\int_0^\tau d\sigma\, a_{m-1}(\varphi+\sigma)\right] . \tag{3.7.24}$$

To solve the inhomogeneous equation (3.7.22) we make a hypothesis familiar from ordinary differential equations, namely

$$\tilde{q}^{(m-1)}(t,\varphi) = \exp\left[\int_0^t d\sigma\, a_{m-1}(\varphi-\sigma)\right] g^{(m-1)}(t,\varphi) , \tag{3.7.25}$$

where $g^{(m-1)}$ is still unknown. Inserting (3.7.25) into (3.7.22) we immediately obtain

$$\exp\left[\int_0^{t+\tau} d\sigma\, a_{m-1}(\varphi+\tau-\sigma)\right] g^{(m-1)}(t+\tau,\varphi+\tau)$$

$$= \exp\left[\int_0^{\tau} d\sigma\, a_{m-1}(\varphi+\sigma)\right] \exp\left[\int_0^{t} d\sigma\, a_{m-1}(\varphi-\sigma)\right]$$

$$\cdot\, g^{(m-1)}(t,\varphi) + G_{m-1,m}(\tau,\varphi)\tilde{q}^{(m)}(t,\varphi)\,. \tag{3.7.26}$$

After dividing (3.7.26) by the exponential function on the lhs of (3.7.26) we obtain

$$g^{(m-1)}(t+\tau,\varphi+\tau) = g^{(m-1)}(t,\varphi) + f(t,\tau,\varphi)\,, \tag{3.7.27}$$

where we have used the abbreviation

$$f(t,\tau,\varphi) = G_{m-1,m}(\tau,\varphi) \exp\left[-\int_0^{t+\tau} d\sigma\, a_{m-1}(\varphi+\tau-\sigma)\right] \tilde{q}^{(m)}(t,\varphi)\,. \tag{3.7.28}$$

In order to determine the form of $G_{m-1,m}$ for finite τ, we first start for infinitesimal τ's because G is differentiable with respect to τ and because $G_{m-1,m}$ must vanish for $\tau = 0$. According to (3.7.22) we may write

$$\tau \to 0\,, \quad G_{m-1,m} = \tau b(\varphi)\,. \tag{3.7.29}$$

To save some indices we write for the time being

$$g^{(m-1)}(t,\varphi) = y(t,\varphi) \tag{3.7.30}$$

and consider (3.7.27) for the sequence $\tau, 2\tau, \ldots$. We write this sequence as follows

$$y(t+\tau,\varphi+\tau) - y(t,\varphi) = f(t,\tau,\varphi)\,, \tag{3.7.31}$$

$$y(t+2\tau,\varphi+2\tau) - y(t+\tau,\varphi+\tau) = f(t+\tau,\tau,\varphi+\tau)\,, \tag{3.7.32}$$

$$y(t+N\tau,\varphi+N\tau) - y(t+(N-1)\tau,\varphi+(N-1)\tau)$$
$$= f(t+(N-1)\tau,\tau,\varphi+(N-1)\tau)\,. \tag{3.7.33}$$

Summing up over all these equations we readily obtain

$$y(t+\underbrace{N\tau}_{t_0},\varphi+\underbrace{N\tau}_{t_0}) = y(t,\varphi) + \sum_{l=0}^{N-1} f(t+l\tau,\tau,\varphi+l\tau)\,, \tag{3.7.34}$$

or taking the limit to infinitesimal τ's

$$y(t+t_0, \varphi+t_0) = y(t, \varphi) + \int_0^{t_0} ds f(t+s, \tau, \varphi+s) . \tag{3.7.35}$$

Using the definition of f (3.7.28) and reintroducing $g^{(m-1)}$ according to (3.7.30) we obtain

$$g^{(m-1)}(t+t_0, \varphi+t_0) = g^{(m-1)}(t, \varphi) + K(t, t_0, \varphi) , \tag{3.7.36}$$

where we have used the abbreviation

$$K(t, t_0, \varphi) = \int_0^{t_0} ds\, b(\varphi+s) \exp\left[-\int_0^{t+s} d\sigma\, a_{m-1}(\varphi+s-\sigma) + \int_{-s}^{0} a_m(\varphi-\sigma)\, d\sigma\right] \tilde{q}^{(m)}(t, \varphi) . \tag{3.7.37}$$

Choosing the initial conditions appropriately we can write the solution of (3.7.36) in the form

$$g^{(m-1)}(t, \varphi) = \int_{\infty}^{0} ds\, b(\varphi+s) \exp\left[-\int_{-s}^{t} d\sigma\, a_{m-1}(\varphi-\sigma) + \int_{-s}^{0} a_m(\varphi-\sigma)\, d\sigma\right] \cdot \tilde{q}^{(m)}(t, \varphi) . \tag{3.7.38}$$

Inserting (3.7.38) into (3.7.36) and making some simple shifts of integration variables s and σ on the resulting lhs of (3.7.36), one can readily convince oneself that (3.7.38) indeed fulfills (3.7.36). Then (3.7.38) can be written in a more practical form as

$$g^{(m-1)}(t, \varphi) = \exp\left[-\int_0^t a_{m-1}(\varphi-\sigma)\, d\sigma\right] J \tilde{q}^{(m)}(t, \varphi) \tag{3.7.39}$$

with

$$J = \int_{\infty}^{0} ds\, b(\varphi+s) \exp\left\{-\int_0^s [a_{m-1}(\varphi+\sigma) - a_m(\varphi+\sigma)]\, d\sigma\right\} . \tag{3.7.40}$$

To discuss the properties of the integral J we make the decomposition

$$a_j(\varphi-\sigma) = \lambda_j + \hat{a}_j(\varphi-\sigma) . \tag{3.7.41}$$

It follows from (3.7.23) and our results on $G(\tau, \varphi)$ that $\hat{a}_{m-1}(\varphi)$ is T_j-periodic and C^k with respect to φ. If in addition we invoke assumption (6) of Theorem 3.1.1, the differentiability C^k with respect to φ and the KAM condition secure that the integrals over $\hat{a}_j$ in (3.7.38) remain finite for $\sigma \to \pm\infty$ and, in particular, that they are quasiperiodic in s. Since $\boldsymbol{b}$ is also a quasiperiodic function which is bounded and

$$\lambda_m - \lambda_{m-1} < 0 , \tag{3.7.42}$$

we may readily convince ourselves that the integral J converges.

We now recall our original hypothesis (3.7.25) for the solution of the inhomogeneous equation (3.7.22). To obtain (3.7.25) we have to multiply (3.7.38) by

$$\exp\left[\int_0^t d\sigma\, a_{m-1}(\varphi-\sigma)\right], \tag{3.7.43}$$

which gives us

$$\tilde{q}^{(m-1)}(t,\varphi) = J(\varphi)\tilde{q}^{(m)}(t,\varphi) . \tag{3.7.44}$$

We immediately see that (3.7.44) has precisely the form required in (3.7.19) where we may identify the second term on the rhs with the rhs of (3.7.44) and J with h. Thus we can indeed choose $h(\varphi)$ or $J(\varphi)$ as a function of φ alone, which makes it possible to cast (3.7.22) into diagonal form.

We now want to study the explicit form of

$$G_{m-1,m}\tilde{q}^{(m)} . \tag{3.7.45}$$

A comparison between (3.7.22) and (3.7.36) with help of (3.7.26) indicates that therefore we have to multiply (3.7.37) by

$$\exp\left[\int_0^{t+\tau} d\sigma\, a_{m-1}(\varphi+\tau-\sigma)\right]. \tag{3.7.46}$$

After a few elementary transformations of the integrals we obtain

$$G_{m-1,m}(\tau,\varphi)$$
$$= \exp\left[\int_0^\tau d\sigma\, a_{m-1}(\varphi+\sigma)\right]\int_0^\tau ds\, b(\varphi+s)\exp\left\{\int_0^s d\sigma[a_m(\varphi+\sigma) - a_{m-1}(\varphi+\sigma)]\right\}. \tag{3.7.47}$$

We recognize that $G_{m-1,m}$ is indeed a function of τ and φ (but not a function of t) as it has to be.

Our procedure can be generalized to the set of all equations (3.7.1) in an obvious way, namely, by making the hypothesis

$$\tilde{q}^{(j)}(t,\varphi) = \hat{q}^{(j)}(t,\varphi) + \sum_{k>j} h_k(\varphi)\tilde{q}^{(k)}(t,\varphi) \tag{3.7.48}$$

we may determine the coefficients $h_k(\varphi)$ in such a way that we obtain a diagonal matrix G for $\hat{q}$, i.e., the solution vectors of (3.7.1) can be chosen so that

$$\hat{q}^{(j)}(t+\tau, \varphi+\tau) = G_{jj}(\tau, \varphi)\hat{q}^{(j)}(t, \varphi) \tag{3.7.49}$$

holds. Equation (3.7.49) can be considered as a functional equation for $\hat{q}^{(j)}$. We have solved this equation above [cf. (3.7.16)]. Its solution reads

$$\hat{q}^{(j)}(t, \varphi) = \exp\left[\int_0^t d\sigma\, a_j(\varphi-\sigma)\right] w^{(j)}(0, \varphi+t)\,. \tag{3.7.50}$$

Provided assumptions $(1-6)$ of Theorem 3.1.1 are fulfilled, (3.7.50) can be written

$$\hat{q}^{(j)}(t, \varphi) = \exp(\hat{\lambda}_j t)\, v^{(j)}(\varphi+t)\,, \tag{3.7.51}$$

where $v^{(j)}$ is T_j-periodic and C^k with respect to φ and therefore quasiperiodic in t. Equation (3.7.51) is precisely of the form which was asserted in (3.1.20). (Clearly, $\mathrm{Re}\{\hat{\lambda}_j\}$ is the generalized characteristic exponent λ_j.)

3.8 The General Case: Some of the Generalized Characteristic Exponents Coincide

In this section we shall present two theorems. The first one deals with a reduction of the matrix C (3.1.18) if some of the generalized characteristic exponents coincide. The second theorem shows that C can even be diagonalized if all generalized characteristic exponents coincide, and an additional assumption on the growth rate of $|q^{(j)}|$ is made. The first of these theorems is formulated as follows.

Theorem 3.8.1. Under assumptions (1, 2, 4) of Theorem 3.1.1 and by a suitable choice of the solution vectors $q^{(j)}(t, \varphi)$, the matrix C (3.1.18) can be brought to the triangular form [cf. (3.7.1)]

$$C^T = G = \begin{pmatrix} \square & & & \\ & \square & & \\ & & \square & \\ & 0 & & \ddots \end{pmatrix}, \tag{3.8.1}$$

where the boxes correspond to generalized characteristic exponents λ_j which are different. Each box has at maximum the size m_j, where m_j is the degree of λ_j. The coefficients of the matrix $C(\tau, \varphi)$ are differentiable with respect to τ, T_j-periodic in φ and (at least) C^0 with respect to φ. If in addition assumption (5) of Theorem 3.1.1 is valid (for $\lambda_j \neq \lambda_k$), the coefficients of $C(\tau, \varphi)$ are (at least) C^1 with respect to τ and T_j-periodic and C^k with respect to φ.

We now give a sketch of the construction of C.

We assume that the generalized characteristic exponents λ_j are labeled in such a way that

$$\lambda_1 \geqslant \lambda_2 \geqslant \lambda_3 \geqslant \ldots \geqslant \lambda_m . \tag{3.8.2}$$

If

$$\lambda_1 > \lambda_2 > \lambda_3 > \ldots > \lambda_k \geqslant \lambda_{k+1} , \tag{3.8.3}$$

we may apply our previous reduction scheme (Sects. 3.1 – 7) at least up to λ_{k-1}.

In order to treat the case in which several of the λ's coincide, we therefore assume that λ_1 is l times degenerate, i.e.,

$$\lambda_1 = \lambda_2 \ldots = \lambda_l . \tag{3.8.4}$$

We assume again that all $\boldsymbol{q}$'s are linearly independent with finite angles even if $t \to +\infty$. We define a normalization factor by [cf. (3.5.3, 4)]

$$\mathcal{N} = \left[\sum_j |\boldsymbol{q}^{(j)}(t, \varphi_0 + \tau)|^2 \right]^{-1/2} \tag{3.8.5}$$

and form

$$\boldsymbol{\chi}^{(j)}(t, \varphi_0 + \tau) = \mathcal{N} \boldsymbol{q}^{(j)}(t, \varphi_0 + \tau) . \tag{3.8.6}$$

We define

$$\operatorname*{Lim\,Sup}_{t \to \infty} \tag{3.8.7}$$

by the following description. Take a sequence $t = t_n$, $t_n \to \infty$, such that at least for one of the $\boldsymbol{q}^{(j)}$'s and some positive ε

$$\frac{1}{t_n} \ln |\boldsymbol{q}^{(j)}(t_n, \varphi_0 + \tau)| \geqslant \lambda_1 - \varepsilon > \lambda_{l+1} \tag{3.8.8}$$

holds. Because the present procedure is a rather straightforward generalization of the case of nondegenerate λ's, we shall give only a sketch of our procedure. Loosely speaking, for such $t_n > t_{n_0}$ we may neglect

$$\mathcal{N} \sum_{k=l+1}^{m} D_{jk} \boldsymbol{q}^{(k)}(t + \tau, \varphi_0) \exp[-z_k(\tau)] \tag{3.8.9}$$

against

$$\mathcal{N} \sum_{k=1}^{l} D_{jk} \boldsymbol{q}^{(k)}(t + \tau, \varphi_0) \exp[-z_k(\tau)] , \tag{3.8.10}$$

or, more precisely, we have

$$\operatorname*{Lim\,Sup}_{t\to\infty} \{\boldsymbol{\chi}^{(j)}(t, \varphi_0+\tau)\}$$

$$= \sum_{k=1}^{l} D_{jk}(\tau) \operatorname*{Lim\,Sup}_{t\to\infty} \{\mathcal{N}(t+\tau) \exp [z_k(t+\tau) - z_k(\tau)]\, \boldsymbol{u}^{(k)}(t+\tau)\}\,. \quad (3.8.11)$$

In Lim Sup we take all such sequences t_n for which at least one $\boldsymbol{q}^{(j)}$ fulfills (3.8.8). Note that such sequences may depend on τ. We now form

$$\sum_{j=1}^{m} \alpha_j(\tau) \operatorname*{Lim\,Sup}_{t\to\infty} \{\boldsymbol{\chi}^{(j)}(t, \varphi_0+\tau)\} \quad (3.8.12)$$

and require that the α_j's are determined in such a way that (3.8.12) vanishes.

Using the explicit form of $\boldsymbol{\chi}^{(j)}$ and (3.8.12) we thus obtain

$$\sum_{k=1}^{l} \sum_{j=1}^{m} \alpha_j(\tau) D_{jk}(\tau) \operatorname*{Lim\,Sup}_{t\to\infty} \{\mathcal{N}(t+\tau) \exp [z_k(t+\tau) - z_k(\tau)]\, \boldsymbol{u}^{(k)}(t+\tau)\} = 0 \quad (3.8.13)$$

and because the $\boldsymbol{u}^{(k)}$'s are linearly independent of each other we obtain

$$\sum_{j=1}^{m} \alpha_j(\tau) D_{jk}(\tau) \operatorname*{Lim\,Sup}_{t\to\infty} \{\mathcal{N}(t+\tau) \exp [z_k(t+\tau) - z_k(\tau)]\, \boldsymbol{u}^{(k)}(t+\tau)\} = 0\,. \quad (3.8.14)$$

In it k runs from 1 till l.

We thus find l equations for the m unknown α_j's. This can be seen most clearly from the fact that (3.8.14) can be fulfilled if and only if

$$\sum_{j=1}^{m} \alpha_j(\tau) D_{jk}(\tau) = 0 \quad \text{for} \quad k = 1, \ldots, l \quad (3.8.15)$$

holds. From (3.8.15) it follows that α_j can be chosen independently of t. On the other hand, the differentiability properties of $D(\tau)$ are not known. Therefore it is advisable to go back to (3.8.12) where the differentiability properties of the coefficients of $\alpha_j(\tau)$ are known. There are (at least) $m-l$ linearly independent solution vectors $\boldsymbol{\alpha} = (\alpha_1, \ldots, \alpha_m)$ of (3.8.15).

To secure the solvability condition even for finite t_n we may just take l rows of (3.8.12) provided not all coefficients of α_j vanish identically. If this happens, however, we may go over to another selection of l rows and apply the procedure described in Sect. 3.6.2.

Furthermore, we may choose l' $(l' \leqq l)$ vectors

$$(\alpha_1, \alpha_2, \ldots, \alpha_m)\,, \quad (3.8.16)$$

such that (3.8.12) does not vanish. Let us treat the case in which the least reduction is possible, $l' = l$. If we label the vectors (3.8.16) for which (3.8.12) remains finite by

$$\alpha^{(k)} \quad \text{with} \quad k = 1, \ldots, l, \tag{3.8.17}$$

and those for which (3.8.12) vanishes by

$$\alpha^{(k)} \quad \text{with} \quad k = l+1, \ldots, m, \tag{3.8.18}$$

we may quite generally form new solutions of (3.1.5) by the definition

$$\tilde{q}^{(k)} = \sum \alpha_j^{(k)} q^{(j)}. \tag{3.8.19}$$

On account of our construction, the new solutions $\tilde{q}^{(k)}$, $k = l+1, \ldots, m$, possess generalized characteristic exponents λ_2 or smaller. When we use some arguments of Sects. 3.1 – 7, we recognize that the matrix $C^T = (G_{jk})$, [cf. (3.7.1)], is reduced to the form

$$\begin{array}{c|c|c|} & l & m-l \\ \hline l & & \\ \hline m-l & 0 & \\ \hline \end{array} \, . \tag{3.8.20}$$

Again one may show that the $\alpha_j(\tau)$ have the same differentiability properties as their coefficients in (3.8.12). If the coefficients in (3.8.12) are not considered as a function of τ but rather of φ (compare the previous sections), the α's acquire the same differentiability properties. Therefore $\tilde{q}^{(k)}$ (3.8.19) has again the same differentiability properties as the original solution vectors $q^{(j)}(t, \varphi)$. After having reached (3.8.20), we may continue our procedure so that we can reduce the scheme of coefficients in (3.8.20) to one of the triangular form

$$\begin{pmatrix} \square & & & \\ & \square & & \\ & & \square & \\ & 0 & & \ddots \end{pmatrix} . \tag{3.8.21}$$

The question arises whether we may reduce the scheme (3.8.21) further to one in which we have nonvanishing matrices along the main diagonal only

$$\begin{pmatrix} \square & & & \\ & \square & & 0 \\ & & \square & \\ & 0 & & \ddots \end{pmatrix} . \tag{3.8.22}$$

There are various circumstances under which such reduction can be reached. This can be achieved, for instance, if for $t \to -\infty$ the generalized characteristic exponents λ_j' just obey the inverse of the corresponding relations (3.8.2). It is, however, also possible to reduce scheme (3.8.21) to one of the form (3.8.22) if it can be shown that the squares in (3.8.21) can be further reduced individually to diagonal form. Even if that is not possible, further general statements can be made on the form of the solution matrix $Q(t, \varphi)$. Namely, if the solution matrices belonging to the Q_j-squares in (3.8.21) are known, the total matrix $Q(t, \varphi)$ can be found consecutively by the method of variation of the constant. The procedure is analogous to that in Sect. 3.7, but $\tilde{q}^{(j)}$ must be replaced by submatrices $\tilde{Q}^{(j)}$, etc.

In the final part of this section we want to present a theorem in which all λ_k's are equal.

Theorem 3.8.2. Let us make the following assumptions:

1) M (3.1.4) is at least C^1 with respect to φ;
2) $\lambda_k = \lambda$, $k = 1, \ldots, m$;
3) $e^{-\lambda n \tau} \| T_\tau^n q \|$ and $e^{\lambda n \tau} \| T_\tau^{-n} q \|$ are bounded for $n \to \infty$ and τ arbitrary, real, and for all vectors of the solution space of $\dot{Q} = M(t, \varphi) Q, q$, which are C^1 with respect to φ. ($\| \ldots \|$ denotes the Hilbert space norm.)

Then we may find a new basis $\tilde{q}$ so that $T_\tau \to \tilde{T}_\tau$ becomes diagonal,

$$\tilde{T}_\tau \tilde{q} = e^{\hat{\lambda} \tau} \tilde{q} , \quad \hat{\lambda} \text{ imaginary} , \tag{3.8.23}$$

provided the spectrum [of T_τ acting in the space $q(t, \varphi)$] is a point spectrum.

In the case of continuous spectrum, T_τ is equivalent to an operator $\hat{T}_\tau$ of "scalar type", i.e., $\hat{T}_\tau$ can be decomposed according to

$$\hat{T}_\tau = \int e^{\hat{\lambda} \tau} E(d\hat{\lambda}) , \tag{3.8.24}$$

where E is the resolution of the identity for $\hat{T}_\tau$.

The proof of this theorem can be achieved by using several theorems of the theory of linear operators. We shall quote these latter theorems only (cf. references) and indicate how they allow us to derive Theorem 3.8.2. Because of assumption (1) of Theorem 3.8.2 and Lemma 3.2.2, $q(t, \varphi)$ is (at least) C^1 with respect to φ for $-\infty < t < \infty$. Then the solutions form a Hilbert space for $-\infty < t < \infty$.

The operators T_τ form an Abelian group. As indicated in Sect. 3.7, we can decompose the group representation G of T_τ for $\tau \to 0$ into $1 + \tau A(\varphi)$. Correspondingly, we decompose $T_\tau = 1 + A\tau$, $\tau \to 0$, where A is the infinitesimal generator of the group. With the help of A, for finite τ, we can represent T_τ in the form $T_\tau = \exp(A\tau)$. We now invoke the following lemma: Let $\hat{G}$ be a bounded Abelian group of operators in Hilbert space H. Then there is a bounded self-adjoint operator in H with a bounded inverse defined everywhere such that for every $\hat{A}$ in $\hat{G}$ the operator $B T_\tau B^{-1}$ is unitary.

Consequently, the bounded group $\exp(A\tau)$ introduced above is equivalent to a group of unitary operators. By Stone's theorem, the latter group has an infinitesimal generator $\mathrm{i}\tilde{A}$, where $\tilde{A}$ is self-adjoint. Thus A is equivalent to $\mathrm{i}\tilde{A}$, the transformation matrix being given by B. Because in our present case A and thus $\tilde{A}$ are independent of t and τ, also B is independent of t and τ. Thus B describes a t- and τ-independent transformation of $\boldsymbol{q}$ to another set of $\boldsymbol{q}$'s which therefore must be again solutions to (3.1.5).

In the last step of our analysis we recall that T_τ becomes unitary and therefore, according to the theory of linear operators, a spectral operator of scalar type, so $\tilde{A}$ and $\hat{T}_\tau$ satisfy the equations

$$\tilde{A} = -\int \hat{\lambda} E(d\hat{\lambda}) \tag{3.8.25}$$

and

$$T_\tau \rightarrow \hat{T}_\tau = \int \mathrm{e}^{\hat{\lambda}\tau} E(d\hat{\lambda}) \,, \tag{3.8.26}$$

respectively, where E is the resolution of the identity for $\tilde{A}$ (and $\hat{T}_\tau$). If the spectrum $\hat{\lambda}$ is a point spectrum, it follows from (3.8.26) that

$$\hat{T}_\tau \tilde{q} = \mathrm{e}^{\hat{\lambda}\tau} \tilde{q} \,.$$

This equation can now be treated in the same manner as (3.7.9).

In conclusion we mention that a further class of solutions can be identified if $\| T_\tau^{\pm n} \boldsymbol{q} \|$ goes with n^m. The solutions are then polynomials in t (up to order m) with quasiperiodic coefficients.

3.9 Explicit Solution of (3.1.1) by an Iteration Procedure

In Sect. 2.9, where we treated differential equations with periodic coefficients, we developed a procedure to calculate the solutions explicitly by perturbation theory. Unfortunately there are difficulties with respect to convergence if one tries to extend this procedure in a straightforward manner to the present case with quasiperiodic coefficients. It is possible, however, to devise a rapidly converging iteration scheme whose convergence can be rigorously proven. The basic idea consists in a decomposition of the matrix M of the equation $\dot{\tilde{Q}} = M\tilde{Q}$ into a constant matrix A and a matrix M which contains the quasiperiodic time dependence, $M = A + \tilde{M}$. The matrix $\tilde{M}$ is considered as a small perturbation.

We first formulate the basic theorem and then show how the solution can be constructed explicitly.

Theorem 3.9.1. Let us assume that A and $\tilde{M}$ satisfy the following conditions:

1) The matrix $\tilde{M}$ is periodic in φ_j and analytic on the domain

$$|\mathrm{Im}\{\varphi\}| \equiv \sup_j \{|\mathrm{Im}\{\varphi_j\}|\} \leqslant \rho_0, \quad \rho_0 > 0 \tag{3.9.1}$$

and real for real φ_j;

2) The matrix $\tilde{M}$ does not contain any constant terms;

3) For some positive ω_j and d the inequality

$$|(\boldsymbol{n}, \boldsymbol{\omega})| \geqslant \varepsilon \|\boldsymbol{n}\|^{-(m+1)}, \quad \|\boldsymbol{n}\| \neq 0, \tag{3.9.2}$$

i.e., the Kolmogorov-Arnold-Moser condition holds for every vector

$$\boldsymbol{n} = (n_1, n_2, \ldots, n_N) \tag{3.9.3}$$

with integer components, n_j;

4) The eigenvalues λ of the matrix A have distinct real parts.

Then the assertion is: a sufficiently small positive constant K can be found such that for

$$\sum_{j,k=1}^{m} |\tilde{M}_{jk}| \leqslant K \tag{3.9.4}$$

a solution can be constructed which has the form

$$V(\omega t + \varphi_0) e^{\Lambda t}. \tag{3.9.5}$$

In it the matrix V is a quasiperiodic matrix which is analytic and analytically invertible in the domain

$$|\text{Im}\{\varphi\}| \leqslant \rho_0/2, \tag{3.9.6}$$

Λ is a constant matrix.

To show how V and Λ can be constructed explicitly, we start from the equation

$$\dot{\tilde{Q}}(t) = [A + \tilde{M}(t)]\, \tilde{Q}(t) \tag{3.9.7}$$

where A and $\tilde{M}$ have been specified above. By means of a constant matrix C we transform A into its Jordan normal form

$$J = C^{-1} A C. \tag{3.9.8}$$

Due to our assumption that the real parts of the eigenvalues of A are distinct, all eigenvalues of A are distinct and therefore we may assume that J is of purely diagonal form. We now put

$$\tilde{Q}(t) = C Q(t) \tag{3.9.9}$$

and insert it into (3.9.7). After multiplication of (3.9.7) by C^{-1} from the left we obtain

$$\dot{Q} = [J + M_0(t)]\, Q, \tag{3.9.10}$$

where we have used the abbreviation

$$C^{-1}\tilde{M}C = M_0 . \tag{3.9.11}$$

In order to solve (3.9.10) we make the hypothesis

$$Q(t) = [1 + U_1(t)]\, Q_1(t) , \tag{3.9.12}$$

where 1 is the unity matrix. Since we want to construct a solution of the form (3.9.5), it would be desirable to choose U_1 as a quasiperiodic matrix but Q_1 as an exponential function of time.

For a practical calculation it turns out, however, that we cannot construct U_1 exactly. Therefore we resort to an iteration scheme in which we aim at calculating U_1 approximately but retaining the requirement that U_1 is quasiperiodic. We then derive a new equation for Q_1. It will turn out that by an adequate choice of the equation for U_1, the equation for Q_1 takes the same shape as equation (3.9.10) with one major difference. In this new equation M_0 is replaced by a new quasiperiodic matrix M_1 whose matrix elements are smaller than those of M_0. Thus the basic idea of the present approach is to introduce repeatedly substitutions of the form (3.9.12).

In this way the original equation (3.9.10) is reduced to a form in which the significance of the quasiperiodic matrix $M(t)$ becomes smaller and smaller. These ideas will become clearer when we perform the individual steps explicitly. We insert hypothesis (3.9.12) into (3.9.10) which immediately yields

$$\dot{U}_1 Q_1 + [1 + U_1(t)]\, \dot{Q}_1 = J Q_1 + J U_1 Q_1 + M_0 Q_1 + M_0 U_1 Q_1 . \tag{3.9.13}$$

For reasons which will transpire below we add on both sides of (3.9.13)

$$U_1 J Q_1 - J U_1 Q_1 \tag{3.9.14}$$

and we add on the rhs 0 specifically as

$$0 \equiv D_1 Q_1 - D_1 Q_1 . \tag{3.9.15}$$

In general, we shall define D_1 as the constant part of the main diagonal of the matrix M_0. We do this here only for formal reasons because M_0 was anyway constructed in such a way that its constant part vanishes. However, in our subsequent iteration procedure such terms can arise. Next, D_1 can be found explicitly via the formula

$$D_1 = \frac{1}{(2\pi)^N} \int_0^{2\pi} \cdots \int_0^{2\pi} \begin{pmatrix} M_{0,11} & & 0 \\ & M_{0,22} & \\ 0 & & \ddots\, M_{0,mm} \end{pmatrix} d\varphi_1 \ldots d\varphi_N . \tag{3.9.16}$$

With aid of the just-described manipulations, (3.9.13) is transformed into

$$\underline{\dot{U}_1 Q_1 + U_1 J Q_1 - J U_1 Q_1} + (1 + U_1)\dot{Q}_1$$

$$= J Q_1 + U_1 J Q_1 + \underline{M_0 Q_1} + M_0 U_1 Q_1 + D_1 Q_1 - \underline{D_1 Q_1} \,. \tag{3.9.17}$$

To simplify this equation we assume that U_1 can be chosen in such a way that the underlined expressions in (3.9.17) cancel for an arbitrary matrix Q. Before writing down the resulting equation for U_1, we make explicit use of the assumption that U_1 is a quasiperiodic function, i.e., we write

$$U_1 = \sum U_{1,n} \exp\left[in(\omega t + \varphi)\right] . \tag{3.9.18}$$

By means of it we can immediately cast the time derivative of U_1 in the form

$$\dot{U}_1 = \sum_{l=1}^{N} \frac{\partial U_1}{\partial \varphi_l} \omega_l = \left(\frac{\partial U_1}{\partial \varphi}, \omega\right) . \tag{3.9.19}$$

Using this specific form, the equation for U_1 then reads

$$\left(\frac{\partial U_1}{\partial \varphi}, \omega\right) + U_1 J - J U_1 = M_0 - D_1 \,. \tag{3.9.20}$$

Because of (3.9.20), (3.9.17) reduces to the following equation for Q_1

$$(1 + U_1)\dot{Q}_1 = \underbrace{J Q_1 + U_1 J Q_1}_{(1 + U_1) J Q_1} + M_0 U_1 Q_1 + \underline{D_1 Q_1} \,. \tag{3.9.21}$$

On the rhs we add 0 in the form

$$\underline{U_1 D_1 Q_1} - U_1 D_1 Q_1 \tag{3.9.22}$$

so that we can cast (3.9.21) into the form

$$(1 + U_1)\dot{Q}_1 = (1 + U_1) J Q_1 + \underline{(1 + U_1) D_1 Q_1} + (M_0 U_1 - U_1 D_1) Q_1 \,. \tag{3.9.23}$$

After multiplication of this equation by

$$(1 + U_1)^{-1} \tag{3.9.24}$$

from the left, we obtain

$$\dot{Q}_1 = J Q_1 + D_1 Q_1 + (1 + U_1)^{-1}(M_0 U_1 - U_1 D_1) Q_1 \,. \tag{3.9.25}$$

We now introduce the abbreviations

$$J + D_1 = J_1 \quad \text{and} \tag{3.9.26}$$

$$(1 + U_1)^{-1}(M_0 U_1 - U_1 D_1) = M_1 , \tag{3.9.27}$$

which leaves us with

$$\dot{Q}_1 = J_1 Q_1 + M_1 Q_1 . \tag{3.9.28}$$

Here J_1 is evidently again a diagonal time-independent matrix, whereas M_1 is a quasiperiodic function. So at a first sight it appears that nothing is gained because we are again led to an equation for Q_1 which is of precisely the same form as the original equation (3.9.10). However, a rough estimate will immediately convince us that M_1 is reduced by an order of magnitude. To this end let us introduce a small parameter ε so that

$$M_0 \propto \varepsilon . \tag{3.9.29}$$

Because D_1 is obtained from M_1 by means of (3.9.16), then

$$D_1 \propto \varepsilon \tag{3.9.30}$$

holds. On the other hand, we assume that J is of order unity as compared to M_0

$$J \propto 1 . \tag{3.9.31}$$

From (3.9.30, 31) it follows that the rhs of (3.9.20) is proportional to ε. This leads us to the conclusion

$$U_1 \propto \varepsilon . \tag{3.9.32}$$

An inspection of (3.9.27) reveals that M_1 must be of order

$$M_1 \propto \varepsilon^2 , \tag{3.9.33}$$

whereas J_1 is given by the order of magnitude

$$J_1 \propto 1 + \varepsilon . \tag{3.9.34}$$

Thus expressed in ε's (3.9.28) is of the form

$$\dot{Q}_1 = J_1 Q_1 + \varepsilon^2 \ldots Q_1 . \tag{3.9.35}$$

That means that we have indeed achieved an appreciable reduction of the size of the quasiperiodic part on the rhs. It should be noted that our estimate is a superficial one and that the convergence of the procedure has been proven mathematically rigorously (see below). Our further procedure is now obvious. We shall continue the iteration by making in a second step the hypothesis

$$Q_1 = (1 + U_2) Q_2 \tag{3.9.36}$$

and continue with the general term

$$Q_\nu = (1 + U_{\nu+1}) Q_{\nu+1} \,. \tag{3.9.37}$$

In each of these expressions Q_ν is assumed to be quasiperiodic. Simultaneously we put

$$J_{\nu+1} = J_\nu + D_{\nu+1} \,, \quad \text{where} \tag{3.9.38}$$

$$D_{\nu+1} = \text{main diagonal of } \overline{M}_{\nu+1}, \quad \overline{M}_{\nu+1} = \frac{1}{(2\pi)^N} \int \cdots \int M_{\nu+1} d\varphi_1 \ldots d\varphi_N \,. \tag{3.9.39}$$

Introducing the further abbreviation

$$M_{\nu+1} = (1 + U_{\nu+1})^{-1} (M_\nu U_{\nu+1} - U_{\nu+1} D_{\nu+1}) \,, \tag{3.9.40}$$

we may define the equation for $Q_{\nu+1}$ by

$$\dot{Q}_{\nu+1} = J_{\nu+1} Q_{\nu+1} + M_{\nu+1} Q_{\nu+1} \,. \tag{3.9.41}$$

In order to obtain the relations (3.9.38 – 41), it is only necessary to substitute in the previous relations (3.9.26 – 28) the index 0 or no index by ν and the index 1 by $\nu + 1$. Putting the transformations (3.9.36, 37) one after the other we obtain the explicit form of the solution as

$$Q = (1 + U_1)(1 + U_2) \ldots (1 + U_l) Q_l \,. \tag{3.9.42}$$

Evidently in the limit that l goes to infinity we have the relation

$$\dot{Q}_l = J_l Q_l + O(\varepsilon^l) Q_l \Rightarrow J_l Q_l \,. \tag{3.9.43}$$

This equation can now be solved explicitly. Its general solution reads

$$Q_l = e^{J_l t} Q_l(0) \,, \tag{3.9.44}$$

where J_l is a diagonal matrix.

As a specific choice of initial conditions we can take

$$Q_l(0) = 1 \,. \tag{3.9.45}$$

We are now left with the explicit construction of $U_{\nu+1}$, $\nu = 0, 1, \ldots$, which must incorporate the proof that U can be chosen as a quasiperiodic function. According to (3.9.20) the equation for U reads

$$\left(\frac{\partial U_{\nu+1}}{\partial \varphi}, \omega\right) + U_{\nu+1} J_\nu - J_\nu U_{\nu+1} = M_\nu - D_{\nu+1} . \tag{3.9.46}$$

For simplicity we shall drop the indices ν and $\nu + 1$. We further put

$$M - D = M' , \tag{3.9.47}$$

which can be expanded into a multiple Fourier series

$$M'(\varphi) = \sum_n M'_n \exp(\mathrm{i}\boldsymbol{n} \cdot \varphi) . \tag{3.9.48}$$

In order to solve (3.9.46) we expand $U(\varphi)$ into such a series, too.

$$U(\varphi) = \sum_n U_n \exp(\mathrm{i}\boldsymbol{n} \cdot \varphi) . \tag{3.9.49}$$

Inserting (3.9.48, 49) into (3.9.46), we find for the Fourier coefficients the following relations

$$U_n(J + \mathrm{i}(\boldsymbol{n}, \omega)) - J U_n = M'_n . \tag{3.9.50}$$

We now introduce the matrix elements of the matrices U_n, J and M'_n by means of

$$U_n = (U_{jk}^{(n)}) , \tag{3.9.51}$$

$$J = \begin{pmatrix} J_1 & & \\ & \ddots & \\ & & J_m \end{pmatrix} , \tag{3.9.52}$$

$$M'_n = (M'^{(n)}_{jk}) . \tag{3.9.53}$$

This allows us to write the matrix equation (3.9.50) in the form

$$U_{jk}^{(n)}(J_k + \mathrm{i}(\boldsymbol{n}, \omega)) - J_j U_{jk}^{(n)} = M'^{(n)}_{jk} . \tag{3.9.54}$$

Its solution can be immediately found

$$U_{jk}^{(n)} = [J_k - J_j + \mathrm{i}(\boldsymbol{n}, \omega)]^{-1} M'^{(n)}_{jk} \tag{3.9.55}$$

provided the denominator does not vanish. We can easily convince ourselves that the denominator does not vanish. For $k \neq j$ it was assumed that the real parts of the eigenvalues of the original matrix A (or J) are distinct. Provided M is small enough, $D_{\nu+1}$ are so small that subsequent shifts of J according to (3.9.38) are so small that (3.9.56) is fulfilled for all iterative steps. Therefore we obtain for $k \neq j$

$$\mathrm{Re}\{J_k\} \neq \mathrm{Re}\{J_j\} \tag{3.9.56}$$

$$|J_k - J_j + \mathrm{i}(\boldsymbol{n}, \boldsymbol{\omega})| > 0 . \tag{3.9.57}$$

For $k = j$ we find

$$|\boldsymbol{n}, \boldsymbol{\omega}| > 0 \quad \text{for} \quad |\boldsymbol{n}| > 0 , \tag{3.9.58}$$

but for $\boldsymbol{n} = 0$ we know that according to the construction (3.9.47)

$$M_{jj}^{\prime(0)} = 0 . \tag{3.9.59}$$

To make the solution matrix U_n unique we put $U_{jj}^0 = 0$.

Inserting (3.9.55) into (3.9.49) we thus obtain as the unique and explicit solution of the step ν (where we suppress the index ν of J_k, J_j)

$$U_{\nu+1,jk}(\boldsymbol{\varphi}) = \sum_{\boldsymbol{n}} (J_k - J_j + \mathrm{i}\boldsymbol{n} \cdot \boldsymbol{\omega})^{-1} M_{\nu,jk}^{\prime(\boldsymbol{n})} \exp(\mathrm{i}\boldsymbol{n} \cdot \boldsymbol{\varphi}) . \tag{3.9.60}$$

Equation (3.9.60) can also be expressed as

$$U_{\nu+1} = -\sigma \int_0^t \exp[(J_k - J_j)\tau] M_\nu^\prime(\boldsymbol{\varphi} + \tau) d\tau . \tag{3.9.61}$$

The sign $\sigma \int_0^t \ldots$ means that $t = -\infty$ if $J_k - J_j > 0$ and $t = \infty$ if $J_k - J_j < 0$ are taken as lower limit of the integral.

We may summarize our results as follows. We have devised an iteration procedure by which we can construct the quasiperiodic solution matrix

$$V = \prod_{\nu=1}^{l} (1 + U_\nu) , \quad l \to \infty \tag{3.9.62}$$

explicitly, whereas Λ is given by

$$\Lambda = \lim_{\nu \to \infty} J_\nu . \tag{3.9.63}$$

In practical applications for a sufficiently small quasiperiodic matrix $\tilde{M}$ very few steps may be sufficient. In the literature an important part of the mathematical treatment deals with the proof of the convergence of the above procedure. Since in the present context this representation does not give us any deeper insights, we refer the interested reader to Sect. 5.3, Chap. 6, and Appendix A where the question of convergence is treated in detail. Rather, we make a comment which qualitatively elucidates the range of validity of convergence. As it transpires from (3.9.60) U is obtained from M essentially by dividing M by $J_k - J_j$. In order to secure the convergence of the infinite product in (3.9.62), U_ν must go to 0 sufficiently rapidly. This is at least partly achieved if the original M's are sufficiently small and the differences between the original eigenvalues J sufficiently large.

4. Stochastic Nonlinear Differential Equations

In this chapter we shall present some of the most essential features of stochastic differential equations. Readers interested in learning more about this subject are referred to the book by Gardiner (cf. references).

In many problems of the natural sciences, but also in other fields, we deal with macroscopic features of systems, e.g., with fluids, with electrical networks, with macroscopic brain activities measured by EEG, etc. It is quite natural to describe these macroscopic features by macroscopic quantities, for instance, in an electrical network such variables are macroscopic electric charges and currents. However, one must not forget that all these macroscopic processes are the result of many, more or less coherent, microscopic processes. For instance, in an electrical network the current is ultimately carried by the individual electrons, or electrical brain waves are ultimately generated by individual neurons. These microscopic degrees of freedom manifest themselves in the form of fluctuations which can be described by adding terms to otherwise deterministic equations for the macroscopic quantities. Because in general microscopic processes occur on a much shorter time scale than macroscopic processes, the fluctuations representing the underworld of the individual parts of the system take place on time scales much shorter than the macroscopic process. The theory of stochastic differential equations treats these fluctuations in a certain mathematical idealization which we shall discuss below.

In practical applications one must not forget to check whether such an idealization remains meaningful, and if results occur which contradict those results one would obtain by physical reasoning, one should carefully check whether the idealized assumptions are valid.

4.1 An Example

Let us first consider an example to illustrate the main ideas of this approach. Let us treat a single variable q, which changes during time t. Out of the continuous time sequence we choose a discrete set of times t_i, where we assume for simplicity that the time intervals between t_i and t_{i-1} are equal to each other. The change of q when the system goes from one state t_{i-1} to another one at time t_i will be denoted by

$$\Delta q(t_i) \equiv q(t_i) - q(t_{i-1}) . \tag{4.1.1}$$

This change of the state variable q is generally caused by macroscopic or coherent forces K which may depend on that variable, and by additional fluctuations which occur within the time interval under discussion

$$\Delta t = t_i - t_{i-1} . \tag{4.1.2}$$

This impact of coherent driving forces and fluctuating forces can be described by

$$\Delta q(t_i) = K(q(t_{i-1}))\Delta t + \Delta w(t_i) , \tag{4.1.3}$$

where $\Delta w(t_i) = w(t_i) - w(t_{i-1})$. Because Δw represents the impact of microscopic processes on the macroscopic process described by q, we can expect that Δw depends on very many degrees of freedom of the underworld. As usual in statistical mechanics we treat these many variables by means of statistics. Therefore we introduce a statistical average which we characterize by means of the following two properties.

1) We assume that the average of Δw vanishes

$$\langle \Delta w(t_i)\rangle = 0 . \tag{4.1.4}$$

Otherwise Δw would contain a part which acts in a coherent fashion and could be taken care of by means of K.

2) We assume that the fluctuations occur on a very short time scale. Therefore when we consider the correlation function between Δw at different times t_i and t_j we shall assume that the corresponding fluctuations are uncorrelated. Therefore we may postulate

$$\langle \Delta w(t_i)\Delta w(t_j)\rangle = \delta_{t_i t_j} Q \Delta t . \tag{4.1.5}$$

The quantity Q is a measure for the size of the fluctuations; $\delta_{t_i t_j}$ expresses the statistical independence of Δw at different times t_i, t_j. The time interval Δt enters into (4.1.5) because we wish to define the fluctuating forces in such a way that Brownian motion is covered as a special case. To substantiate this we treat a very simple example, namely

$$K = 0 . \tag{4.1.6}$$

In this case we may solve (4.1.3) immediately by summing both sides over t_i, which yields

$$q(t) - q(t_0) = \sum_{j=1}^{N} \Delta w(t_j) , \quad t - t_0 = N\Delta t . \tag{4.1.7}$$

We have denoted the final time by t. We choose $q(t_0) = 0$. It then follows from (4.1.4) that

$$\langle q(t)\rangle = 0 . \tag{4.1.8}$$

In order to get an insight as to how big the deviations of $q(t)$ may be on the average due to the fluctuations, we form the mean square

$$\langle q^2(t)\rangle = \sum_{i=1}^{N}\sum_{j=1}^{N}\langle \Delta w(t_i)\Delta w(t_j)\rangle\,, \tag{4.1.9}$$

which can be reduced on account of (4.1.5) to

$$Q\sum_{j=1}^{N}\Delta t = Qt\,. \tag{4.1.10}$$

This result is well known from Brownian motion. The mean square of the coordinate of a particle undergoing Brownian motion increases linearly with time t. This result is a consequence of the postulate (4.1.5).

We now return to (4.1.3). As it will transpire from later applications, very often the fluctuations Δw occur jointly with the variable q, i.e., one is led to consider equations of the form

$$\Delta q(t_i) = K(q(t_{i-1}))\Delta t + g(q)\Delta w(t_i)\,. \tag{4.1.11}$$

An important question arises concerning at which time the variable q in g must be taken, and it turns out that the specific choice determines different kinds of processes. That means in particular that this time cannot be chosen arbitrarily. Two main definitions are known in the literature. The one is from Îto, according to which q is taken at time t_{i-1}. This means we put in (4.1.11)

$$g(q)\Delta w(t_i) = g(q(t_{i-1}))\Delta w(t_i)\,. \tag{4.1.12}$$

First the system has reached $q(t_{i-1})$, then fluctuations take place and carry the system to a new state $q(t_i)$. Because the fluctuations Δw occur only after the state q at time t_{i-1} has been reached, $q(t_{i-1})$ and $\Delta w(t_i)$ are uncorrelated

$$\langle g(q(t_{i-1}))\Delta w(t_i)\rangle = \langle g(q(t_{i-1}))\rangle \underbrace{\langle \Delta w(t_i)\rangle}_{=0} = 0\,. \tag{4.1.13}$$

As we shall see below, this feature is very useful when making mathematical transformations. On the other hand, it has turned out that in many applications this choice (4.1.12) does not represent the actual process well. Rather, the fluctuations go on all the time, especially when the system moves on from one time to the next time. Therefore the second choice, introduced by Stratonovich, requires that in (4.1.12) q is taken at the midpoint between the times t_{i-1} and t_i. Therefore the Stratonovich rule reads

$$g(q)\Delta w(t_i) = g\left(q\left(\frac{t_i + t_{i-1}}{2}\right)\right)\Delta w(t_i)\,. \tag{4.1.14}$$

The advantage of this rule is that it allows variables to be transformed according to the usual calculus of differential equations.

Because we have considered here the variable q at the discrete times t_i, it may seem surprising that we introduce now new times inbetween the old sequence. At this moment we should mention that in both the Îto and Stratonovich procedures we have a limiting procedure in mind in which Δt (4.1.2) tends to 0. Consequently, midpoints are also then included in the whole approximation procedure.

4.2 The Îto Differential Equation and the Îto-Fokker-Planck Equation

In this section we consider the general case in which the state of the system is described by a vector $\boldsymbol{q}$ having the components q_l. We wish to study a stochastic process in which, in contrast to (4.1.3 or 11), the time interval tends to zero. In the generalization of (4.1.11) we consider a stochastic equation in the form

$$dq_l(t) = K_l(\boldsymbol{q}(t))\,dt + \sum_m g_{lm}(\boldsymbol{q}(t))\,dw_m(t)\,. \tag{4.2.1}$$

We postulate

$$\langle dw_m \rangle = 0 \quad \text{and} \tag{4.2.2}$$

$$\langle dw_m(t)\,dw_l(t)\rangle = \delta_{lm}\,dt\,. \tag{4.2.3}$$

In contrast to (4.1.5) we make $Q = 1$ because any $Q \neq 1$ could be taken care of by an appropriate choice of g_{lm}. For a discussion of orders of magnitude the following observation is useful. From (4.2.3) one may guess that

$$dw_m \sim \sqrt{dt}\,. \tag{4.2.4}$$

Though this is not mathematically rigorous, it is a most useful tool in determining the correct orders of magnitude of powers of dw. From (4.2.1) we may deduce a mean value equation for q_l by taking a statistical average on both sides of (4.2.1). So we get for the sum on the rhs

$$\sum_m \langle g_{lm}(\boldsymbol{q}(t))\,dw_m(t)\rangle = \sum_m \langle g_{lm}(\boldsymbol{q}(t))\rangle \langle dw_m(t)\rangle\,. \tag{4.2.5}$$

Working from the Îto assumption that $\boldsymbol{q}$ occurring in g, and dw_m in (4.2.1) are statistically uncorrelated, we have therefore split the average of the last term into a product of two averages. From (4.2.2) we find

$$d\langle q_l(t)\rangle = \langle K_l(\boldsymbol{q}(t))\rangle\,dt\,, \tag{4.2.6}$$

or, after a formal division of (4.2.6) by dt, we obtain the mean value equation of Îto in the form

$$\frac{d}{dt}\langle q_l(t)\rangle = \langle K_l(\boldsymbol{q}(t))\rangle . \tag{4.2.7}$$

Instead of following the individual paths of $q_l(t)$ of each member of the statistical ensemble, we may also ask for the probability to find the variable q_l within a given interval $q_l \ldots q_l + dq_l$ at a given time t. We wish to derive an equation for the corresponding (probability) distribution function. To this end we introduce an arbitrary function $\boldsymbol{u}(\boldsymbol{q})$

$$\boldsymbol{u} = \boldsymbol{u}(\boldsymbol{q}) , \tag{4.2.8}$$

which does not explicitly depend on time. Furthermore, we form the differential of u_j, i.e., du_j, up to terms linear in dt. To this end we have to remember that dq_l contains two parts

$$dq_l = dq_{l,1} + dq_{l,2} \quad \text{with} \quad dq_{l,1} = O(dt) \quad \text{and} \quad dq_{l,2} = O(\sqrt{dt}) , \tag{4.2.9}$$

where the first part stems from the coherent force $\boldsymbol{K}$ and the second part from the fluctuating force. Therefore we must calculate du_j up to second order

$$du_j = \sum_k \frac{\partial u_j}{\partial q_k} dq_k + \frac{1}{2}\sum_{kl} \frac{\partial^2 u_j}{\partial q_k \partial q_l} dq_k dq_l . \tag{4.2.10}$$

Inserting (4.2.1) in (4.2.10) we readily obtain

$$\begin{aligned} du_j = {} & \sum_k \frac{\partial u_j}{\partial q_k}\Big[K_k(\boldsymbol{q})\, dt + \sum_m g_{km} dw_m(t)\Big] \\ & + \frac{1}{2}\sum_{kl} \frac{\partial^2 u_j}{\partial q_k \partial q_l}\Big[\sum_{mn} g_{km} g_{ln} dw_m dw_n + \underbrace{O(dw\, dt)}_{=O((dt)^{3/2})}\Big] , \end{aligned} \tag{4.2.11}$$

where we omit terms of order $(dt)^{3/2}$. Note that

$$g_{km} \equiv g_{km}(\boldsymbol{q}) , \tag{4.2.12}$$

where $\boldsymbol{q}$ is taken such that it is not correlated with $d\boldsymbol{w}$. We now turn to the determination of the distribution function. To elucidate our procedure we assume that time is taken at discrete values t_i. We consider an individual path described by the sequence of variables $\boldsymbol{q}_j$, $\boldsymbol{w}_j$ at time t_j, $j = 0, 1, \ldots, i$. The probability (density) of finding this path is quite generally given by the joint probability

$$P(\boldsymbol{q}_i, \boldsymbol{w}_i - \boldsymbol{w}_{i-1}, t_i; \boldsymbol{q}_{i-1}, \boldsymbol{w}_{i-1} - \boldsymbol{w}_{i-2}, t_{i-1}; \ldots \boldsymbol{q}_0, \boldsymbol{w}_0, t_0) . \tag{4.2.13}$$

When we integrate P over all variables q_j and w_j at all intermediate times $k = 1, \dots, i-1$ and over w_i, w_0, we obtain the probability distribution $f(q, t_i | q_0, t_0)$, $q \equiv q_i$. This is the function we are looking for and we shall derive an equation for it. To this end we take the average $\langle \dots \rangle$ of (4.2.11). Taking the average means that we multiply both sides of (4.2.11) by (4.2.13) and integrate over all variables q_k, w_k except q_0. This multiple integration can be largely simplified when we use the fact that we are dealing here with a Markov process, as it transpires from the form of (4.2.1) in connection with the Îto rule. Indeed the values of q_i are determined by those of q_{i-1} and dw_i alone. This Markov property allows us to write P (4.2.13) in the form [1].

$$P = P_q(q_i, t_i | w_i - w_{i-1}, q_{i-1}, t_{i-1}) P_w(w_i - w_{i-1}, t_{i-1}) \cdot P(q_{i-1}, w_{i-1} - w_{i-2}, t_{i-1}, \dots q_0, w_0, t_0) \,. \tag{4.2.14}$$

In this relation P_q is the conditional probability of finding $q = q_i$ at time t_i, provided $\Delta w = \Delta w_i$ and $q = q_{i-1}$ at time t_{i-1}, and $P_w(\Delta w_i, t_i)$ is the probability distribution of Δw at time t_i. The last factor is again the joint probability distribution. We need further the usual normalization properties, namely

$$\int P_q(q_k, t_k | \Delta w_k, q_{k-1}, t_{k-1}) d^n q_k = 1 \tag{4.2.15}$$

and

$$\int P_w(\Delta w_k, t_k) d^n w_k = 1 \,, \tag{4.2.16}$$

where n is the dimension of the vectors q and w. We are now in a position to derive and discuss the averages $\langle \dots \rangle$ over the individual terms of (4.2.11). Because u_j depends on $q \equiv q_i$ only, the multiple integral over all q_k, dw_k (except q_0) reduces to

$$\langle u_j \rangle = \int d^n q \, u_j(q) f(q, t | q_0, t_0) \,, \tag{4.2.17}$$

where we used the definition of f introduced above. Interchanging the averaging procedure and the time evolution of the system described by the differential operator d and using (4.2.11) we readily find

$$\underbrace{d\langle u_j \rangle}_{1} = \underbrace{\sum_k \left\langle \frac{\partial u_j}{\partial q_k} K_k(q) \right\rangle dt}_{2} + \underbrace{\sum_{km} \left\langle \frac{\partial u_j}{\partial q_k} g_{km} \right\rangle \langle dw_m \rangle}_{3}$$

$$\underbrace{+ \frac{1}{2} \sum_{kl} \left\langle \frac{\partial^2 u_j}{\partial q_k \partial q_l} \sum_{mn} g_{km} g_{ln} \right\rangle \langle dw_m dw_n \rangle}_{4} \,, \tag{4.2.18}$$

where we have numerated the corresponding terms to be discussed. Because of the statistical independence of $\boldsymbol{q}_{i-1}$ and $\Delta \boldsymbol{w}_i$, which is also reflected by the decomposition (4.2.14), we have been able to split the average $\langle \ldots \rangle$ on the rhs into a product (cf. terms 3 and 4). The averages containing $\boldsymbol{q}$ are of precisely the same structure as (4.2.17) in that they contain the variable $\boldsymbol{q}$ only at a single time $i = 1$. A short, straightforward analysis shows that we may again reduce the average to an integral over $\boldsymbol{q}$. Taking the differential quotient of (4.2.18) with respect to time t we then obtain from (4.2.17)

$$\frac{d\langle u_j \rangle}{dt} = \int d^n q u_j(\boldsymbol{q}) \frac{\partial}{\partial t} f(\boldsymbol{q}, t | \boldsymbol{q}_0, t_0) . \tag{4.2.19}$$

The expressions under the sum of term 2 in (4.2.18) read

$$\left\langle \frac{\partial u_j}{\partial q_k} K_k(\boldsymbol{q}) \right\rangle = \int d^n q \left[\frac{\partial u_j}{\partial q_k} K_k(\boldsymbol{q}) \right] f(\boldsymbol{q}, t | \boldsymbol{q}_0, t_0) . \tag{4.2.20}$$

We perform a partial integration with respect to the coordinate q_k by which we may transform (4.2.20) into

$$- \int d^n q u_j(\boldsymbol{q}) \frac{\partial}{\partial q_k} [K_k(\boldsymbol{q}) f(\boldsymbol{q}, t | \boldsymbol{q}_0, t_0)] , \tag{4.2.21}$$

where we assumed that f (and its derivatives) vanishes at the boundaries. Term 3 vanishes because the average factorizes since $\boldsymbol{q}$ and $d\boldsymbol{w}$ are uncorrelated and because of (4.2.2). The last term 4 of (4.2.18) can be transformed in analogy to our procedure applied to the term (4.2.20), but using two consecutive partial integrations. This yields for a single term under the sums

$$\left\langle \frac{1}{2} \frac{\partial^2 u_j}{\partial q_k \partial q_l} g_{km} g_{lm} \right\rangle = \frac{1}{2} \int d^n q \, u_j(\boldsymbol{q}) \frac{\partial^2}{\partial q_k \partial q_l} [g_{km} g_{lm} f(\boldsymbol{q}, t | \boldsymbol{q}_0, t_0)] . \tag{4.2.22}$$

When we consider the resulting terms 1, 2 and 4, which stem from (4.2.18), we note that all of them are of the same type, namely of the form

$$\int d^n q u_j(\boldsymbol{q}) [\ldots] f(\boldsymbol{q}, t | \boldsymbol{q}_0, t_0) , \tag{4.2.23}$$

where the bracket may contain differential operators. Because these expressions occur on both sides of (4.2.18) and because u_j was an arbitrary function, we conclude that (4.2.18), if expressed by the corresponding terms 1, 2 and 4, must be fulfilled, even if the integration over $d^n q$ and the factor $u_j(\boldsymbol{q})$ are dropped. This leads us to the equation

$$\frac{\partial f}{\partial t} = - \sum_k \frac{\partial}{\partial q_k} [K_k(\boldsymbol{q}) f] + \frac{1}{2} \sum_{kl} \frac{\partial^2}{\partial q_k \partial q_l} \left[\sum_m g_{km} g_{lm} f \right] . \tag{4.2.24}$$

This Fokker-Planck type equation is related to the Îto stochastic equation (4.2.1).

4.3 The Stratonovich Calculus

In this calculus $\boldsymbol{q}$, which occurs in the functions g, is taken at the midpoint of the time interval $t_{i-1} \dots t_i$, i.e., we have to consider expressions of the form

$$g_{lm}\left(\boldsymbol{q}\left(\frac{t_i + t_{i-1}}{2}\right)\right) dw_m(t_i) \,. \tag{4.3.1}$$

(Here and in the following summation will be taken over dummy indices.)

This rule can be easily extended to the case in which g depends explicitly on time t, where t must be replaced according to (4.3.1) by the midpoint rule. Two things must be observed.

1) If we introduce the definition (4.3.1) into the original Îto equation (4.2.1), a new stochastic process is defined.

2) Nevertheless, the Îto and Stratonovich processes are closely connected with each other and one may go from one definition to the other by a simple transformation. To this end we integrate the Îto equation (4.2.1) in a formal manner by summing up the individual time steps and then going over to the integral. Thus we obtain from (4.2.1)

$$q_l(t) = q_l(t_0) + \int_{t_0}^{t} K_l(\boldsymbol{q})\, dt' + \int\!\!\!\!\!I_{t_0}^{t} g_{lm}(\boldsymbol{q})\, dw_m(t') \,. \tag{4.3.2}$$

In it the Îto integral is defined by taking $\boldsymbol{q}$ in g_{lm} at a time just before the fluctuation dw_m occurs. We have denoted this integral by the sign

$$\int\!\!\!\!\!I \,. \tag{4.3.3}$$

We now consider a process which leads to precisely the same $q_l(t)$ as in (4.3.2) but by means of the Stratonovich definition

$$q_l(t) = q_l(t_0) + \int_{t_0}^{t} \tilde{K}_l(\boldsymbol{q})\, dt' + \int\!\!\!\!\!S_{t_0}^{t} \tilde{g}_{lm}(\boldsymbol{q})\, dw_m(t') \,. \tag{4.3.4}$$

We shall show that this can be achieved by an appropriate choice of a new driving force $\tilde{\boldsymbol{K}}$ and new factors $\tilde{g}_{lm}$. In fact, we shall show that we need just a new force $\tilde{\boldsymbol{K}}$, but that we may choose $\tilde{g} = g$. Here the Stratonovich integral

$$\int\!\!\!\!\!S \tag{4.3.5}$$

is to be evaluated as the limit of the sum with the individual terms (4.3.1). Therefore we consider

$$\mathop{\mathrm{S}\!\!\!\int} \tilde{g}_{lm}(q)\,dw_m(t') = \lim_{\substack{\Delta t \to 0 \\ N \to \infty}} \sum_i \tilde{g}_{lm}\left(q\left(\frac{t_{i-1}+t_i}{2}\right)\right)[w_m(t_i) - w_m(t_{i-1})]\,, \tag{4.3.6}$$

where $\Delta t \to 0$ and $N \to \infty$ such that $N\Delta t = t - t_0$.

Inserting an additional term containing

$$w_m\left(\frac{t_{i-1}+t_i}{2}\right), \tag{4.3.7}$$

we obtain for the rhs of (4.3.6) (before performing "lim")

$$\begin{aligned}&\sum_{i=1}^{N} \tilde{g}_{lm}\left(q\left(\frac{t_{i-1}+t_i}{2}\right)\right)\left[w_m(t_i) - w_m\left(\frac{t_{i-1}+t_i}{2}\right)\right] \\ &\quad + \sum_{i=1}^{N} \tilde{g}_{lm}\left(q\left(\frac{t_{i-1}+t_i}{2}\right)\right)\left[w_m\left(\frac{t_{i-1}+t_i}{2}\right) - w_m(t_{i-1})\right].\end{aligned} \tag{4.3.8}$$

We put

$$q\left(\frac{t_{i-1}+t_i}{2}\right) = q\left(t_{i-1} + \frac{t_i - t_{i-1}}{2}\right) = q(t_{i-1}) + dq(t_{i-1}) \tag{4.3.9}$$

and

$$dt = \frac{t_i - t_{i-1}}{2}. \tag{4.3.10}$$

Because $q(t)$ obeys the Îto equation we find for (4.3.9)

$$\begin{aligned}q\left(\frac{t_{i-1}+t_i}{2}\right) &= q(t_{i-1}) + K(q(t_{i-1}))\frac{t_i - t_{i-1}}{2} \\ &\quad + g[q(t_{i-1})]\left[w\left(\frac{t_{i-1}+t_i}{2}\right) - w(t_{i-1})\right].\end{aligned} \tag{4.3.11}$$

We expand $\tilde{g}_{lm}$ with repect to (4.3.10) and obtain, taking into account the terms of order dt,

$$\tilde{g}_{lm}\left(q\left(\frac{t_{i-1}+t_i}{2}\right)\right) = \tilde{g}_{lm}(q(t_{i-1})) + \frac{\partial \tilde{g}_{lm}(q(t_{i-1}))}{\partial t}\,\frac{t_i - t_{i-1}}{2} +$$

$$+\left[\frac{\partial\tilde{g}_{lm}(\boldsymbol{q}(t_{i-1}))}{\partial q_k}K_k(\boldsymbol{q}(t_{i-1}))\right.$$

$$\left.+\frac{1}{2}\frac{\partial^2 g_{lm}(\boldsymbol{q}(t_{i-1}))}{\partial q_k\partial q_l}g_{kp}(\boldsymbol{q}(t_{i-1}))\,g_{lp}(\boldsymbol{q}(t_{i-1}))\right]\frac{t_i-t_{i-1}}{2}$$

$$+\frac{\partial\tilde{g}_{lm}(\boldsymbol{q}(t_{i-1}))}{\partial q_k}g_{kp}(\boldsymbol{q}(t_{i-1}))\left[w_p\left(\frac{t_{i-1}+t_i}{2}\right)-w_p(t_{i-1})\right]. \tag{4.3.12}$$

The second term on the rhs occurs if $\tilde{g}_{lm}$ explicitly depends on time. By means of (4.3.12), the integral (4.3.6) acquires the form

$$\$\ldots=\sum_{i=1}^{N}\tilde{g}_{lm}\left(\boldsymbol{q}\left(\frac{t_{i-1}+t_i}{2}\right)\right)\left[w_m(t_i)-w_m\left(\frac{t_{i-1}+t_i}{2}\right)\right]$$

$$+\sum_{i=1}^{N}\tilde{g}_{lm}(\boldsymbol{q}(t_{i-1}))\left[w_m\left(\frac{t_{i-1}+t_i}{2}\right)-w_m(t_{i-1})\right]$$

$$+\sum_{i=1}^{N}\frac{\partial\tilde{g}_{lm}(\boldsymbol{q}(t_{i-1}))}{\partial q_k}g_{kp}(\boldsymbol{q}(t_{i-1}))\left[w_m\left(\frac{t_{i-1}+t_i}{2}\right)-w_m(t_{i-1})\right]^2\delta_{mp}$$

$$+O(\Delta t\,\Delta w)\,. \tag{4.3.13}$$

An inspection of the rhs of (4.3.13) reveals that the first two sums jointly define the Îto integral. This can be seen as follows. The first sum contains a contribution from dw_m over the time interval from $(t_{i-1}+t_i)/2$ till t_i, whereas the second sum contains the time interval from t_{i-1} till $(t_{i-1}+t_i)/2$. Thus taking both time intervals together and summing up over i the sums cover the total time interval. The third sum over i is again of the Îto type. Here we must use a result from stochastic theory according to which the square bracket squared converges against $dt/2$ in probability. Thus the whole sum converges towards an ordinary integral. These results allow us to write the connection between the Stratonovich and Îto integrals in the form

$$\$\tilde{g}_{lm}dw_m(t')=\int\tilde{g}_{lm}dw_m(t')+\int\frac{\partial\tilde{g}_{lm}}{\partial q_k}g_{km}\,dt\,. \tag{4.3.14}$$

We are now in a position to compare the Îto and Stratonovich processes described by (4.3.2 and 4) explicitly. We see that both processes lead to the same result, as required if we make the following identifications:

Stratonovich		Îto
$\tilde{g}_{lm}$	$=$	g_{lm}
$\tilde{\boldsymbol{K}}_l$	$=$	$K_l-\dfrac{1}{2}\dfrac{\partial g_{lm}}{\partial q_k}g_{km}\,.$

(4.3.15)

In other words, we find exactly the same result if we use a Stratonovich stochastic equation, but use in it $\tilde{g}$ and $\tilde{\boldsymbol{K}}$ instead of g and $\boldsymbol{K}$ of the Îto equation in the way indicated by (4.3.15). These results allow us in particular to establish a Stratonovich-Fokker-Planck equation. If the Stratonovich stochastic equation is written in the form

$$dq_l = K_l(q)\,dt + g_{lm}(q)\,dw_m(t) \tag{4.3.16}$$

and $g_{lm} \cdot dw_m$ is evaluated according to (4.3.1), the corresponding Fokker-Planck equation reads

$$\frac{\partial f}{\partial t} = -\frac{\partial}{\partial q_l}\left\{\left[K_l(q) + \frac{1}{2}\,\frac{\partial g_{lj}}{\partial q_k}\,g_{kj}\right]f\right\} + \frac{1}{2}\,\frac{\partial^2}{\partial q_l \partial q_m}(g_{li}g_{mi}f)\,. \tag{4.3.17}$$

We remind the reader that we used the convention of summing up over dummy indices. Because in other parts of this book we do not use this convention, we write the Fokker-Planck equation in the usual way by explicitly denoting the sums

$$\frac{\partial f}{\partial t} = -\sum_l \frac{\partial}{\partial q_l}\left\{\left[K_l(q) + \frac{1}{2}\sum_{kj}\frac{\partial g_{lj}}{\partial q_k}\,g_{kj}\right]f\right\} + \frac{1}{2}\sum_{lm}\frac{\partial^2}{\partial q_l \partial q_m}\left(\sum_i g_{li}g_{mi}f\right). \tag{4.3.18}$$

4.4 Langevin Equations and Fokker-Planck Equation

For sake of completeness we mention the Langevin equations which are just special cases of the Îto or Stratonovich equations, because their fluctuating forces are independent of the variable $\boldsymbol{q}$ and of time t. Therefore the corresponding Fokker-Planck equation is the same in the Îto and Stratonovich calculus.

Exercises. 1) Why do we need relations (4.2.14, 15) when evaluating the rhs of (4.2.18)? *Hint:* while u_j depends on $\boldsymbol{q}$ at time t_i, K_k and g_{km} depend on $\boldsymbol{q}$ at time t_{i-1}.

2) How does the explicit expression for the joint probability (4.2.13) read? *Hint:* the conditional probability $P_w(w, t\,|\,w' = \boldsymbol{0}, 0)$ is given by

$$P_w(w, t\,|\,\boldsymbol{0}, 0) = \frac{1}{(2\pi t)^{n/2}}\exp\left(-\frac{w^T \cdot w}{2t}\right). \tag{4.4.1}$$

5. The World of Coupled Nonlinear Oscillators

When speaking of oscillators, probably most of us first think of mechanical oscillators such as springs. Another example from mechanics is provided by the pendulum. It can be treated as a linear oscillator if its amplitude of oscillation is small enough, but otherwise it represents a nonlinear oscillator. In many cases of practical importance we have to deal with coupled oscillators. For instance, think of a piece of elastic material which for mathematical treatment is treated as a set of coupled finite elements each of which can be represented as an oscillator. Such methods are of great importance in mechanical engineering, for instance when dealing with vibrations of engines or towers, or with flutter of wings of airplanes. Of course, in a number of cases we consider the limiting case in which the finite elements approach a continuous distribution which corresponds to our original picture of an elastic medium. Oscillations occur equally well in electrical and radio engineering. Here we deal not only with the old radio tube oscillator but with modern circuits using transistors and other electronic devices.

In the field of optics the laser can be considered as being built up of many coupled quantum-mechanical oscillators, namely the electrons of the laser atoms. Through the cooperation of these oscillators, laser light in the form of a coherent oscillation of the electromagnetic field is produced. In a number of experiments done with fluids, the observed phenomena can be interpreted as if some specific oscillators representing complicated motions of the fluid interact with each other. As mentioned in the introduction, chemical reactions can also show oscillations and can serve as examples for the behavior of coupled oscillators. Even in elementary particle physics we have to bear in mind that we are dealing with fields which in one way or another can be considered as a set of homogeneously distributed coupled oscillators.

As is well known, the brain can show macroscopic electric oscillations stemming from a coherent firing of neurons. For these reasons and many others the problem of dealing with the behavior of coupled oscillators is of fundamental importance.

5.1 Linear Oscillators Coupled Together

We now turn to a more mathematically orientated discussion of coupled oscillators where we shall be interested in the mathematical features of the solutions irrespective of the physical nature of the oscillators. For the following it will be important to distinguish between linear and nonlinear cases because their behavior may be entirely different.

5.1.1 Linear Oscillators with Linear Coupling

A linear oscillator may be described by an equation of the form

$$\ddot{x}_1 + \omega_1^2 x_1 = 0\,, \tag{5.1.1}$$

where x_1 is a time-dependent variable, whereas ω_1 is the eigenfrequency. Such an oscillator may be coupled linearly to a second one, i.e., by additional terms which are linear in x_2 and x_1, respectively,

$$\ddot{x}_1 + \omega_1^2 x_1 = \alpha x_2 \tag{5.1.2}$$

$$\ddot{x}_2 + \omega_2^2 x_2 = \beta x_1\,. \tag{5.1.3}$$

Introducing additional variables, which in Hamiltonian mechanics are called momenta, p_1 and p_2 by

$$p_1 = \dot{x}_1/\omega_1 \quad \text{and} \tag{5.1.4}$$

$$p_2 = \dot{x}_2/\omega_2\,, \tag{5.1.5}$$

we may cast (5.1.2) in the form of first-order differential equations

$$\dot{p}_1 = -\omega_1 x_1 + (\alpha/\omega_1)x_2 \tag{5.1.6}$$

$$\dot{x}_1 = \omega_1 p_1\,, \tag{5.1.7}$$

and the same can be done with (5.1.3). Writing the variables x_1, p_1, x_2, p_2 in the form of a vector

$$\begin{pmatrix} p_1 \\ x_1 \\ p_2 \\ x_2 \end{pmatrix} = q\,, \tag{5.1.8}$$

we may cast the set of equations (5.1.6) and (5.1.7) and their corresponding equations with the index 2 into the form

$$\dot{q} = Lq\,, \tag{5.1.9}$$

where L is a matrix with time-independent coefficients. Clearly any number of linearly coupled linear oscillators can be cast into the form (5.1.9). The solutions of this type of equation were derived in Sect. 2.6 so that the problem of these oscillators is entirely solved. Later we shall show that the same kind of solutions as those derived in Sect. 2.6 holds if the variables x_j are considered as continuously distributed in space.

5.1.2 Linear Oscillators with Nonlinear Coupling. An Example. Frequency Shifts

An example of this type is provided by

$$\ddot{x}_1 + \omega_1^2 x_1 = \alpha x_1 x_2 , \tag{5.1.10}$$

$$\ddot{x}_2 + \omega_2^2 x_2 = \beta x_1 x_2 , \tag{5.1.11}$$

where the terms on the right-hand side of these equations provide the coupling.

In order to discuss some features of the solutions of these equations we introduce new variables. We make

$$p_j = \dot{x}_j / \omega_j \tag{5.1.12}$$

so that we may replace (5.1.10) by

$$\dot{p}_1 = \ddot{x}_1 / \omega_1 = -\omega_1 x_1 + (\alpha/\omega_1) x_1 x_2 \quad \text{and} \tag{5.1.13}$$

$$\dot{x}_1 = \omega_1 p_1 , \tag{5.1.14}$$

and similar equations can be obtained for the second oscillator. Introducing

$$b_j = x_j + \mathrm{i} p_j , \quad b_j^* = x_j - \mathrm{i} p_j , \tag{5.1.15}$$

we may cast (5.1.13 and 14) into a single equation. To this end we multiply (5.1.13) by i and add it to (5.1.14). With use of (5.1.15) we then obtain

$$\dot{b}_1 = -\mathrm{i}\omega_1 b_1 + \mathrm{i}\gamma(b_1 + b_1^*)(b_2 + b_2^*) , \quad \gamma = \alpha/(4\omega_1) . \tag{5.1.16}$$

Inserting the hypothesis

$$b_j = r_j \exp(-\mathrm{i}\varphi_j) , \tag{5.1.17}$$

r_j, φ_j being real, into (5.1.16), dividing both sides by $\exp(-\mathrm{i}\varphi_j)$, and separating the real from the imaginary part, we obtain the two equations

$$\dot{r}_1 = -2\gamma r_1 r_2 \sin(2\varphi_1)\cos(\varphi_2) \quad \text{and} \tag{5.1.18}$$

$$\dot{\varphi}_1 = \omega_1 - 4\gamma r_2 \cos^2(\varphi_1)\cos(\varphi_2) . \tag{5.1.19}$$

Similar equations can be obtained for r_2 and φ_2. These equations represent equations of motion for the radii r_j and phase angles φ_j. Collecting variables in the form of vectors

$$\begin{pmatrix} \varphi_1 \\ \varphi_2 \end{pmatrix} = \boldsymbol{\varphi} , \tag{5.1.20}$$

$$\begin{pmatrix} \omega_1 \\ \omega_2 \end{pmatrix} = \boldsymbol{\omega} , \tag{5.1.21}$$

$$\begin{pmatrix} r_1 \\ r_2 \end{pmatrix} = \boldsymbol{r} , \tag{5.1.22}$$

and writing the rhs of the equations for φ in the form

$$\begin{pmatrix} f_1 \\ f_2 \end{pmatrix} = \boldsymbol{f} , \tag{5.1.23}$$

and those for r in the form

$$\begin{pmatrix} g_1 \\ g_2 \end{pmatrix} = \boldsymbol{g} , \tag{5.1.24}$$

we find equations of the general form

$$\dot{\boldsymbol{\varphi}} = \boldsymbol{\omega} + \boldsymbol{f}(\boldsymbol{r}, \boldsymbol{\varphi}) \quad \text{and} \tag{5.1.25}$$

$$\dot{\boldsymbol{r}} = \boldsymbol{g}(\boldsymbol{r}, \boldsymbol{\varphi}) . \tag{5.1.26}$$

Again it is obvious how we can derive similar equations for a number n of linear oscillators which are coupled nonlinearly. Nonlinear coupling between linear oscillators can cause behavior of the solutions which differs qualitatively from those of linearly coupled linear oscillators. As is evident from (5.1.19), $\varphi_1 = \omega_1 t$ is no longer a solution, i.e., the question arises immediately as to how we can still speak of a periodic oscillation.

A still rather simple but nevertheless fundamental effect can be easily observed when we change the kind of coupling to a somewhat different one, e.g., into the form

$$\ddot{x}_1 + \omega_1^2 x_1 = \alpha x_1 x_2^2 . \tag{5.1.27}$$

The equation for the phase angle φ_1 then acquires the form

$$\dot{\varphi}_1 = \omega_1 - 4\gamma r_2^2 \cos^2(\varphi_1)\cos^2(\varphi_2) . \tag{5.1.28}$$

Let us now assume in the way of a model that r_2 is a time-independent constant. In order to take into account the effect of the term containing φ_1 and φ_2 on rhs of (5.1.28), we may average that term in a first approximation over a long time interval. Because for such a time average we obtain

$$\cos^2(\varphi_1)\cos^2(\varphi_2) = 1/4 , \tag{5.1.29}$$

we obtain instead of (5.1.28) a new equation for φ_1 in lowest approximation, i.e., $\varphi_1^{(0)}$,

$$\dot{\varphi}_1^{(0)} = \omega_1 - \gamma r_2^2 . \tag{5.1.30}$$

It indicates that in lowest approximation the effect of the nonlinear terms of the rhs of (5.1.28) consists in a shift of frequencies. Therefore whenever nonlinear coupling between oscillators is present we have to expect such frequency shifts and possibly further effects. In the following we shall deal with various classes of nonlinearly coupled oscillators. In one class we shall study the conditions under which the behavior of the coupled oscillators can be represented by a set of, possibly, different oscillators which are uncoupled. We shall see that this class plays an important role in many practical cases. On the other hand, other large classes of quite different behavior have been found more recently. One important class consists of solutions which describe chaotic behavior (Sect. 8.11.2)

5.2 Perturbations of Quasiperiodic Motion for Time-Independent Amplitudes (Quasiperiodic Motion Shall Persist)

In order to elucidate and overcome some of the major difficulties arising when we deal with nonlinearly coupled oscillators, we shall consider a special case, namely equations of the form (5.1.25) where we assume that $\boldsymbol{r}$ is a time-independent constant.

If there is no coupling between oscillators their phase angles obey equations of the form

$$\frac{d\varphi}{dt} = \omega . \tag{5.2.1}$$

Since these oscillators generally oscillate at different frequencies $\omega_1, \ldots, \omega_n$, we may call their total motion quasiperiodic. Now we switch on a coupling between these oscillators and let it grow to a certain amount. In such a case we have to deal with equations of the form

$$\frac{d\varphi}{dt} = \omega + \varepsilon f(\varphi) . \tag{5.2.2}$$

It is assumed that f is periodic in $\varphi_1, \dots, \varphi_n$ and that it can be represented as a Fourier series in the form

$$f(\varphi) = \sum_m f_m e^{i m \cdot \varphi}, \quad \text{where} \tag{5.2.3}$$

$$m\varphi = m_1\varphi_1 + \dots + m_1\varphi_n . \tag{5.2.4}$$

The small quantity ε in (5.2.2) indicates the smallness of the additional force f.

Let us assume that under the influence of coupling the solutions do not lose their quasiperiodic character. As we have seen in the preceding section by means of the explicit examples (5.1.28, 30) the additional term f can cause a frequency shift. On the other hand, we shall see below that we can treat the perturbation εf in an adequate way only if the frequencies ω_j remain fixed. The situation will turn out to be quite analogous to that in Sect. 2.1.3, where the convergence of certain Fourier series depended crucially on certain irrationality conditions between frequencies ω_j. Indeed, it shall turn out that we must resort precisely to those same irrationality conditions. To this end we must secure that all the time the basic frequencies ω_j remain unaltered. This can be achieved by a formal trick, namely we not only introduce the perturbation f into (5.2.1) but at the same time a counterterm $\Delta(\varepsilon)$ which in each order of ε just cancels the effect of the frequency shift caused by f. Therefore we consider instead of (5.2.2) the equation

$$\frac{d\varphi}{dt} = \omega + \Delta(\varepsilon) + \varepsilon f(\varphi) , \tag{5.2.5}$$

where $\Delta(\varepsilon)$ is a function of ε still to be determined. (The introduction of such counterterms is, by the way, not new to physicists who have been dealing with quantum electrodynamics or quantum field theory. Here the ω's correspond to the observed energies of a particle, but due to the coupling of that particle with a field (corresponding to f) an energy shift is caused. In order to obtain the observed energies, adequate counterterms are introduced into the Hamiltonian. The corresponding procedure is called renormalization of the mass, electric charge, etc.)

To simplify the notation in the following we shall make the substitution

$$\varepsilon f(\varphi) \to f(\varphi) . \tag{5.2.6}$$

We shall now devise an iteration procedure to solve (5.2.5). It will turn out that this procedure, developed by Kolmogorov and Arnold and further elaborated by Bogolyubov and others, converges very rapidly. We first proceed in a heuristic fashion. In lowest approximation we average equation (5.2.5) over a sufficiently long time interval so that we obtain

$$\frac{d\bar{\varphi}}{dt} = \omega + \Delta + \bar{f}\,. \tag{5.2.7}$$

We now choose in this approximation

$$\Delta = \Delta_0 = -\bar{f} \tag{5.2.8}$$

so that the solution, which we also denote by

$$\bar{\varphi} \equiv \varphi^{(0)}\,, \tag{5.2.9}$$

acquires the form

$$\varphi^{(0)} = \omega t + \varphi_0\,. \tag{5.2.10}$$

Obviously we have chosen the counterterm Δ in such a way that $\varphi^{(0)}$ retains the old frequencies ω_j. In ordinary perturbation theory we should now insert (5.2.10) into rhs of (5.2.5) in order to obtain an improved solution $\hat{\varphi}^{(1)}$, i.e.,

$$\frac{d\hat{\varphi}^{(1)}}{dt} = (\omega + \Delta + \bar{f}) + \sum_{m \neq 0} f_m \exp\left[\mathrm{i}\boldsymbol{m}(\omega t + \varphi_0)\right]. \tag{5.2.11}$$

This solution can be found directly by integrating the rhs in (5.2.11), yielding

$$\hat{\varphi}^{(1)} = \omega t + \varphi_0 + \sum_{m \neq 0} (f_m/(\mathrm{i}\boldsymbol{m} \cdot \omega)) \exp\left[\mathrm{i}\boldsymbol{m}(\omega t + \varphi_0)\right]. \tag{5.2.12}$$

The expression on the rhs, namely

$$\tilde{f}(\Psi) = \sum_{m \neq 0} (f_m/(\mathrm{i}\boldsymbol{m} \cdot \omega)) \exp(\mathrm{i}\boldsymbol{m} \cdot \Psi)\,, \quad \Psi \equiv \omega t + \varphi_0\,, \tag{5.2.13}$$

is familiar from Sect. 2.1.3. There we have seen that (5.2.13) may cause serious convergence difficulties, but these convergence difficulties do not occur if the ω's fulfill a Kolmogorov-Arnold-Moser (KAM) condition which we shall take in the form

$$\begin{aligned} &|\boldsymbol{m} \cdot \omega| \geqslant K \|\boldsymbol{m}\|^{-(n+1)}\,, \\ &\|\boldsymbol{m}\| = |m_1| + |m_2| + \ldots |m_n|. \end{aligned} \tag{5.2.14}$$

Below a condition on f will be imposed (namely that of being analytic in a certain domain) to ensure that (5.2.13) converges when (5.2.14) is fulfilled. The solution (5.2.12) can then be cast into the form ($\varphi = \hat{\varphi}^{(1)}$)

$$\varphi = \omega t + \varphi_0 + f(\omega t + \varphi_0)\,. \tag{5.2.15}$$

Within conventional perturbation theory we should now continue our approximation scheme by inserting (5.2.15) into the rhs of (5.2.5), thus hoping to improve the approximation. However, the convergence of this procedure has been questioned and the following rapidly converging procedure has been devised. The basic idea is as follows.

We shall use (5.2.15) as a starting point for a sequence of transformations of variables, namely we shall introduce instead of the wanted variable φ yet an other unknown variable $\varphi^{(1)}$ by the hypothesis

$$\varphi = \varphi^{(1)} + \tilde{f}(\varphi^{(1)}) , \tag{5.2.16}$$

where $\tilde{f}$ is identical with that occurring in (5.2.15). When we insert (5.2.16) into (5.2.5) we find a rather complicated equation for $\varphi^{(1)}$ so that nothing seems to be gained. But when rearranging the individual terms of this equation it transpires that it has precisely the same form as (5.2.5), with one major difference. On the rhs terms occur which can be shown to be smaller than those of equation (5.2.5), i.e., the smallness parameter ε is now replaced by ε^2. Continuing this procedure, we find a new equation in which the smallness parameter is ε^4, then we find ε^8, etc. This indicates, at least from a heuristic point of view, that the iteration procedure is rapidly converging, as shall be proved in this section.

We insert the hypothesis (5.2.16) into both sides of (5.2.5). Of course, $\varphi^{(1)}$ is a function of time. Therefore, when differentiating $\tilde{f}$ with respect to time, by applying the chain rule we first differentiate $\tilde{f}$ with respect to $\varphi^{(1)}$ and then $\varphi^{(1)}$ with respect to time t.

In this connection it will be advantageous to use a notation which is not too clumsy. If $\boldsymbol{v}$ is an arbitrary vector with components $v_1, \ldots, v_n$, we introduce the following notation

$$\frac{\partial \tilde{f}(\varphi^{(1)})}{\partial \varphi^{(1)}} \boldsymbol{v} = \begin{pmatrix} \sum_{l=1}^{n} \frac{\partial \tilde{f}_1}{\partial \varphi_l^{(1)}} v_l \\ \sum_{l=1}^{n} \frac{\partial \tilde{f}_2}{\partial \varphi_l^{(1)}} v_l \\ \vdots \\ \sum_{l=1}^{n} \frac{\partial \tilde{f}_N}{\partial \varphi_l^{(1)}} v_l \end{pmatrix} , \tag{5.2.17}$$

where lhs is defined by rhs. When we choose $\boldsymbol{v} = \boldsymbol{\omega}$ we readily find from (5.2.13) the identity

$$\frac{\partial \tilde{f}(\varphi^{(1)})}{\partial \varphi^{(1)}} \omega = f(\varphi^{(1)}) - \bar{f}(\varphi^{(1)}) . \tag{5.2.18}$$

It just tells us that by differentiating (5.2.13) with respect to $\varphi^{(1)}$ and multiplying it by ω, we obtain f except for the constant term $(\bar{f}(\varphi^{(1)}) = \bar{f} \equiv f_0)$.

After these preliminaries we may immediately write the result we obtain when inserting (5.2.16) into (5.2.5):

$$\frac{d\varphi^{(1)}}{dt}+\frac{\partial\tilde{f}(\varphi^{(1)})}{\partial\varphi^{(1)}}\frac{d\varphi^{(1)}}{dt}=\omega+\Delta+f[\varphi^{(1)}+\tilde{f}(\varphi^{(1)})]\,. \tag{5.2.19}$$

This equation can be given a somewhat different form as one can check directly. The following equation

$$\left(1+\frac{\partial\tilde{f}(\varphi^{(1)})}{\partial\varphi^{(1)}}\right)\left(\frac{d\varphi^{(1)}}{dt}-\omega\right)=\Delta+\bar{f}+f[\varphi^{(1)}+\tilde{f}(\varphi^{(1)})]-f(\varphi^{(1)}) \tag{5.2.20}$$

is identical with (5.2.19) provided we make use of (5.2.18). Dividing both sides of (5.2.20) by

$$\left(1+\frac{\partial\tilde{f}(\varphi^{(1)})}{\partial\varphi^{(1)}}\right), \tag{5.2.21}$$

we obtain

$$\begin{aligned}\frac{d\varphi}{dt}=\omega+\Delta+\bar{f}+&\left[\left(1+\frac{df(\varphi)}{\partial\varphi^{(1)}}\right)^{-1}-1\right](\Delta+\bar{f})\\&+\left(1+\frac{\partial\tilde{f}(\varphi^{(1)})}{\partial\varphi^{(1)}}\right)^{-1}\{f[\varphi^{(1)}+\tilde{f}(\varphi^{(1)})]-f(\varphi^{(1)})\}\,.\end{aligned} \tag{5.2.22}$$

We now wish to show that (5.2.22) has the same shape as (5.2.5). To this end we introduce the abbreviation

$$\Delta^{(1)}=\Delta+\bar{f}(\varphi) \tag{5.2.23}$$

and abbreviate the remaining terms on the rhs of (5.2.22) by $f^{(1)}$ so that

$$\begin{aligned}f^{(1)}(\varphi^{(1)},\Delta^{(1)})=&\left[\left(1+\frac{\partial\tilde{f}(\varphi^{(1)})}{\partial\varphi^{(1)}}\right)^{-1}-1\right]\Delta^{(1)}\\&+\left(1+\frac{\partial\tilde{f}(\varphi^{(1)})}{\partial\varphi^{(1)}}\right)^{-1}\{f[\varphi^{(1)}+f(\varphi^{(1)})]-f(\varphi^{(1)})\}\,.\end{aligned} \tag{5.2.24}$$

In this way (5.2.22) acquires the form

$$\frac{d\varphi^{(1)}}{dt}=\omega+\Delta^{(1)}+f^{(1)}(\varphi^{(1)},\Delta^{(1)})\,, \tag{5.2.25}$$

which, as predicted, has the same form as (5.2.5), but where $\Delta^{(1)}$ now replaces Δ and $f^{(1)}$ replaces f (or εf).

We now want to convince ourselves, at least for heuristic purposes, that $f^{(1)}$ and $\Delta^{(1)}$ are smaller by an order of ε than the terms in (5.2.5). Since f is proportional to ε [compare (5.2.6)] we have immediately

$$\frac{\partial \tilde{f}}{\partial \varphi^{(1)}} \propto \varepsilon, \tag{5.2.26}$$

and from (5.2.23) we obtain

$$\Delta^{(1)} \propto \varepsilon. \tag{5.2.27}$$

Due to (5.2.26) we have for small enough ε

$$\left(1 + \frac{\partial \tilde{f}}{\partial \varphi^{(1)}}\right)^{-1} - 1 \propto \varepsilon \tag{5.2.28}$$

and because $\tilde{f} \propto \varepsilon$ and $f \propto \varepsilon$ we conclude

$$f(\varphi^{(1)} + \tilde{f}) - f(\varphi^{(1)}) \propto \varepsilon^2. \tag{5.2.29}$$

Taking all these estimates together we immediately arrive at the result that $f^{(1)}$ goes with ε^2. Now, the idea is to continue this iteration procedure by making the following substitutions

$$f^{(1)} \to f^{(2)}, \tag{5.2.30}$$

$$f \equiv f^{(0)} \to f^{(1)}, \tag{5.2.31}$$

$$\Delta^{(1)} \to \Delta^{(2)}, \tag{5.2.32}$$

and in particular

$$\varphi^{(1)} = \varphi^{(2)} + f^{(1)}(\varphi^{(2)}). \tag{5.2.33}$$

It is obvious how to continue this iteration procedure:

$$\varphi^{(s)} = \varphi^{(s+1)} + f^{(s)}(\varphi^{(s+1)}). \tag{5.2.34}$$

Because the last two terms on the rhs of (5.2.25), etc., become smaller and smaller, we expect that eventually we obtain

$$\lim_{s \to \infty} \left(\frac{d\varphi^{(s)}}{dt}\right) = \omega \tag{5.2.35}$$

as solution of our starting equation (5.2.5).

For readers interested in applications we show how the iteration procedure works explicitly. To this end let us write down the first three steps

$$\varphi = \varphi^{(1)} + \tilde{f}(\varphi^{(1)}) , \tag{5.2.36}$$

$$\varphi^{(1)} = \varphi^{(2)} + \tilde{f}^{(1)}(\varphi^{(2)}) , \tag{5.2.37}$$

$$\varphi^{(2)} = \varphi^{(3)} + \tilde{f}^{(2)}(\varphi^{(3)}) . \tag{5.2.38}$$

If $\tilde{f}^{(2)}$ is small enough we may assume

$$\frac{d\varphi^{(2)}}{dt} \equiv \omega \tag{5.2.39}$$

with the solution

$$\varphi^{(2)} = \omega t + \varphi_0 , \tag{5.2.40}$$

where we may take $\varphi_0 = \mathbf{0}$ for simplicity.

Inserting (5.2.40) into (5.2.37) we obtain as an explicit solution to the starting equation (5.2.5)

$$\varphi = \omega t + \tilde{f}^{(1)}(\omega t) + \tilde{f}[\omega t + \tilde{f}^{(1)}(\omega t)] . \tag{5.2.41}$$

If $\tilde{f}^{(2)}$ is not sufficiently small but $\tilde{f}^{(3)}$ is instead, we may have

$$\frac{d\varphi^{(3)}}{dt} = \omega \tag{5.2.42}$$

and therefore

$$\varphi^{(3)} = \omega t , \tag{5.2.43}$$

from which we obtain

$$\varphi^{(2)} = \omega t + \tilde{f}^{(2)}(\omega t) . \tag{5.2.44}$$

Inserting this into $\varphi^{(1)}$ we obtain

$$\varphi^{(1)} = \omega t + \tilde{f}^{(2)}(\omega t) + \tilde{f}^{(1)}[(\omega t + \tilde{f}^{(2)}(\omega t)] , \tag{5.2.45}$$

and repeating this step once again to come back to φ we obtain our final result

$$\begin{aligned} \varphi = {} & \omega t + \tilde{f}^{(2)}(\omega t) + \tilde{f}^{(1)}[\omega t + \tilde{f}^{(2)}(\omega t)] \\ & + \tilde{f}(\omega t + \tilde{f}^{(2)}(\omega t) + \tilde{f}^{(1)}[\omega t + \tilde{f}^{(2)}(\omega t)]) . \end{aligned} \tag{5.2.46}$$

This result clearly indicates the construction of φ. It contains terms which increase linearly with t. The next terms are quasiperiodic functions of time where, for instance, the second term $\tilde{f}^{(1)}$ depends on a frequency which itself is again a periodic function of time. Evidently, by continuing this kind of substitution we obtain a representation of φ in form of a series of terms $\tilde{f}^{(s)}$. In addition, the arguments of these $\tilde{f}$'s appear themselves as a series. A proof of the convergence of the procedure must show that the $\tilde{f}$'s converge provided the initial $\tilde{f}$ is small enough. In the subsequent section we shall study some of the most interesting aspects of the convergence properties of $\tilde{f}^{(s)}$.

5.3 Some Considerations on the Convergence of the Procedure*

A complete proof of the convergence of the procedure described in the preceding section is not given here since a detailed proof of a more general problem is presented in the Appendix.

Here we wish to give the reader an idea what the specific underlying ideas of the whole procedure are. We shall see that the procedure contains a sophisticated balance between two competing processes which tend to create and destroy the convergence, respectively. We start with an auxiliary theorem.

Lemma 5.3.1. We assume that $f(\varphi)$ is an analytic function on the domain

$$|\mathrm{Im}\{\varphi_j\}| \leqslant \rho\,, \tag{5.3.1}$$

where ρ is a given positive constant. We further assume that f is bounded in this domain with

$$|f(\varphi)| \leqslant M\,. \tag{5.3.2}$$

Then, for the function

$$\tilde{f}(\varphi) = \sum_{m \neq 0} (f_m / \mathrm{i}m \cdot \omega) \exp(\mathrm{i}m \cdot \varphi) \tag{5.3.3}$$

on the domain

$$|\mathrm{Im}\{\varphi_j\}| \leqslant \rho - 2\delta\,, \tag{5.3.4}$$

the inequality

$$|\tilde{f}(\varphi)| \leqslant \frac{MC}{K}\,\frac{1}{\delta^{2n+1}} \tag{5.3.5}$$

holds. In it C is a numerical constant which depends on the dimension n of the vector φ

$$C = \left(\frac{n+1}{\mathrm{e}}\right)^{n+1} (1+\mathrm{e})^n\,. \tag{5.3.6}$$

The function (5.3.3) is analytic on the domain (5.3.4). While the constant M is introduced in (5.3.2), the constant K occurs in the inequality of the KAM condition (5.2.14). In (5.3.4), δ is a positive constant which must be suitably chosen as is explained below. In a similar way one finds the inequality

$$\sum_k \left| \frac{\partial \tilde{f}(\varphi)}{\partial \varphi_k} \right| \leqslant \frac{MC'}{K} \frac{1}{\delta^{2n+2}}, \tag{5.3.7}$$

where the constant C' is given by

$$C' = \left(\frac{n+2}{\mathrm{e}}\right)^{n+2} (1+\mathrm{e})^n . \tag{5.3.8}$$

We now want to show how the inequality (5.3.5, 6) can be proven. The reader who is not so interested in these details can proceed to formula (5.3.30). We decompose $f(\varphi)$ into its Fourier series. The Fourier coefficients are given by the well-known formula

$$f_m = \frac{1}{(2\pi)^n} \int_0^{2\pi} \dots \int_0^{2\pi} f(\Theta)\, \mathrm{e}^{-\mathrm{i}m \cdot \Theta} d\Theta_1 \dots d\Theta_n . \tag{5.3.9}$$

Integration over the angles Θ can be interpreted as an integration over the complex plane where $z = r \exp(\mathrm{i}\Theta)$, $r = 1$.

Because f is an analytic function, according to the theory of functions we may deform the path of integration as long as f remains analytic on the domain over which the integration runs. This allows us to replace (5.3.9) by

$$\frac{1}{(2\pi)^n} \int_0^{2\pi} \dots \int_0^{2\pi} f(\Theta + \mathrm{i}\Phi) \exp[-\mathrm{i}m(\Theta + \mathrm{i}\Phi)]\, d\Theta_1 \dots d\Theta_n . \tag{5.3.10}$$

We now choose the individual components φ_j of Φ by

$$\varphi_j = -\rho \operatorname{sign}\{m_j\} . \tag{5.3.11}$$

We then immediately obtain

$$m\varphi = -\rho \|m\| , \tag{5.3.12}$$

where

$$\|m\| = |m_1| + |m_2| + \dots + |m_n| . \tag{5.3.13}$$

By use of (5.3.2), (5.3.9) can be estimated by

$$|f_m| \leqslant M \exp(-\rho \|m\|) . \tag{5.3.14}$$

We choose the imaginary part of φ_j according to

$$|\mathrm{Im}\{\varphi_j\}| < \rho - 2\delta. \tag{5.3.15}$$

Starting from the definition (5.3.3) for $\tilde{f}$ we obtain the estimate

$$|\tilde{f}(\varphi)| \leqslant \sum_{m \neq 0} \frac{|f_m|}{|\boldsymbol{m} \cdot \omega|} \exp\left[\|\boldsymbol{m}\|(\rho - 2\delta)\right] \tag{5.3.16}$$

by taking in each term of the series the absolute value and using (5.3.15). Using the KAM condition (5.2.14) and (5.3.14) we obtain

$$|\tilde{f}(\varphi)| \leqslant \frac{M}{K} \sum_{m} \|\boldsymbol{m}\|^{n+1} \exp(-2\delta\|\boldsymbol{m}\|). \tag{5.3.17}$$

It is now rather easy to estimate the sum over $\boldsymbol{m}$ in the following way. We write

$$\sum_{m} \|\boldsymbol{m}\|^{\nu} \exp(-2\delta\|\boldsymbol{m}\|) \quad \text{with} \tag{5.3.18}$$

$$0 < \delta < 1, \quad \nu > 1 \tag{5.3.19}$$

and introduce the abbreviation

$$\|\boldsymbol{m}\| = z. \tag{5.3.20}$$

The maximum of

$$\nu \ln z - \delta z \tag{5.3.21}$$

for $z \geqslant 1$ (as $\|\boldsymbol{m}\| \geqslant 1$) lies at

$$z = \frac{\nu}{\delta} \geqslant 1. \tag{5.3.22}$$

Using (5.3.22) we may estimate (5.3.21) by

$$\nu \ln z - \delta z \leqslant \nu\left(\ln \frac{\nu}{\delta} - 1\right). \tag{5.3.23}$$

Taking the exponential of both sides of (5.3.23), we find after a slight rearrangement

$$z^{\nu} \leqslant \mathrm{e}^{\delta z} \mathrm{e}^{\nu(\ln \nu/\delta - 1)} = \left(\frac{\nu}{\mathrm{e}}\right)^{\nu} \frac{\mathrm{e}^{\delta z}}{\delta^{\nu}}. \tag{5.3.24}$$

Inserting this relation into (5.3.18), we readily obtain

$$\sum_m \exp(-2\|\boldsymbol{m}\|\delta)\|\boldsymbol{m}\|^\nu \leqslant \left(\frac{\nu}{\mathrm{e}}\right)^\nu \frac{1}{\delta^\nu}$$

$$\cdot \sum_m \exp(-|m_1|\delta - |m_2|\delta - \ldots - |m_n|\delta) \equiv S\,, \tag{5.3.25}$$

where the sum over $\boldsymbol{m}$ can be written as the nth power of the infinite sum over a single variable $q = 1, 2, 3, \ldots$ so that the rhs of (5.3.25) is

$$S = \left(\frac{\nu}{\mathrm{e}}\right)^\nu \frac{1}{\delta^\nu}\left(1 + 2\sum_{q=1}^{\infty} \mathrm{e}^{-\delta q}\right)^n. \tag{5.3.26}$$

Summing up the geometric series yields

$$S = \left(\frac{\nu}{\mathrm{e}}\right)^\nu \frac{1}{\delta^\nu}\left(\frac{1+\mathrm{e}^{-\delta}}{1-\mathrm{e}^{-\delta}}\right)^n. \tag{5.3.27}$$

Using

$$1 - \mathrm{e}^{-\delta} > \delta \mathrm{e}^{-\delta} \tag{5.3.28}$$

and thus

$$\frac{1+\mathrm{e}^{-\delta}}{1-\mathrm{e}^{-\delta}} \leqslant \frac{1}{\delta}(1+\mathrm{e}^{\delta}) < \frac{1+\mathrm{e}}{\delta}\,, \tag{5.3.29}$$

we obtain the final result

$$\sum_m \exp(-2\|\boldsymbol{m}\|\delta)\|\boldsymbol{m}\|^\nu < \left(\frac{\nu}{\mathrm{e}}\right)^\nu \frac{1}{\delta^{\nu+n}}(1+\mathrm{e})^n\,. \tag{5.3.30}$$

Using (5.3.30, 17) we obtain the desired result (5.3.5, 6). As each member of the infinite sum (5.3.3) is analytic and this sum (5.3.3) is bounded [cf. (5.3.5)] on the domain (5.3.4), so (5.3.3) is analytic on the domain (5.3.4).

After this small exercise we resume the main theme, namely, we wish to consider the iteration procedure described in Sect. 5.2 and its convergence. Since $\boldsymbol{f}^{(1)}$ in (5.2.24) is built up of functions $\boldsymbol{f}, \tilde{\boldsymbol{f}}$ and derivatives as they appear in (5.3.2, 5, 7) estimates for $\boldsymbol{f}^{(1)}$ can now be derived. Since these estimates are somewhat lengthy and we shall come to this question in a more general frame in the Appendix, we shall first give the result of the estimate. One finds that

$$\boldsymbol{f}^{(1)}(\boldsymbol{\varphi}^{(1)}, \boldsymbol{\Delta}^{(1)}) \tag{5.3.31}$$

is analytic in the domain

$$|\mathrm{Im}\{\varphi_j^{(1)}\}| \leqslant \rho - 2\delta \tag{5.3.32}$$

and obeys the inequality

$$|f^{(1)}| \leqslant M_1 \quad \text{where} \tag{5.3.33}$$

$$M_1 = \frac{C''}{K} \frac{M^2}{\delta^{2n+2}} . \tag{5.3.34}$$

Here C'' is a numerical constant of a similar structure to (5.3.6, 8), M is introduced in (5.3.2), K is the constant occurring in the KAM condition and δ occurs in (5.3.4 or 32).

We now come to the discussion of the competition between two different processes. Namely, when we wish to continue our iteration procedure, we have to proceed from M_1 to another M_2 and so on. In order to obtain convergence we have to choose δ big enough. On the other hand, we are limited in choosing the size of δ, because each time we go one step further we must replace the estimate (5.3.4) correspondingly, i.e., in the first step we have

$$\rho_1 = \rho - 2\delta , \tag{5.3.35}$$

in the second step,

$$\rho_2 = \rho_1 - 2\delta_1 = \rho - 2\delta - 2\delta_1 , \tag{5.3.36}$$

and so on.

Because each time ρ_k must be bigger than 0 the sum over the δ's must not become arbitrarily big. Rather, we have to require that ρ_k must be bounded from below so that we can perform the estimates safely in each step. Therefore we have two conflicting requirements. On the one hand, the δ's must be big enough to make M_j [cf. (5.3.34)] converging, on the other hand, the δ's must not become too big but rather must converge to 0. The attractive feature of the whole approach is that δ can be chosen such that in spite of this conflict a convergent approach can bc constructed. To this end we put ($\gamma < 1$)

$$\delta = \gamma, \delta_1 = \gamma^2, \ldots, \delta_s = \gamma^{s+1} \tag{5.3.37}$$

and choose γ such that $\rho_k \to \rho/2$ for $k \to \infty$, e.g.,

$$2(\gamma + \gamma^2 + \ldots + \gamma^s + \ldots) = \frac{\rho}{2} \tag{5.3.38}$$

holds. Summing up the geometric series on lhs of (5.3.38) we find

$$\delta = \gamma = \frac{\rho}{4+\rho} (<1) . \tag{5.3.39}$$

We are now in a position to give explicit estimates for the subsequent M_j's.

We introduce the abbreviation $P = C''/K$ so that (5.3.33, 34) reads

$$|f^{(1)}| \leqslant M_1 = P \frac{M^2}{\delta^{2n+2}} . \tag{5.3.40}$$

When we go from M_1 to M_2 we have to replace M by M_1 and Δ by Δ_1, thus obtaining

$$|f^{(2)}| \leqslant M_2 = P \frac{M_1^2}{\delta_1^{2n+2}} . \tag{5.3.41}$$

It is obvious how to continue:

$$|f^{(3)}| \leqslant M_3 = P \frac{M_2^2}{\delta_2^{2n+2}} , \tag{5.3.42}$$

$$\vdots$$

$$|f^{(s+1)}| \leqslant M_{s+1} = P \frac{M_s^2}{\delta_s^{2n+2}} . \tag{5.3.43}$$

Using (5.3.37) we may write (5.3.43) in the form

$$M_{s+1} = P \frac{M_s^2}{\gamma^{(s+1)(2n+2)}} . \tag{5.3.44}$$

Now we introduce as an abbreviation

$$N = \frac{1}{\gamma^{2n+2}} , \tag{5.3.45}$$

which allows us to write (5.3.44) in the form

$$M_{s+1} = P N^{s+1} M_s^2 . \tag{5.3.46}$$

We choose M sufficiently small so that

$$P N^2 M \leqslant r_0 < 1 . \tag{5.3.47}$$

Now we require $P \geqslant 1$, i.e., the constant K appearing in the KAM condition has to be chosen small enough as follows from (5.3.40). It is then easy to derive for the recursion formulas (5.3.40 – 43) the relation

$$M_s \leqslant \frac{r_0^{2s}}{N^{s+2}} . \tag{5.3.48}$$

It tells us the following. If we choose r_0 small enough, M_s converges rapidly, namely with powers of r_0^2. Since M is related with r_0 by (5.3.47), we have shown that the whole procedure converges like a geometric series provided M is chosen small enough. This concludes our present considerations.

6. Nonlinear Coupling of Oscillators: The Case of Persistence of Quasiperiodic Motion

In this chapter we shall present a theorem developed by *Moser* which extends former work by Kolmogorov and Arnold. The problem to be treated contains those of Sects. 3.9 and 5.2, 3 as special cases. As we have seen before, a set of linearly coupled harmonic oscillators can oscillate at certain basic frequencies so that their total motion is a special case of a quasiperiodic motion. In this chapter we shall deal with the important problem whether *nonlinearly* coupled oscillators can also perform quasiperiodic motion. We also include in this consideration oscillators which by themselves are nonlinear.

6.1 The Problem

To illustrate the basic idea let us start with the most simple case of a single oscillator described by an amplitude r and a phase angle φ. In the case of a linear (harmonic) oscillator the corresponding equations read

$$\dot{\varphi} = \omega \, , \tag{6.1.1}$$

$$\dot{r} = 0 \, . \tag{6.1.2}$$

In many practical applications, e.g., radio engineering or fluid dynamics, we deal with a self-sustained oscillator whose amplitude r obeys a nonlinear equation, e.g.,

$$\dot{r} = \alpha r - \beta r^3 \, . \tag{6.1.3}$$

While (6.1.2) allows for solutions with arbitrary time-independent amplitude r, (6.1.3) allows only for a definite solution with time-independent amplitude, namely

$$r = r_0 = \sqrt{\frac{\alpha}{\beta}} \, , \quad \alpha, \beta > 0 \, , \tag{6.1.4}$$

(besides the unstable solution $r = 0$).

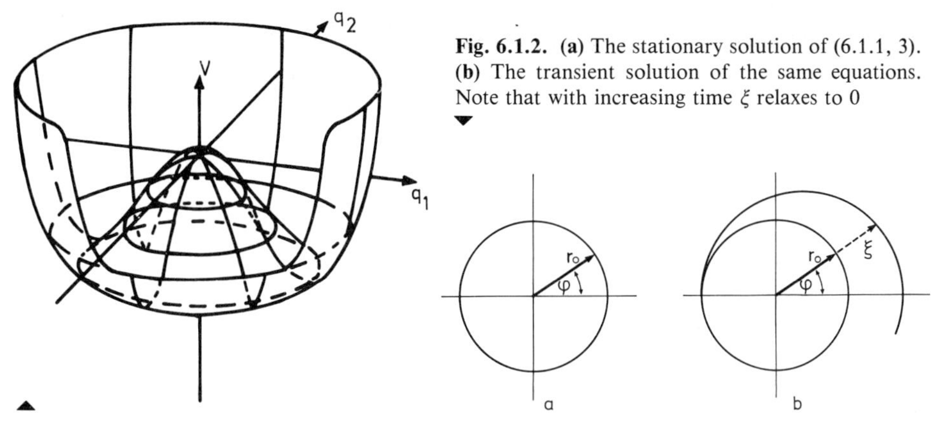

Fig. 6.1.2. (a) The stationary solution of (6.1.1, 3). (b) The transient solution of the same equations. Note that with increasing time ξ relaxes to 0

Fig. 6.1.1. Visualization of the solution of (6.1.1, 3), namely by a particle which moves in an overdamped way in a potential shown in this figure, and which rotates with angular frequency ω

As is well known [1] equations (6.1.1 and 3) can describe a particle moving in a potential which rotates at a frequency ω and whose motion in r-direction is overdamped. As is obvious from Fig. 6.1.1, the particle will always relax with its amplitude $r(t)$ towards r_0 (6.1.4).

If we focus our attention on small deviations of r from r_0 we may put

$$r = r_0 + \xi \tag{6.1.5}$$

and derive a linearized equation for ξ in the form

$$\dot{\xi} = -2\alpha\xi + O(\xi^2) . \tag{6.1.6}$$

In the following discussion we shall neglect the term $O(\xi^2)$ which indicates a function of ξ of order ξ^2 or still higher order. The solutions of the equations (6.1.1) and (6.1.3, 6) can be easily visualized by means of the potential picture and can be plotted according to Fig. 6.1.2, where Fig. 6.1.2a shows stationary motion, while Fig. 6.1.2b shows how the system relaxes when initially ξ was chosen unequal to zero.

These considerations can be extended to a set of oscillators and to the more general case in which the number of frequencies $\omega_1, \omega_2, \ldots, \omega_n$ differs from the number of amplitude displacements $\xi_1, \ldots, \xi_m$. In such a case equations (6.1.1, 6) can be generalized to equations

$$\dot{\varphi}^{(0)} = \omega , \tag{6.1.7}$$

$$\dot{\xi}^{(0)} = \Lambda\xi^{(0)} , \tag{6.1.8}$$

where we use an obvious vector notation. Here Λ is a matrix which we assume to be diagonizable. Because we shall assume that $\xi^{(0)}$ is real, in general we cannot assume that Λ has been fully diagonalized. The reader may note that (6.1.7) has the same form as the unperturbed ($\varepsilon = 0$) equation (5.2.1), whereas (6.1.8) has the same form as the unperturbed equation (3.9.10) with $M_0 = 0$.

We now introduce a nonlinear coupling between the ω's and ξ's, i.e., we are interested in equations of the form

$$\dot{\varphi}^{(1)} = \omega + \varepsilon \boldsymbol{f}(\varphi^{(1)}, \xi^{(1)}, \varepsilon)\,, \tag{6.1.9}$$

$$\dot{\xi}^{(1)} = \Lambda \xi^{(1)} + \varepsilon \boldsymbol{g}(\varphi^{(1)}, \xi^{(1)}, \varepsilon)\,, \tag{6.1.10}$$

where $\boldsymbol{f}$ and $\boldsymbol{g}$ are functions which are periodic in $\varphi_j^{(1)}$ with period 2π. For mathematical reasons we shall further assume that they are real analytic in $\varphi^{(1)}$, $\xi^{(1)}$, ε, where ε is a small parameter.

As we know from the discussion in Sect. 5.2, the additional term $\boldsymbol{f}$ may cause a change of the basic periods $T_j = 2\pi/\omega_j$. For reasons identical with those of Sect. 5.2, we must introduce a counterterm in (6.1.9) which is a constant vector and will guarantee that the basic periods remain the same, even if the interaction term $\boldsymbol{f}$ is taken into account. This leads us to consider equations of the form

$$\dot{\varphi} = \omega + \varepsilon \boldsymbol{f}(\varphi, \xi, \varepsilon) + \Delta\,, \tag{6.1.11}$$

where the counterterm Δ has been added.

In addition, we wish that the matrix Λ remains unchanged even if $\varepsilon \boldsymbol{g}$ is taken into account. We shall see below what this sentence "Λ remains unchanged" means precisely. As it will turn out again, we have to introduce another counterterm in (6.1.10) which must be chosen in the form $\boldsymbol{d} + D\xi$, where $\boldsymbol{d}$ is a constant vector while D is a constant matrix.

In this way, (6.1.10) is transformed into

$$\dot{\xi} = \Lambda \xi + \varepsilon \boldsymbol{g}(\varphi, \xi, \varepsilon) + \boldsymbol{d} + D\xi\,. \tag{6.1.12}$$

It will transpire from the subsequent mathematical transformations that we may add the requirements

$$\Lambda \boldsymbol{d} = \boldsymbol{0}\,, \tag{6.1.13}$$

$$\Lambda D = D\Lambda \tag{6.1.14}$$

to $\boldsymbol{d}$ and D. Let us visualize the effect of the additional terms $\boldsymbol{f}$ and $\boldsymbol{g}$ and of the counterterms Δ, $\boldsymbol{d}$ and D. In order to discuss these effects let us start with the case $n = m = 1$.

We wish to consider a situation in which the quasiperiodic motion, or, in the present special case, the periodic motion, persists. According to (6.1.7 and 8), the unperturbed equations are given by

$$\dot{\varphi}^{(0)} = \omega\,, \quad \text{and} \tag{6.1.15}$$

$$\dot{\xi}^{(0)} = -\lambda \xi^{(0)}\,. \tag{6.1.16}$$

They describe a uniform rotation of the angle $\varphi^{(0)}$, $\varphi^{(0)} = \omega t$ and a relaxation of $\xi^{(0)}$ towards zero so that the spiral motion of Fig. 6.1.3 emerges. As a special case

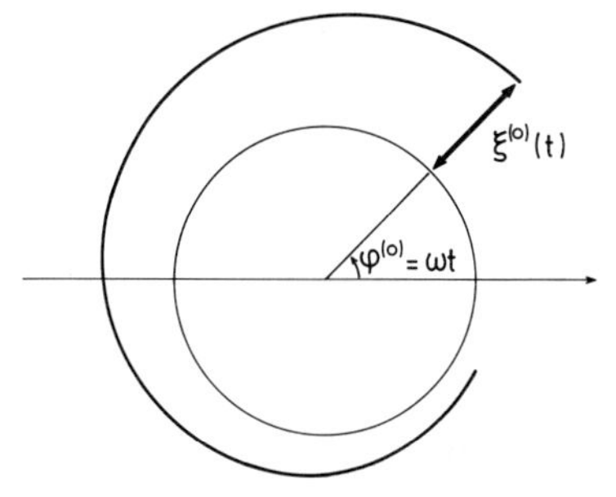

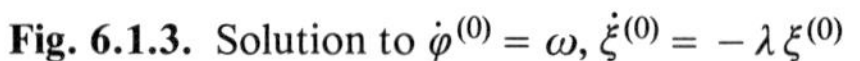
Fig. 6.1.3. Solution to $\dot{\varphi}^{(0)} = \omega$, $\dot{\xi}^{(0)} = -\lambda\xi^{(0)}$

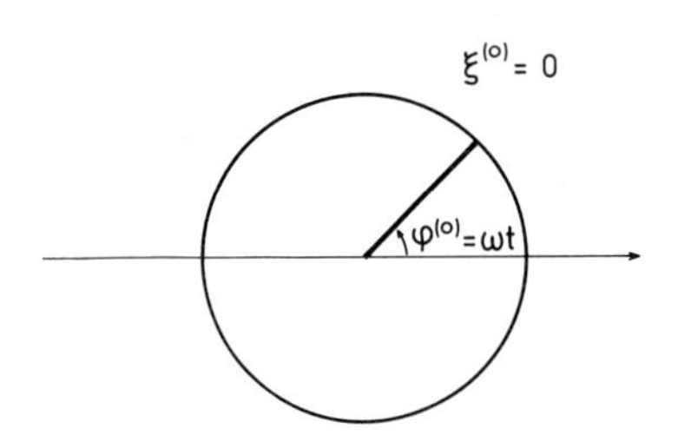

Fig. 6.1.4. The steady-state solution of Fig. 6.1.3

we obtain $\varphi^{(0)} = \omega t$, $\xi^{(0)} = 0$, which is the motion on the circle of Fig. 6.1.4. These motions are perturbed when we take into account the additional terms $\boldsymbol{f}$ and $\boldsymbol{g}$ of (6.1.9, 10) in which case we find

$$\dot{\varphi}^{(1)} = \omega + \varepsilon f(\varphi^{(1)}, \xi^{(1)}, \varepsilon)\,, \qquad (6.1.17)$$

$$\dot{\xi}^{(1)} = -\lambda\xi^{(1)} + \varepsilon g(\varphi^{(1)}, \xi^{(1)}, \varepsilon)\,, \qquad (6.1.18)$$

where f and g are 2π-periodic in $\varphi^{(1)}$. Let us first consider the case which corresponds to $\xi^{(0)} = 0$, i.e., the circle of Fig. 6.1.4. To visualize what happens we adopt the adiabatic approximation from Sect. 1.13. We put $\dot{\xi}^{(1)} = 0$ in (6.1.18). For small enough ε we may express $\xi^{(1)}$ uniquely by means of $\varphi^{(1)}$, i.e.,

$$\xi^{(1)} \equiv \xi_0^{(1)} = F(\varphi^{(1)})\,. \qquad (6.1.19)$$

(In fact this relation may be derived even without the adiabatic approximation as we shall demonstrate below, but for our present purpose this is not important.) Relation (6.1.19) tells us that the circle of Fig. 6.1.4 is deformed into some other closed curve (Fig. 6.1.5). Thus the old orbit $\xi^{(0)} = 0$ is deformed into the new one (6.1.19). Inserting (6.1.19) into (6.1.17) we obtain an equation of the form

$$\dot{\varphi}^{(1)} = \omega + \varepsilon \hat{f}(\varphi^{(1)}, \varepsilon)\,. \qquad (6.1.20)$$

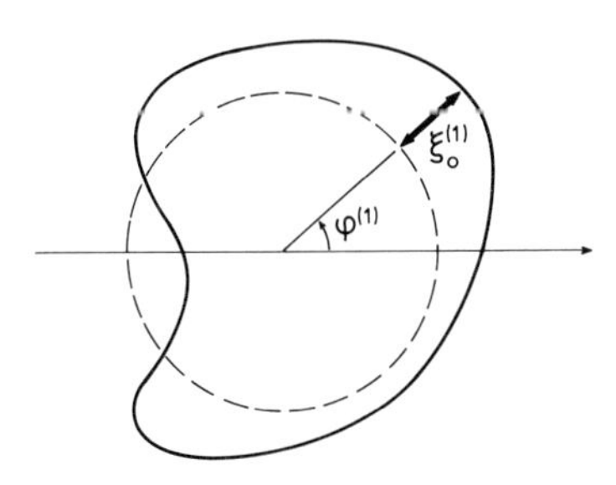

Fig. 6.1.5. Stationary solution to (6.1.17, 18), qualitative plot

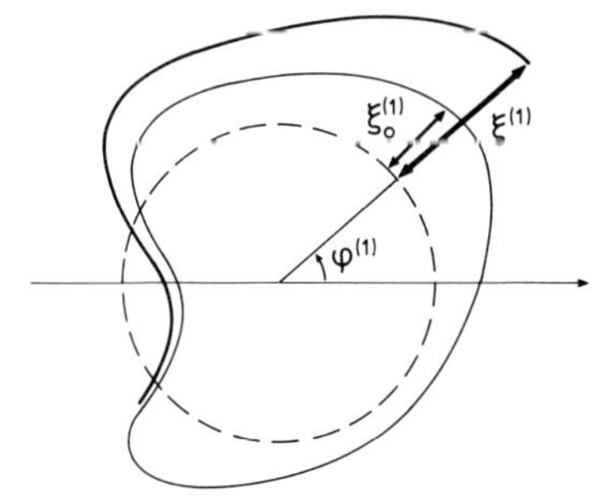

Fig. 6.1.6. Transient solution of (6.1.17, 18), qualitative plot

It tells us that the rotation speed $\dot{\varphi}^{(1)}$ depends on $\varphi^{(1)}$. Therefore, at least in general, we must expect that the length of the period of motion around the closed orbit of Fig. 6.1.5 is different from that of Fig. 6.1.4. When we introduce a suitable counterterm Δ in the equation

$$\dot{\varphi} = \omega + \varepsilon \hat{f}(\varphi, \varepsilon) + \Delta \,, \tag{6.1.21}$$

we may secure that the period remains that of (6.1.15) (Fig. 6.1.4).

Let us now consider the analog of Fig. 6.1.3. There we considered an initial state which is somewhat elongated from the stationary state $\xi^{(0)} = 0$ and which tends to $\xi^{(0)} = 0$ with increasing time. If ε is small, the additional term g in (6.1.18) will deform the spiral of Fig. 6.1.3 to that of Fig. 6.1.6, but qualitatively $\xi^{(1)}$ will behave the same way as $\xi^{(0)}$, i.e., it relaxes towards its stationary state. However, the relaxation speed may change. If we wish that $\xi^{(1)}$ relaxes with the same speed as $\xi^{(0)}$, at least for small enough $\xi^{(1)}$'s, we have to add the above-mentioned counterterm $D\xi$.

The basic idea of the procedure we want to describe is now as follows. We wish to introduce instead of φ and ξ new variables ψ and χ, respectively, so that the Figs. 6.1.5, 6 of φ and ξ can be transformed back into those of Figs. 6.1.4, 3, respectively. Before we discuss how to construct such a transformation, we indicate what happens in dimensions where $n \geqslant 2$ and/or $m \geqslant 2$. All the essential features can be seen with help of the case $n = 2$ and, depending on an adequate interpretation, $m = 2$ (or 3). Let us start again with the unperturbed motion described by (6.1.7 and 8). In the steady state, $\dot{\boldsymbol{\xi}}^{(0)} = \boldsymbol{\xi}^{(0)} = \mathbf{0}$ and we may visualize the solutions of (6.1.7) by plotting them with coordinates φ_1, φ_2. The variables can be visualized by means of a local coordinate system on the torus (Fig. 6.1.7).

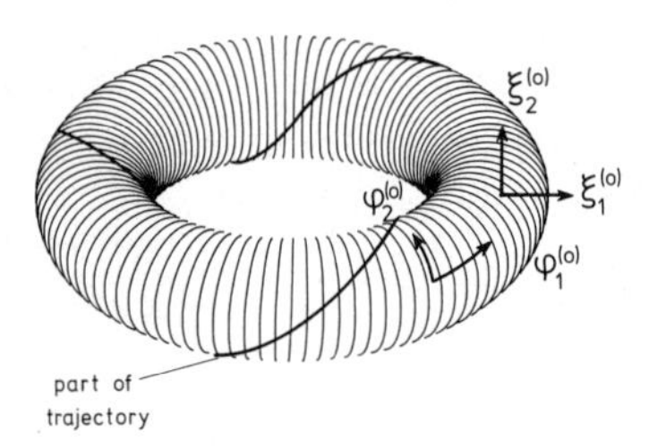

Fig. 6.1.7. Local coordinates with respect to a torus (schematic). We distinguish between two systems: 1) Local coordinates on the torus represented by $\varphi_j^{(0)}$; 2) Coordinates pointing away from the torus represented by $\xi_j^{(0)}$. In the figure representing a two-dimensional torus in three-dimensional space a single ξ would be sufficient. In this figure we try to visualize what happens in high-dimensional space where two linearly indirect directions point away from the torus

In the case $\dot{\boldsymbol{\xi}}^{(0)} \neq \mathbf{0}$, provided Λ is diagonal,

$$\Lambda = \begin{pmatrix} -\lambda_1 & 0 \\ 0 & -\lambda_2 \end{pmatrix}, \quad \lambda_j > 0 \,, \tag{6.1.22}$$

(6.1.8) describes a relaxation of the system's motion towards the torus. To grasp the essential changes caused by the additional term $\boldsymbol{g}$ in (6.1.10), let us again adopt the adiabatic approximation so that $\xi_1^{(1)}$, $\xi_2^{(1)}$ may be expressed by $\varphi_1^{(1)}$, $\varphi_2^{(1)}$ through

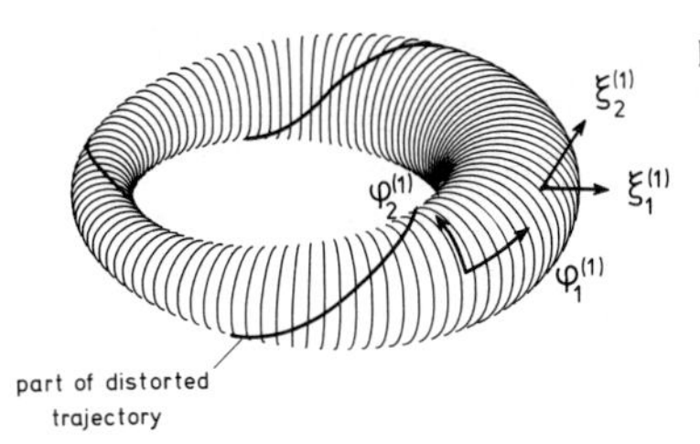

Fig. 6.1.8. Deformed torus with its new coordinate system

$$\begin{aligned} \xi_1^{(1)} &\equiv \xi_{1,0}^{(1)} = F_1(\varphi_1^{(1)}, \varphi_2^{(1)}) , \\ \xi_2^{(1)} &\equiv \xi_{2,0}^{(1)} = F_2(\varphi_1^{(1)}, \varphi_2^{(1)}) . \end{aligned} \tag{6.1.23}$$

Clearly, the original torus is now distorted into a new surface (Fig. 6.1.8). In it, the uniform rotation of $\varphi_1^{(0)}$, $\varphi_2^{(0)}$ is replaced by a nonuniform rotation of $\varphi_1^{(1)}$, $\varphi_2^{(1)}$ due to the additional term $\boldsymbol{f}$ in (6.1.9). Consequently, in general the corresponding periods T_1 and T_2 will be changed. To retain the original periods we must introduce the counterterms Δ_1, Δ_2. Let us now consider the relaxation of $\xi_1^{(1)}$, $\xi_2^{(1)}$ towards the deformed surface. While originally according to our construction (or visualization) the rectangular coordinate system $\xi_1^{(0)}$, $\xi_2^{(0)}$ was rotating with $\varphi_1^{(0)}$, $\varphi_2^{(0)}$ uniformly along the torus, $\xi_1^{(1)}$, $\xi_2^{(1)}$ may form a new coordinate system which is no longer orthogonal and in which the relaxation rate is different from λ_1, λ_2 due to the additional term $\boldsymbol{g}$ in (6.1.10). The counterterm $D\xi$ allows us to correct for the locally distorted coordinate system $\xi_1^{(1)}$, $\xi_2^{(1)}$ by making it orthogonal again and by restoring the original relaxation rates. If the eigenvalues of Λ are complex (or purely imaginary), D makes it possible to keep in particular the original Λ fixed while $\boldsymbol{d}$ may correct (in the sense of an average) for the distortion of the torus. Therefore, due to $\boldsymbol{f}$, $\boldsymbol{g}$ and the counterterms, the trajectories are deformed in three ways:

1) The local coordinate system $\boldsymbol{\varphi}$ on the torus is changed (see Fig. 6.1.9a).
2) The radial displacement has been changed or, more generally, the position of the surface elements of the torus has been changed (Fig. 6.1.9b).
3) The orientation of the coordinate system ξ has been changed (Fig. 6.1.9c).

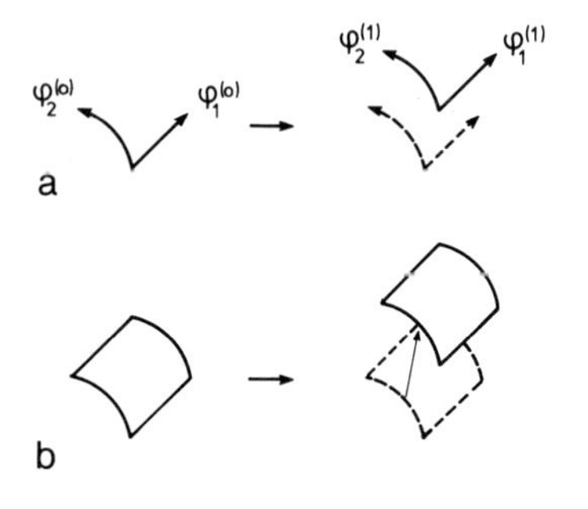

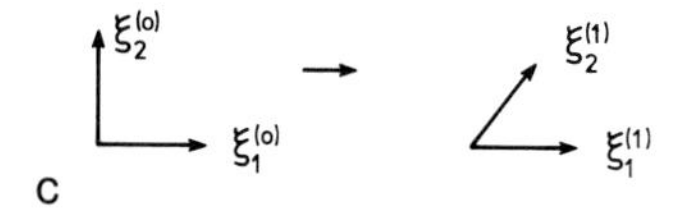

Fig. 6.1.9 a – c. Local changes due to the deformation of a torus. **(a)** New coordinates on the torus. **(b)** A displacement of elements of the torus. **(c)** Change of the coordinate system pointing away from the torus

This suggests the introduction of coordinate transformations that lead us back from the deformed trajectories described by (6.1.11, 12) to the undeformed trajectories described by (6.1.7, 8) by a coordinate transformation taking these three aspects into account. To this end we express φ and ξ by new coordinates ψ and χ, respectively. When treating the motion along ξ_0 we make for φ the ansatz

$$\varphi = \psi + \varepsilon \boldsymbol{u}(\psi, \varepsilon) \,, \tag{6.1.24}$$

which takes into account that the rotation speed is not uniform but changes during the rotation. The factor ε in front of $\boldsymbol{u}$ indicates explicitly that this term tends to $\mathbf{0}$ and therefore φ coincides with ψ when we let ε tend to 0 in (6.1.11, 12). With respect to ξ, we have to take into account that the deformed trajectory differs from the undeformed trajectory (Fig. 6.1.7) by dilation in radial direction and by rotation of the local coordinate system. Therefore we make the following hypothesis

$$\xi = \chi + \varepsilon \boldsymbol{v}(\psi, \varepsilon) + \varepsilon V(\psi, \varepsilon) \chi \,, \tag{6.1.25}$$

where $\boldsymbol{v}$ takes care of the dilation in radial direction, whereas the second term $V \cdot \chi$ takes care of the rotation of the local coordinate system. Because ψ is an angular variable, we require that $\boldsymbol{u}$, $\boldsymbol{v}$, V are periodic in ψ with period 2π. Furthermore, these functions must be analytic in ψ, ξ and ε. We wish to show that by a suitable choice of $\boldsymbol{u}$, $\boldsymbol{v}$, V and the counterterms $\boldsymbol{\Delta}$, $\boldsymbol{d}$, D, we may transform (6.1.11, 12) into equations for the new variables ψ and χ which describe "undistorted trajectories" (Fig. 6.1.7) and which obey the equations

$$\dot{\psi} = \omega + O(\chi) \,, \tag{6.1.26}$$

$$\dot{\chi} = \Lambda \chi + O(\chi^2) \,. \tag{6.1.27}$$

The terms O describe those terms which are neglected and $O(\chi^k)$ denotes an analytic function of ψ, χ, ε which vanishes with χ-derivatives up to order $k - 1 \geqslant 0$ for $\chi = \mathbf{0}$. In other words, we are interested in a transformation from (6.1.11, 12) to (6.1.26, 27), neglecting terms denoted by O. When we specifically choose the solution

$$\chi = \mathbf{0} \,, \tag{6.1.28}$$

the solution of (6.1.26) reads

$$\psi = \omega t \tag{6.1.29}$$

(up to an arbitrary constant ψ_0). Inserting (6.1.28, 29) into (6.1.24, 25), we find

$$\varphi = \omega t + \varepsilon \boldsymbol{u}(\omega t, \varepsilon) \quad \text{and} \tag{6.1.30}$$

$$\xi = \varepsilon \boldsymbol{v}(\omega t, \varepsilon) \,, \tag{6.1.31}$$

indicating that we are dealing with a quasiperiodic solution.

6.2 Moser's Theorem[1] (Theorem 6.2.1)

So far we have given an outline of what we wish to prove. In order to be able to perform the proof we need a number of assumptions which we now list. The object of our study is (6.1.11, 12). We assume that the matrix Λ in (6.1.12) is diagonalizable, and we denote its eigenvalues by Λ_μ.

1) Let us consider the expressions

$$\mathrm{i} \sum_{\nu=1}^{n} j_\nu \omega_\nu + \sum_{\mu=1}^{m} k_\mu \Lambda_\mu \quad \text{for} \tag{6.2.1}$$

$$\sum |k_\mu| \leqq 2 \quad \text{and} \tag{6.2.2}$$

$$|\sum k_\mu| \leqq 1\,, \tag{6.2.3}$$

with integer coefficients

$$(j_1, \ldots, j_n) = \boldsymbol{j}\,, \quad (k_1, \ldots, k_m) = \boldsymbol{k}\,. \tag{6.2.4}$$

We require that of the expressions (6.2.1) only finitely many vanish, namely only for $\boldsymbol{j} = \boldsymbol{0}$. Obviously this implies that $\omega_1, \ldots, \omega_n$ are rationally independent.

2) For some constants K and τ with

$$0 < K \leqslant 1\,, \quad \tau > 0\,, \tag{6.2.5}$$

we require

$$\left| \mathrm{i} \sum_{\nu=1}^{n} j_\nu \omega_\nu + \sum_{\mu=1}^{m} k_\mu \Lambda_\mu \right| \geqq K(\|\boldsymbol{j}\|^\tau + 1)^{-1} \tag{6.2.6}$$

for all integers mentioned under point (1), for which lhs of (6.2.6) does not vanish. Condition (6.2.6) may be considered as a generalized version of the KAM condition. Note that $\|\boldsymbol{j}\|$ is defined in the present context by

$$\|\boldsymbol{j}\| = |j_1| + |j_2| + \ldots + |j_n|.$$

3) We require that $\boldsymbol{f}$ and $\boldsymbol{g}$ in (6.1.11 and 12) have the period 2π in φ and are real analytic in φ, ξ, ε. We now formulate Moser's theorem.

Theorem 6.2.1. Under conditions (1 – 3) there exist unique analytic power series $\Delta(\varepsilon)$, $\boldsymbol{d}(\varepsilon)$, $D(\varepsilon)$ satisfying (6.1.13, 14) such that the system (6.1.11, 12) possesses a quasiperiodic solution with the same characteristic numbers $\omega_1, \ldots, \omega_n$, $\Lambda_1, \ldots, \Lambda_m$ as the unperturbed solution. More explicitly speaking, a coordinate transformation of the form (6.1.24, 25) exists, analytic in ψ, χ and ε which transforms (6.1.11, 12) into a system of the form (6.1.26, 27). In particular, (6.1.30, 31) represent a quasiperiodic solution with the above-mentioned characteristic

[1] Our formulation given here slightly differs from that of Moser's original publication.

numbers. All series of Δ, $\boldsymbol{d}$, D have a positive radius of convergence in ε. While Δ, $\boldsymbol{d}$, D are determined uniquely, $\boldsymbol{u}$, $\boldsymbol{v}$, V are determined only up to a certain class of transformations (which shall be discussed in Appendix A).

We shall proceed as follows. In the next section we shall show how the counterterms Δ, $\boldsymbol{d}$, D, as well as $\boldsymbol{u}$, $\boldsymbol{v}$ and V can be determined by a well-defined iteration procedure. This section is of interest for those readers who wish to apply this formalism to practical cases. In Appendix A we present the rigorous proof of Moser's theorem. The crux of the problem consists in the proof that the iteration procedure described in Sect. 6.3 converges.

6.3 The Iteration Procedure*

Our starting point is (6.1.11, 12), repeated here for convenience

$$\dot{\varphi} = \omega + \varepsilon f(\varphi, \xi, \varepsilon) + \Delta\,, \tag{6.3.1}$$

$$\dot{\xi} = \Lambda\xi + \varepsilon \boldsymbol{g}(\varphi, \xi, \varepsilon) + \boldsymbol{d} + D\xi\,. \tag{6.3.2}$$

We wish to transform these equations by means of the transformations

$$\varphi = \psi + \varepsilon \boldsymbol{u}(\psi, \varepsilon)\,, \tag{6.3.3}$$

$$\xi = \chi + \varepsilon(\boldsymbol{v}(\psi, \varepsilon) + V(\psi, \varepsilon)\chi) \tag{6.3.4}$$

into (6.1.26, 27).

We expand $\boldsymbol{u}$, $\boldsymbol{v}$, V, Δ, $\boldsymbol{d}$, D into a power series in ε and equate the coefficients of the same powers of ε on the left- and right-hand sides of (6.3.1, 2). It is sufficient to discuss the resulting equations for the coefficients of ε to the first power as the higher coefficients are determined by equations of the same form. As is well known from iteration procedures, higher-order equations contain coefficients of lower order.

Expressing the various quantities $\boldsymbol{u}, \ldots, D$ up to order ε we write

$$\begin{aligned}
\varphi &= \psi + \varepsilon \boldsymbol{u}' + O(\varepsilon^2)\\
\Delta &= \varepsilon\Delta' + O(\varepsilon^2)\\
\xi &= \chi + \varepsilon(\boldsymbol{v}' + V'\chi) + O(\varepsilon^2)\\
\boldsymbol{d} &= \varepsilon \boldsymbol{d}' + O(\varepsilon^2)\\
D &= \varepsilon D' + O(\varepsilon^2)\,.
\end{aligned} \tag{6.3.5}$$

We insert (6.3.3, 4) into (6.3.1), using (6.3.5), and keep only terms up to order ε. Since (6.3.1, 2) are vector equations and we have to differentiate for instance $\boldsymbol{u}$ with respect to φ, which itself is a vector, we must be careful with respect to the notation. Therefore, we first use vector components. Thus we obtain

$$\dot{\psi}_\mu + \varepsilon \sum_\nu (\partial u'_\mu / \partial \psi_\nu) \dot{\psi}_\nu = \omega_\mu + \varepsilon f_\mu(\psi, \chi, 0) + \varepsilon \Delta'_\mu . \tag{6.3.6}$$

We now require that

$$\dot{\psi}_\mu = \omega_\mu \quad \text{for} \quad \varepsilon = 0 \tag{6.3.7}$$

and use the following abbreviation. We consider

$$(\partial u'_\mu / \partial \psi_\nu) \tag{6.3.8}$$

as an element with indices μ, ν of a matrix which we write in the form

$$\boldsymbol{u}'_\psi . \tag{6.3.9}$$

Furthermore, we specialize (6.3.6) to $\boldsymbol{\chi} = \boldsymbol{0}$. In agreement with (6.1.26, 27) we require that $\dot{\psi}_\mu = \omega_\mu$ for $\boldsymbol{\chi} = \boldsymbol{0}$ is fulfilled up to order ε. It is now easy to cast (6.3.6) again in the form of a vector equation, namely

$$\boldsymbol{u}'_\psi \boldsymbol{\omega} - \boldsymbol{\Delta}' = \boldsymbol{f}(\boldsymbol{\psi}, \boldsymbol{0}, 0) . \tag{6.3.10}$$

We now insert (6.3.3, 4) using (6.3.5) into (6.3.2) and keep only terms up to order ε, where we consider again the individual components μ. Using the chain rule of differential calculus we obtain for lhs of (6.3.2)

$$\dot{\chi}_\mu + \varepsilon \sum_\nu (\partial v'_\mu(\psi) / \partial \psi_\nu) \dot{\psi}_\nu + \sum_{\nu,\lambda} (\partial V'_{\mu\nu}(\psi) / \partial \psi_\lambda) \dot{\psi}_\lambda \chi_\nu + \sum_\nu V'_{\mu\nu}(\psi) \dot{\chi}_\nu . \tag{6.3.11}$$

For the rhs of (6.3.2) we obtain

$$\begin{aligned} &\varepsilon g_\mu(\psi, \chi, 0) + \varepsilon \sum_\nu (\partial g_\mu(\psi, \chi, 0) / \partial \chi_\nu) \chi_\nu + \sum_\nu \Lambda_{\mu\nu} \chi_\nu + \varepsilon \sum_\nu \Lambda_{\mu\nu} v'_\nu(\psi) \\ &\quad + \varepsilon \sum_\nu \Lambda_{\mu\nu} \sum_{\nu'} V'_{\nu\nu'} \chi_{\nu'} + \varepsilon d'_\mu + \varepsilon \sum_\nu D'_{\mu\nu} \chi_\nu . \end{aligned} \tag{6.3.12}$$

Since we intend to derive (6.1.27) for $\boldsymbol{\chi}$ we require

$$\dot{\chi}_\mu = \sum_\nu \Lambda_{\mu\nu} \chi_\nu , \tag{6.3.13}$$

or in vector notation

$$\dot{\boldsymbol{\chi}} = \boldsymbol{\Lambda} \boldsymbol{\chi} . \tag{6.3.14}$$

Furthermore we consider

$$\partial v'_\mu(\psi) / \partial \psi_\nu \tag{6.3.15}$$

as an element with indices μ, ν of the matrix

$$v'_{\psi} . \tag{6.3.16}$$

We equate the terms which are independent of χ in (6.3.11, 12). Making use of (6.3.13, 15, 16), we readily obtain (specializing to $\chi = \mathbf{0}$)

$$v'_{\psi}\omega - \Lambda v' = g(\psi, \mathbf{0}, 0) + d' . \tag{6.3.17}$$

In the last step we equate the terms linear in χ which occur in (6.3.11, 12). We denote the matrix, whose elements μ, ν are given by

$$\sum_{\lambda} (\partial V'_{\mu\nu}(\psi)/\partial \psi_{\lambda})\, \omega_{\lambda} , \quad \text{by} \tag{6.3.18}$$

$$V'_{\psi}\omega . \tag{6.3.19}$$

Furthermore we consider

$$\partial g_{\mu}/\partial \chi_{\nu} \tag{6.3.20}$$

as elements with indices μ, ν of the matrix

$$g_{\chi} . \tag{6.3.21}$$

This leads us to the following matrix equation

$$V'_{\psi}\omega + V'\Lambda = g_{\chi} + \Lambda V' + D' . \tag{6.3.22}$$

Collecting the equations (6.3.10, 17, 22) we are led to the following set of equations (setting $\chi = \mathbf{0}$)

$$\left.\begin{aligned} u'_{\psi}\omega &- \Delta' = f(\psi, \mathbf{0}, 0) , \\ v'_{\psi}\omega &- \Lambda v' - d' = g(\psi, \mathbf{0}, 0) , \\ V'_{\psi}\omega &+ V'\Lambda - \Lambda V' - D' = g_{\chi}(\psi, \mathbf{0}, 0) \end{aligned}\right\} . \tag{6.3.23}$$

In these equations the functions f, g, g_{χ} are, of course, given, as are the constants ω and Λ. In the present context, the quantities ψ can be considered as the independent variables, whereas the functions u', v', V' are still to be determined. The equations (6.3.23) can be easily brought into a form used in linear algebra. To this end we rearrange the matrix V' into a vector just by relabeling the indices, e.g., in the case

$$V' = \begin{pmatrix} V'_{11} & V'_{12} \\ V'_{21} & V'_{22} \end{pmatrix} \tag{6.3.24}$$

we may introduce a vector

$$\boldsymbol{V}' = \begin{pmatrix} V'_{11} \\ V'_{12} \\ V'_{21} \\ V'_{22} \end{pmatrix} . \tag{6.3.25}$$

Consequently we have the vector

$$\boldsymbol{U} = \begin{pmatrix} \boldsymbol{u}' \\ \boldsymbol{v}' \\ \boldsymbol{V}' \end{pmatrix} . \tag{6.3.26}$$

We replace the matrix $\boldsymbol{g}_\chi$ in the same way by the vector $\boldsymbol{G}$ and introduce the vector

$$\boldsymbol{F} = \begin{pmatrix} \boldsymbol{f} \\ \boldsymbol{g} \\ \boldsymbol{G} \end{pmatrix} . \tag{6.3.27}$$

Following the same steps with $\boldsymbol{\Delta}'$, $\boldsymbol{d}'$, $\boldsymbol{D}'$, we introduce the vector

$$\hat{\boldsymbol{\Delta}} = \begin{pmatrix} \boldsymbol{\Delta}' \\ \boldsymbol{d}' \\ \boldsymbol{D}' \end{pmatrix} . \tag{6.3.28}$$

Finally we have the matrix

$$L = \begin{pmatrix} L_1 & & 0 \\ & L_2 & \\ 0 & & L_3 \end{pmatrix} , \tag{6.3.29}$$

whose submatrices are matrices of dimension n, m and m^2, respectively. As can be seen from the equations (6.3.23), these matrices must have the form

$$L_1 = \sum_{\nu=1}^{n} \omega_\nu \partial/\partial \psi_\nu , \tag{6.3.30}$$

$$L_2 = \sum_{\nu=1}^{n} \omega_\nu \partial/\partial \psi_\nu - \Lambda , \tag{6.3.31}$$

$$L_3 = \sum_{\nu=1}^{n} \omega_\nu \partial/\partial \psi_\nu - \underbrace{(\Lambda' - \Lambda'')}_{\hat{\Lambda}} , \tag{6.3.32}$$

where Λ' and Λ'' are again matrices stemming from Λ via rearrangement of indices. By means of (6.3.26 – 29) we may cast (6.3.23) in the form

$$LU = F + \hat{\Delta} . \tag{6.3.33}$$

This is an inhomogeneous equation for the vector U. We first determine the eigenvalues of the operator L. This is most easily achieved by expanding the solutions $\tilde{U}$ of the equation

$$L\tilde{U} = \lambda\tilde{U} \tag{6.3.34}$$

into a Fourier series

$$\tilde{U}(\psi) = \sum_j \tilde{U}_j \exp[\mathrm{i}(j, \psi)] . \tag{6.3.35}$$

For the moment, the use of L may seem to be roundabout, but its usefulness will become obvious below. When we insert (6.3.35) into (6.3.34) we readily obtain

$$\mathrm{i}(j, \omega)\tilde{u}_j = \lambda\tilde{u}_j , \tag{6.3.36}$$

$$[\mathrm{i}(j, \omega) - \Lambda]\tilde{v}_j = \lambda\tilde{v}_j , \tag{6.3.37}$$

$$\mathrm{i}(j, \omega)\tilde{V}_j' - \hat{\Lambda}\tilde{V}_j' = \lambda\tilde{V}_j' . \tag{6.3.38}$$

In the notation of the original indices inherent in (6.3.23), the last equation (6.3.38) reads

$$\mathrm{i}(j, \omega)\tilde{V}_j' - \Lambda\tilde{V}_j' + \tilde{V}_j'\Lambda = \lambda\tilde{V}' . \tag{6.3.39}$$

These equations show clearly what the eigenvalues of L are: since the equations for $\tilde{u}_j$, $\tilde{v}_j$, $\tilde{V}_j'$ are uncoupled, we may treat these equations individually. From (6.3.36) we obtain as eigenvalue:

$$\mathrm{i}(j, \omega) , \tag{6.3.40}$$

by diagonalizing (6.3.37)

$$\mathrm{i}(j, \omega) - \Lambda_\mu , \tag{6.3.41}$$

and by diagonalizing (6.3.38), which can be most simply seen from (6.3.39),

$$\mathrm{i}(j, \omega) - \Lambda_\mu + \Lambda_{\mu'} . \tag{6.3.42}$$

According to linear algebra, the null-space is defined by those vectors $\tilde{U}_j$ for which the eigenvalues are zero. But on account of assumption (1) in Sect. 6.2, the eigenvalues (6.3.40 – 42) vanish only for $\boldsymbol{j} = \boldsymbol{0}$. We are now in a position to invoke a well-known theorem of linear algebra on the solubility of the inhomogeneous equations (6.3.33). To this end we expand F into a Fourier series of the form (6.3.35) and equate the coefficients of $\exp(\mathrm{i}j, \psi)$ on both sides of (6.3.33).

For $\boldsymbol{j} \neq \mathbf{0}$ we obtain

$$\mathrm{i}(\boldsymbol{j}, \boldsymbol{\omega})\boldsymbol{u}_j' = \boldsymbol{f}_j \,, \tag{6.3.43}$$

$$[\mathrm{i}(\boldsymbol{j}, \boldsymbol{\omega}) - \Lambda]\,\boldsymbol{v}_j' = \boldsymbol{g}_j \,, \tag{6.3.44}$$

$$[\mathrm{i}(\boldsymbol{j}, \boldsymbol{\omega})]\, V_j' - \Lambda\, V_j' + V_j' \Lambda = \boldsymbol{g}_{\chi,j} \,, \tag{6.3.45}$$

or, writing (6.3.45) in components (provided Λ is diagonal for example),

$$\mathrm{i}(\boldsymbol{j}, \boldsymbol{\omega})\, V_{j(\mu,\nu)}' - (\Lambda_\mu - \Lambda_\nu)\, V_{j(\mu,\nu)}' = \boldsymbol{g}_{\chi,j(\mu,\nu)} \,. \tag{6.3.46}$$

Because the eigenvalues do not vanish, these equations can be solved uniquely for the unknown $\boldsymbol{u}'$, $\boldsymbol{v}'$, V'.

We now turn to the eigenvectors $\tilde{U}_j$ belonging to the vanishing eigenvalues, obtained according to assumption (1) in Sect. 6.2 only for $\boldsymbol{j} = \mathbf{0}$. Thus the eigenvectors of the null-space obey the equations

$$0 \cdot \boldsymbol{u}_0' = \mathbf{0} \,, \tag{6.3.47}$$

$$\Lambda \cdot \boldsymbol{v}_0' = \mathbf{0} \,, \tag{6.3.48}$$

$$\Lambda\, V_0' - V_0' \Lambda = 0 \,, \tag{6.3.49}$$

which follow from (6.3.36 – 38) and the condition that (6.3.40 – 42) vanish.

We now consider the inhomogeneous equations which correspond to (6.3.43 – 45) for $\boldsymbol{j} = \mathbf{0}$. They read

$$-\boldsymbol{\Delta}' = \boldsymbol{f}_0 \,, \tag{6.3.50}$$

$$-\Lambda\, \boldsymbol{v}_0' - \boldsymbol{d}' = \boldsymbol{g}_0 \,, \tag{6.3.51}$$

$$-\Lambda\, V_0' + V_0' \Lambda - D' = \boldsymbol{g}_{\chi,\mathbf{0}} \,. \tag{6.3.52}$$

Obviously (6.3.50) fixes $\boldsymbol{\Delta}'$ uniquely. Because of (6.3.48), i.e.,

$$\Lambda\, \boldsymbol{v}_0' = \mathbf{0} \,, \tag{6.3.53}$$

$\boldsymbol{d}'$ is determined uniquely by (6.3.51). Similarly, from (6.3.49) it follows that D' is determined uniquely by (6.3.52). What happens, however, with the coefficients $\boldsymbol{u}_0'$, $\boldsymbol{v}_0'$, V_0'? According to linear algebra, within the limitation expressed by (6.3.47 – 49), these null-space vectors can be chosen freely. In the following we put them equal to zero, or in other words, we project the null-space out. Let us summarize what we have achieved so far. We have described the first step of an iteration procedure by which we transform (6.3.1, 2) into (6.1.26, 27) by means of the transformations (6.3.3, 4). This was done by determining $\boldsymbol{u}$, $\boldsymbol{v}$, V and the counterterms $\boldsymbol{\Delta}$, $\boldsymbol{d}$, D in lowest approximation. We have shown explicitly how the corresponding equations can formally be solved. This procedure may now be continued so that we may determine the unknown quantities up to any desired order in ε. We shall see in later applications that it is most useful to calculate in particular Δ, $\boldsymbol{d}$, D and the more practically oriented reader may stop here.

From the mathematical point of view, the following questions have to be settled.

1) In each iteration step, the equations were solved by means of Fourier series. Do these series converge? This question is by no means trivial, because, e.g., it follows from (6.3.43) that

$$\boldsymbol{u}_j' = [\mathrm{i}(\boldsymbol{j},\boldsymbol{\omega})]^{-1}\boldsymbol{f}_j\,, \tag{6.3.54}$$

i.e., we have to deal with the problem of small divisors (Sects. 2.1.3 and 5.2). We shall treat it anew in Appendix A.1, making use of our assumption that $\boldsymbol{f}$ and $\boldsymbol{g}$ are analytic functions in $\boldsymbol{\varphi}$ and $\boldsymbol{\xi}$.

2) We have to prove the convergence of the iteration procedure as a whole. This will shed light on the choice of the null-space vectors in each iteration step. As it will turn out, any choice of these vectors will give an admitted solution provided the ε-sum over the null-space vectors converges. In connection with this we shall show that there are classes of possible transformations (6.3.3, 4) which are connected with each other. We turn to a treatment of these problems in Appendix A.

7. Nonlinear Equations. The Slaving Principle

The main objective of this book is the study of dramatic macroscopic changes of systems. As seen in the introduction, this may happen when linear stability is lost. At such a point it becomes possible to eliminate very many degrees of freedom so that the macroscopic behavior of the system is governed by very few degrees of freedom only. In this chapter we wish to show explicitly how to eliminate most of the variables close to points where linear stability is lost. These points will be called critical points. It will be our goal to describe an easily applicable procedure covering most cases of practical importance. To this end the essential ideas are illustrated by a simple example (Sect. 7.1), followed by a presentation of our general procedure for nonlinear differential equations (Sects. 7.2 – 5). While the basic assumptions are stated in Sect. 7.2, the final results of our approach are presented in Sect. 7.4 up to (and including) formula (7.4.5). Section 7.3 and the rest of Sect. 7.4 are of a more technical nature. The rest of this chapter is devoted to an extension of the slaving principle to discrete noisy maps and to stochastic differential equations of the Îto (and Stratonovich) type (Sects. 7.6 – 9).

7.1 An Example

We treat the following nonlinear equations in the two variables u and s

$$\dot{u} = \alpha u - us\,, \tag{7.1.1}$$

$$\dot{s} = -\beta s + u^2\,. \tag{7.1.2}$$

We assume that

$$\alpha \gtrsim 0 \quad \text{and} \tag{7.1.3}$$

$$\beta > 0\,. \tag{7.1.4}$$

When we neglect the nonlinear terms $u \cdot s$ and u^2, respectively, we are left with two uncoupled equations that are familiar from our linear stability analysis. Evidently (7.1.1) represents a mode which is neutral or unstable in a linear stability analysis, whereas s represents a stable mode. This is why we have chosen the notation u (unstable or undamped) and s (stable). We wish to show that s may be expressed by means of u so that we can eliminate s from (7.1.1, 2). Equation (7.1.2) can be immediately solved by integration

$$s(t) = \int_{-\infty}^{t} e^{-\beta(t-\tau)} u^2(\tau) d\tau \,, \tag{7.1.5}$$

where we have chosen the initial condition $s = 0$ for $t = -\infty$. The integral in (7.1.5) exists if $|u(\tau)|^2$ is bounded for all τ or if this quantity diverges for $\tau \to -\infty$ less than $\exp(|\beta\tau|)$. This is, of course, a self-consistency requirement which has to be checked.

We now want to show that the integral in (7.1.5) can be transformed in such a way that s at time t becomes a function of u at the same time t only.

7.1.1 The Adiabatic Approximation

We integrate (7.1.5) by parts according to the rule

$$\int \dot{v} w \, d\tau = v w - \int v \, \dot{w} \, d\tau \,, \tag{7.1.6}$$

where we identify v with $\exp[-\beta(t-\tau)]$ and w with u^2. We thus obtain

$$s(t) = \frac{1}{\beta} u^2(t) - \frac{1}{\beta} \int_{-\infty}^{t} e^{-\beta(t-\tau)} 2(u\dot{u})_\tau d\tau \,. \tag{7.1.7}$$

We now consider the case in which u varies very little so that $\dot{u}$ can be considered as a small quantity. This suggests that we can neglect the integral in (7.1.7). This is the *adiabatic approximation* in which we obtain

$$s(t) = \frac{1}{\beta} u^2(t) \,. \tag{7.1.8}$$

Solution (7.1.8) could have been obtained from (7.1.2) by merely putting

$$\dot{s} = 0 \tag{7.1.9}$$

on lhs of (7.1.2). Let us now consider under which condition we may neglect the integral in (7.1.7) compared to the first term. To this end we consider

$$(|u||\dot{u}|)_{\max} \tag{7.1.10}$$

in (7.1.7). We may then put (7.1.10) in front of the integral which can be evaluated, leaving the condition

$$\frac{(|u||\dot{u}|)_{\max}}{\beta^2} \ll \frac{|u|^2}{\beta} \,. \tag{7.1.11}$$

This condition can be fulfilled provided

$$|\dot{u}|_{\max} \ll \beta |u| \tag{7.1.12}$$

holds. Condition (7.1.12) gives us an idea of the meaning of the adiabatic approximation. We require that u changes slowly enough compared to a change prescribed by the damping constant β.

We now want to show how we may express $s(t)$ by $u(t)$ in a precise manner.

7.1.2 Exact Elimination Procedure

In order to bring out the essential features of our approach we choose $\alpha = 0$, so that (7.1.1) is replaced by

$$\dot{u} = -us. \tag{7.1.13}$$

We now use the still exact relation (7.1.7), substituting $\dot{u}$ in it by $-u \cdot s$ according to (7.1.13)

$$s(t) = \frac{1}{\beta} u^2(t) + \frac{2}{\beta} \int_{-\infty}^{t} e^{-\beta(t-\tau)} (u^2 s)_\tau \, d\tau. \tag{7.1.14}$$

This is an integral equation for $s(t)$ because s occurs again under the integral at times τ. To solve this equation we apply an iteration procedure which amounts to expressing s by powers of u. In the lowest order we may express s on the rhs of (7.1.14) by the approximate relation (7.1.8). In this way, (7.1.14) is transformed into

$$s(t) = \frac{1}{\beta} u^2(t) + \frac{2}{\beta} \int_{-\infty}^{t} e^{-\beta(t-\tau)} \frac{1}{\beta} u^4(\tau) \, d\tau. \tag{7.1.15}$$

To obtain the next approximation we integrate (7.1.15) by parts using formula (7.1.6) in which we again identify v with the exponential function in (7.1.15). We thus obtain

$$s(t) = \frac{1}{\beta} u^2(t) + \frac{2}{\beta^3} u^4(t) - \underbrace{\frac{8}{\beta^3} \int_{-\infty}^{t} e^{-\beta(t-\tau)} (u^3 \dot{u})_\tau \, d\tau}_{I}. \tag{7.1.16}$$

Under the integral we may replace $\dot{u}$ according to (7.1.13), giving

$$I = \frac{8}{\beta^3} \int_{-\infty}^{t} e^{-\beta(t-\tau)} u^4 \left(\frac{1}{\beta} u^2 + \frac{2}{\beta^2} u^4 \right) d\tau. \tag{7.1.17}$$

We split (7.1.17) into an integral containing u^6 and another part containing still higher-order terms

$$I = \frac{8}{\beta^4} \int_{-\infty}^{t} e^{-\beta(t-\tau)} u^6 d\tau + \text{h.o.} \,. \tag{7.1.18}$$

Performing a partial integration as before, $s(t)$ occurs in the form

$$s(t) = \frac{1}{\beta} u^2 + \frac{2}{\beta^3} u^4 + \frac{8u^6}{\beta^5} + \dots \,. \tag{7.1.19}$$

It is now obvious that by this procedure we may again and again perform partial integrations allowing us to express $s(t)$ by a certain power series in $u(t)$ at the same time. Provided u is small enough we may expect to find a very good approximation by keeping few terms in powers of u. Before we study the problem of the convergence of this procedure, some exact relations must be derived. To this end we introduce the following abbreviations into the original equations (7.1.13, 2), namely

$$\dot{u} = -us = Q(u, s) \,, \tag{7.1.20}$$

$$\dot{s} = -\beta s + u^2 = -\beta s + P(u, s) \,. \tag{7.1.21}$$

Note that in our special case P is a function of u alone

$$P = P(u) \,. \tag{7.1.22}$$

The procedure just outlined above indicates that it is possible to express s by u:

$$s = s(u) \,, \tag{7.1.23}$$

so that we henceforth assume that this substitution has been made. We wish to derive a formal series expansion for $s(u)$. We start from (7.1.5), written in the form

$$s(u(t)) = \int_{-\infty}^{t} e^{-\beta(t-\tau)} P(u)_\tau d\tau \,. \tag{7.1.24}$$

As before, we integrate by parts identifying $P(u)$ with w in (7.1.6). The differentiation of $P(u)$ with respect to time can be performed in such a way that we first differentiate P with respect to u and then u with respect to time. But according to (7.1.20) the time derivative of u can be replaced by $Q(u, s)$ where, at least in principle, we can imagine that s is a function of u again. Performing this partial integration again and again, each time using (7.1.20), we are led to the relation

$$s(u) = \frac{1}{\beta} P(u) - \frac{1}{\beta} Q \frac{\partial}{\partial u} \frac{1}{\beta} P + \frac{1}{\beta} Q \frac{\partial}{\partial u} \frac{1}{\beta} Q \frac{\partial}{\partial u} \frac{1}{\beta} P + \dots \,. \tag{7.1.25}$$

In the following section it will turn out to be useful to use the following abbreviation

$$\left(\frac{d}{dt}\right)_\infty = Q(u, s(u)) \frac{\partial}{\partial u} \tag{7.1.26}$$

by which we define the left-hand side. It becomes obvious that (7.1.25) can be considered as a formal geometric series in the operator (7.1.26) so that (7.1.25) can be summed up to give

$$s(t) = \frac{1}{\beta} \frac{1}{\left(1 + \frac{1}{\beta}\left(\frac{d}{dt}\right)_\infty\right)} P \tag{7.1.27}$$

or, in the original notation,

$$s(t) = \frac{1}{\beta} \frac{1}{\left(1 + Q \frac{\partial}{\partial u} \frac{1}{\beta}\right)} P. \tag{7.1.28}$$

Multiplying both sides by the operator

$$\left(1 + Q \frac{\partial}{\partial u} \frac{1}{\beta}\right) \tag{7.1.29}$$

from the left we obtain

$$\left(\beta + Q \frac{\partial}{\partial u}\right) s = P(u, s), \tag{7.1.30}$$

which may be transformed in an obvious way,

$$\frac{\partial s}{\partial u} = \frac{P(u, s) - \beta s}{Q(u, s)}. \tag{7.1.31}$$

Using the explicit form of Q and P, (7.1.31) acquires the form

$$\frac{\partial s}{\partial u} = \frac{u^2 - \beta s}{-us}. \tag{7.1.32}$$

Thus we have found a first-order differential equation for $s(u)$.

In the subsequent section we shall show how all these relations can be considerably generalized to equations much more general than (7.1.1, 2). In practical applications we shall have to evaluate sums as (7.1.25) but, of course, we shall approximate the infinite series by a finite one. In this way we also circumvent difficulties which may arise with respect to the convergence of the formal series equation (7.1.25).

Rather, in practical cases it will be necessary to estimate the rest terms. To this end we split the formal infinite expansion (7.1.25) up in the following way

$$s(t)=\frac{1}{\beta}\sum_{n=0}^{m}\left(-Q\frac{\partial}{\partial u}\frac{1}{\beta}\right)^{n}P(u,s)+\underbrace{\frac{1}{\beta}\sum_{n=m+1}^{\infty}\left(-Q\frac{\partial}{\partial u}\frac{1}{\beta}\right)^{n}P(u,s)}_{r}. \tag{7.1.33}$$

Again it is possible to sum up the rest term

$$r=\frac{1}{\beta}\sum_{n=0}^{\infty}\left(-Q\frac{\partial}{\partial u}\frac{1}{\beta}\right)^{n+m+1}P(u,s(u)), \tag{7.1.34}$$

which yields

$$r=\frac{1}{\beta}\frac{1}{\left(1+Q\dfrac{\partial}{\partial u}\dfrac{1}{\beta}\right)}\left(-Q\frac{\partial}{\partial u}\frac{1}{\beta}\right)^{m+1}P. \tag{7.1.35}$$

This form or the equivalent form

$$\left(\beta+Q\frac{\partial}{\partial u}\right)r=\left(-Q\frac{\partial}{\partial u}\frac{1}{\beta}\right)^{m+1}P \tag{7.1.36}$$

can be used to estimate the rest term. The term Q introduces powers of at least order u^3 while the differentiation repeatedly reduces the order by 1. Therefore, at least, with each power of the bracket in front of (7.1.36), a term u^2 is introduced in our specific example. At the same time corresponding powers $1/\beta$ are introduced. This means that the rest term becomes very small provided u^2/β is much smaller than unity. On the other hand, due to the consecutive differentiations introduced in (7.1.35 or 36), the number of terms increases with $m!$. Therefore the rest term must be chosen in such a way that m is not too big or, in other words, for given m one has to choose u correspondingly sufficiently small. This procedure is somewhat different from the more conventional convergence criteria but it shows that we can determine s as a function of u to any desired accuracy provided we choose u small enough.

Let us now return to our explicit example (7.1.32). We wish to show that (7.1.32) can be chosen as a starting point to express s explicitly by u. We require that rhs of (7.1.32), which we repeat

$$\frac{\partial s}{\partial u}=\frac{u^2-\beta s}{-us}, \tag{7.1.37}$$

remains regular for simultaneous limits $s \to 0$ and $u \to 0$. To solve (7.1.37) we make the hypothesis

$$s = u^2[c_1 + f_1(u)] , \tag{7.1.38}$$

which yields in (7.1.37)

$$2u(c_1+f_1) + u^2 f_1' = \frac{u^2 - \beta u^2(c_1+f_1)}{-uu^2(c_1+f_1)} . \tag{7.1.39}$$

Since rhs must remain regular we have to put

$$c_1 = \frac{1}{\beta} \tag{7.1.40}$$

so that on rhs of (7.1.39)

$$\frac{-u^2 f_1 \beta}{-u^3(c_1+f_1)} \tag{7.1.41}$$

results. In order that $s \to 0$ for $u \to 0$, we put f_1 in the form

$$f_1 = u^2(c_2+f_2) \tag{7.1.42}$$

so that

$$\frac{2u}{\beta} + 2u^3 c_2 + 2u^3 f_2 + u^2 2u(c_2+f_2) + u^4 f_2' = \frac{u(c_2+f_2)\beta}{1/\beta + u^2(c_2+f_2)} \tag{7.1.43}$$

results. A comparison of powers of u on lhs with that of rhs yields

$$c_2 = \frac{2}{\beta^3} . \tag{7.1.44}$$

We are thus led to the expansion

$$s = \frac{u^2}{\beta} + \frac{2u^4}{\beta^3} + \dots \tag{7.1.45}$$

which agrees with our former result (7.1.19).

Starting from (7.1.1, 2) (in which our above treatment with $\alpha = 0$ is included), we can achieve the result (7.1.37) in a much simpler way, namely writing (7.1.1, 2) in the form

$$\lim_{\Delta t \to 0} \left(\frac{\Delta u}{\Delta t}\right) \equiv \dot{u} = \alpha u - us \quad \text{and} \tag{7.1.46}$$

$$\lim_{\Delta t \to 0} \left(\frac{\Delta s}{\Delta t} \right) \equiv \dot{s} = -\beta s + u^2 . \tag{7.1.47}$$

We may divide the corresponding sides of (7.1.47) by those of (7.1.46). In this case Δt drops out, leaving

$$\frac{ds}{du} = \frac{-\beta s + u^2}{\alpha u - us} . \tag{7.1.48}$$

This is a relation very well known from autonomous differential equations referring to trajectories in a plane [1]. Although our series expansion seems roundabout, its usefulness is immediately evident if we are confronted with the question in which way (7.1.48) can be generalized to nonautonomous differential equations in several variables. We are then quickly led to methods we have outlined before.

Thus in the next section we wish to generalize our method in several ways. In practical applications it will be necessary to treat equations in which the control parameter α is unequal zero, so that we have to treat equations of the form

$$\dot{u} = \alpha u - us . \tag{7.1.49}$$

Furthermore, the equations may have a more general structure, for instance those of the slaved variables may have the form

$$\dot{s} = -\beta s + u^2 + us + \dots \tag{7.1.50}$$

or be still more general, containing polynomials of both u and s on rhs. In addition, we wish to treat equations which are nonautonomous, e.g., those in which the coefficients depend on time t

$$\dot{s} = -\beta s + a(t) u^2 . \tag{7.1.51}$$

Also in many applications we need equations which are sets of differential equations in several variables u and s.

Other types of equations must be considered also, for instance

$$\dot{\varphi} = \omega + \Phi(u, s, \varphi) , \tag{7.1.52}$$

where Φ is a function which is periodic in φ. Equations of the type (7.1.52) containing several variables φ may also occur in applications. Last but not least, we must deal with stochastic differential equations

$$\dot{s} = -\beta s + u^2 + F_s(t) \tag{7.1.53}$$

in which $F_s(t)$ is a random force. This random force renders the equations to inhomogeneous equations and an additional difficulty arises from the fact that

$F_s(t)$ is discontinuous. Still more difficulties arise if such a stochastic force depends on variables s and u. In order not to overload our treatment, we shall proceed in two steps. We first consider the fluctuating forces as being approximated by continuous functions.

Then in Sect. 7.9 we shall consider a general class of equations containing discontinuous stochastic forces.

7.2 The General Form of the Slaving Principle. Basic Equations

Consider a set of time dependent variables $\tilde{u}_1, \tilde{u}_2, \ldots, s_1, s_2, \ldots, \tilde{\varphi}_1, \tilde{\varphi}_2, \ldots$, lumped together into the vectors $\tilde{\boldsymbol{u}}$, $\boldsymbol{s}$, $\tilde{\boldsymbol{\varphi}}$, where the sign $\sim$ distinguishes these variables from those used below after some transformations. The notation $\tilde{\boldsymbol{u}}$ and $\boldsymbol{s}$ will become clear later on where $\tilde{\boldsymbol{u}}$ denotes the "*un*stable" or "*un*damped" modes of a system, whereas $\boldsymbol{s}$ refers to "*s*table" modes. The variables $\tilde{\boldsymbol{\varphi}}$ play the role of phase angles, as will transpire from later chapters. We assume that these variables obey the following set of ordinary differential equations[1]

$$\dot{\tilde{\boldsymbol{u}}} = \tilde{\Lambda}_u \tilde{\boldsymbol{u}} + \tilde{\boldsymbol{Q}}(\tilde{\boldsymbol{u}}, \boldsymbol{s}, \tilde{\boldsymbol{\varphi}}, t) + \tilde{\boldsymbol{F}}_u(t) , \tag{7.2.1}$$

$$\dot{\boldsymbol{s}} = \Lambda_s \boldsymbol{s} + \tilde{\boldsymbol{P}}(\tilde{\boldsymbol{u}}, \boldsymbol{s}, \tilde{\boldsymbol{\varphi}}, t) + \tilde{\boldsymbol{F}}_s(t) , \tag{7.2.2}$$

$$\dot{\tilde{\boldsymbol{\varphi}}} = \boldsymbol{\omega} + \tilde{\boldsymbol{R}}(\tilde{\boldsymbol{u}}, \boldsymbol{s}, \tilde{\boldsymbol{\varphi}}, t) + \tilde{\boldsymbol{F}}_\varphi(t) . \tag{7.2.3}$$

In the following, we shall assume that $\tilde{\Lambda}_u$, Λ_s and $\boldsymbol{\omega}$ are time-independent matrices and vectors, respectively. It is not difficult, however, to extend the whole procedure to the case in which these quantities are time dependent. We make the following assumptions. We introduce a smallness parameter δ and assume that $\tilde{\boldsymbol{u}}$ is of the order δ. We assume that the matrix $\tilde{\Lambda}_u$ is in Jordan's normal form and can be decomposed into

$$\tilde{\Lambda}_u = \Lambda_u' + \mathrm{i}\Lambda_u'' + \Lambda_u''' \equiv \Lambda_u' + \hat{\Lambda}_u . \tag{7.2.4}$$

Here Λ_u', Λ_u'' are real diagonal matrices. While the size of the elements of Λ_u'' may be arbitrary, the size of the elements of Λ_u' is assumed to be of order δ. Further, Λ_u''' contains the nondiagonal elements of Jordan's normal form (Sect. 2.4.2). Also the matrix Λ_s is assumed in Jordan's normal form. It can be decomposed into its real and imaginary parts Λ_s', Λ_s'', respectively

$$\Lambda_s = \Lambda_s' + \mathrm{i}\Lambda_s'' . \tag{7.2.5}$$

Here Λ_s'' is a diagonal matrix whose matrix elements may be arbitrary. It is assumed that the diagonal elements γ_i of Λ_s' obey the inequalities

$$\gamma_i \leqslant \beta < 0 . \tag{7.2.6}$$

---

[1] It is actually not difficult to extend our formalism to partial differential equations with a suitable norm (e.g., in a Banach space)

The quantities

$$\tilde{Q}, \tilde{P}, \tilde{R}, \tilde{F}_u, \tilde{F}_s, \tilde{F}_\varphi \tag{7.2.7}$$

are of order δ^2 or smaller and $\tilde{Q}$, $\tilde{P}$, must not contain terms linear in u or s (but may contain $u \cdot s$). The functions

$$\tilde{Q}, \tilde{P}, \tilde{R} \tag{7.2.8}$$

are 2π-periodic in φ, the functions (7.2.7) are assumed to be bounded for $-\infty < t < +\infty$ for any fixed values of $\tilde{u}$, s, $\tilde{\varphi}$. The functions $\tilde{F}$ are driving forces which in particular may mimic stochastic forces. However, we shall assume in this section that these forces are continuous with respect to their arguments.

In subsequent sections we shall then study stochastic forces which can be written in the form

$$\tilde{F}(t) \equiv \tilde{F}(\tilde{u}, s, \tilde{\varphi}, t) = G(\tilde{u}, s, \tilde{\varphi}, t) \cdot F_0(t) , \tag{7.2.9}$$

where F_0 describes a Wiener process. In such a case the matrix G must be treated with care because appropriate limits of the functions $\tilde{u}$, s, $\tilde{\varphi}$ must be taken, e.g., in the form

$$G(\tilde{u}, s, \tilde{\varphi}, t) = \alpha G_{t+0} + \beta G_{t-0} . \tag{7.2.10}$$

We shall leave this problem for subsequent sections.

Let us now rather assume that the functions (7.2.9) can be treated in the same way as $\tilde{Q}$, $\tilde{P}$, $\tilde{R}$. Though a number of our subsequent relations hold also for more general cases, for most applications it is sufficient to assume that $\tilde{P}$, $\tilde{Q}$, $\tilde{R}$ can be expanded into power series of $\tilde{u}$ and s. For instance, we may then write $\tilde{P}$ in the form

$$\begin{aligned} \tilde{P}(\tilde{u}, s, \tilde{\varphi}, t) = {} & \tilde{A}_{suu} : \tilde{u} : \tilde{u} + 2\tilde{A}_{sus} : \tilde{u} : s + \tilde{A}_{sss} : s : s \\ & + \tilde{B}_{suuu} : \tilde{u} : \tilde{u} : \tilde{u} + 3\tilde{B}_{suus} : \tilde{u} : \tilde{u} : s + \dots , \end{aligned} \tag{7.2.11}$$

where the notation introduced in (7.2.11) can be explained as follows. Let us represent $\tilde{P}$ as a vector in the form

$$\tilde{P} = \begin{pmatrix} \tilde{P}_1 \\ \tilde{P}_2 \\ \vdots \end{pmatrix} = (\tilde{P}_j) . \tag{7.2.12}$$

We define

$$\tilde{A}_{suu} : \tilde{u} : \tilde{u} \tag{7.2.13}$$

as a vector whose j-components are given by

$$\sum_{k_1 k_2} \tilde{A}_{j k_1 k_2} \tilde{u}_{k_1} \tilde{u}_{k_2} . \tag{7.2.14}$$

Here $\tilde{A}_{suu}$ may be a 2π-periodic function in φ and may still be a continuous function of time t

$$\tilde{A}_{suu} = \tilde{A}_{suu}(\tilde{\varphi}, t)\,. \tag{7.2.15}$$

A similar decomposition to (7.2.11) is assumed for $\tilde{\boldsymbol{Q}}$. $\tilde{\boldsymbol{R}}$ is assumed in the form

$$\tilde{\boldsymbol{R}}(\tilde{\boldsymbol{u}}, \boldsymbol{s}, \tilde{\varphi}, t) = \boldsymbol{R}_0(\tilde{\varphi}, t) + \boldsymbol{R}_1(\tilde{\varphi}, t):\tilde{\boldsymbol{u}} + \boldsymbol{R}_2(\tilde{\varphi}, t):\boldsymbol{s} + \text{terms of form } \tilde{\boldsymbol{P}}\,. \tag{7.2.16}$$

We shall assume that $\boldsymbol{R}_0$ is of order δ^2 and $\boldsymbol{R}_1$, $\boldsymbol{R}_2$ are of order δ. We now make the transformations[2]

$$\tilde{\boldsymbol{u}} = \exp(\hat{\Lambda}_u t)\boldsymbol{u}\,, \tag{7.2.17}$$

$$\tilde{\varphi} = \omega t + \varphi \tag{7.2.18}$$

to the new time-dependent variables $\boldsymbol{u}$, φ. This transforms the original equations (7.2.1 – 3) into

$$\dot{\boldsymbol{u}} = \Lambda_u \boldsymbol{u} + \bar{\boldsymbol{Q}}(\boldsymbol{u}, \boldsymbol{s}, \varphi, t) + \boldsymbol{F}_u(t) = \boldsymbol{Q}(\boldsymbol{u}, \boldsymbol{s}, \varphi, t)\,, \tag{7.2.19}$$

$$\dot{\boldsymbol{s}} = \Lambda_s \boldsymbol{s} + \bar{\boldsymbol{P}}(\boldsymbol{u}, \boldsymbol{s}, \varphi, t) + \boldsymbol{F}_s(t) = \Lambda_s \boldsymbol{s} + \boldsymbol{P}(\boldsymbol{u}, \boldsymbol{s}, \varphi, t)\,, \tag{7.2.20}$$

$$\dot{\varphi} = \bar{\boldsymbol{R}}(\boldsymbol{u}, \boldsymbol{s}, \varphi, t) + \boldsymbol{F}_\varphi(t) = \boldsymbol{R}(\boldsymbol{u}, \boldsymbol{s}, \varphi, t)\,, \tag{7.2.21}$$

where the last equations in each row are mere abbreviations, whereas the expressions in the middle column of (7.2.19 – 21) are defined as follows:

$$\Lambda_u \equiv \Lambda'_u\,, \tag{7.2.22}$$

$$\bar{\boldsymbol{Q}}(\boldsymbol{u}, \boldsymbol{s}, \varphi, t) = \exp(-\hat{\Lambda}_u t)\tilde{\boldsymbol{Q}}(\exp(\hat{\Lambda}_u t)\boldsymbol{u}, \boldsymbol{s}, \omega t + \varphi, t)\,, \tag{7.2.23}$$

$$\bar{\boldsymbol{P}}(\boldsymbol{u}, \boldsymbol{s}, \varphi, t) = \tilde{\boldsymbol{P}}(\exp(\hat{\Lambda}_u t)\boldsymbol{u}, \boldsymbol{s}, \omega t + \varphi, t)\,, \tag{7.2.24}$$

$$\bar{\boldsymbol{R}}(\boldsymbol{u}, \boldsymbol{s}, \varphi, t) = \tilde{\boldsymbol{R}}(\exp(\hat{\Lambda}_u t)\boldsymbol{u}, \boldsymbol{s}, \omega t + \varphi, t)\,, \tag{7.2.25}$$

$$\boldsymbol{F}_u = \exp(-\hat{\Lambda}_u t)\tilde{\boldsymbol{F}}_u(\exp(\hat{\Lambda}_u t)\boldsymbol{u}, \boldsymbol{s}, \varphi + \omega t, t)\,, \tag{7.2.26}$$

$$\boldsymbol{F}_s = \exp(-\hat{\Lambda}_s t)\tilde{\boldsymbol{F}}_s(\exp(\hat{\Lambda}_u t)\boldsymbol{u}, \boldsymbol{s}, \varphi + \omega t, t)\,, \tag{7.2.27}$$

$$\boldsymbol{F}_\varphi = \tilde{\boldsymbol{F}}_\varphi(\exp(\hat{\Lambda}_u t)\boldsymbol{u}, \boldsymbol{s}, \varphi + \omega t, t)\,. \tag{7.2.28}$$

---

[2] The transformation of a vector by means of $\exp(\hat{\Lambda}_u t) \equiv \exp[(\mathrm{i}\Lambda''_u + \Lambda'''_u)t]$ has the following effects (Sect. 2.6): while the diagonal matrix Λ''_u causes a multiplication of each component of the vector by $\exp(\mathrm{i}\omega_j t)$, where ω_j is an element of Λ''_u, Λ'''_u mixes vector components with coefficients which contain finite powers of t. All integrals occurring in subsequent sections of this chapter exist, because the powers of t are multiplied by damped exponential functions of t.

It will be our goal to express the time-dependent functions $\boldsymbol{s}(t)$ occurring in (7.2.19–21) as functions of $\boldsymbol{u}$ and $\boldsymbol{\varphi}$ (and t) by an explicit construction. Before developing our procedure we mention the following: As we have seen in Chap. 5, a nonlinear coupling between variables can shift frequencies, i.e. in particular the values of the matrix elements of Λ_u'' in (7.2.4) and of $\boldsymbol{\omega}$ in (7.2.3). In order to take care of this effect at a stage as early as possible it is useful to perform the transformations (7.2.17, 18) with the shifted frequencies of $\Lambda_{u,r}''$ and $\boldsymbol{\omega}_r$. In physics, these frequencies are called *renormalized frequencies*. These frequencies are *unknown* and must be determined in a *self-consistent way* at the end of the whole calculation which determines the solutions $\boldsymbol{u}$ and $\boldsymbol{\varphi}$. The advantage of this procedure consists in the fact that in many practical cases few steps of the elimination procedure (with respect to $\boldsymbol{s}$) suffice to give very good results for the variables $\boldsymbol{u}$, $\boldsymbol{s}$, and $\boldsymbol{\varphi}$. We leave it as an exercise to the reader to rederive (7.2.19–21) by means of renormalized frequencies, and to discuss what assumptions on the smallness of $\boldsymbol{\omega}-\boldsymbol{\omega}_r$ and $\Lambda_u''-\Lambda_{u,r}''$ must be made so that the new functions $\boldsymbol{P}$, $\boldsymbol{Q}$, $\boldsymbol{R}$ fulfill the original smallness conditions. The reader is also advised to repeat the initial steps explained in the subsequent chapters using the thus altered equations (7.2.19–21). Our following explicit procedure will be based on the original equations (7.2.19–21).

7.3 Formal Relations

As exemplified in Sect. 7.1, s may be expressed by u at the same time. Now let us assume that such a relation can be established also in the general case of (7.2.19–21) but with suitable generalizations. Therefore, we shall assume that $\boldsymbol{s}$ can be expressed by $\boldsymbol{u}$ and $\boldsymbol{\varphi}$ and may still explicitly depend on time

$$\boldsymbol{s}=\boldsymbol{s}(\boldsymbol{u}(t),\boldsymbol{\varphi}(t),t)\,. \tag{7.3.1}$$

Later it will be shown how $\boldsymbol{s}$ may be constructed in the form (7.3.1) explicitly, but for the moment we assume that (7.3.1) holds, allowing us to establish a number of important exact relations. Since $\boldsymbol{s}$ depends on time via $\boldsymbol{u}$ and $\boldsymbol{\varphi}$ but also directly on time t, we find by differentiation of (7.3.1)

$$\frac{d}{dt}\boldsymbol{s}=\frac{\partial}{\partial t}\boldsymbol{s}+\dot{\boldsymbol{u}}\frac{\partial \boldsymbol{s}}{\partial \boldsymbol{u}}+\dot{\boldsymbol{\varphi}}\frac{\partial \boldsymbol{s}}{\partial \boldsymbol{\varphi}}\,, \tag{7.3.2}$$

where we have used the notation

$$\frac{\partial \boldsymbol{s}}{\partial \boldsymbol{u}}=\begin{pmatrix}\nabla_u s_1\\ \nabla_u s_2\\ \vdots\\ \nabla_u s_n\end{pmatrix} \tag{7.3.3}$$

and a similar one for $\partial\boldsymbol{s}/\partial\boldsymbol{\varphi}$.

According to (7.2.20) the lhs of (7.3.2) is equal to

$$\Lambda_s s + \boldsymbol{P}(\boldsymbol{u}, \boldsymbol{s}, \boldsymbol{\varphi}, t) . \tag{7.3.4}$$

Furthermore, we express $\dot{\boldsymbol{u}}$ and $\dot{\boldsymbol{\varphi}}$ by the corresponding rhs of (7.2.19, 21). This enables (7.3.2, 4) to be simplified to

$$\left(\frac{\partial}{\partial t} - \Lambda_s + \boldsymbol{Q}\frac{\partial}{\partial \boldsymbol{u}} + \boldsymbol{R}\frac{\partial}{\partial \boldsymbol{\varphi}}\right) s = \boldsymbol{P}(\boldsymbol{u}, \boldsymbol{s}(\boldsymbol{u}, \boldsymbol{\varphi}, t), \boldsymbol{\varphi}, t) . \tag{7.3.5}$$

The bracket on the lhs represents an operator acting on s. The formal solution of (7.3.5) can be written as

$$s = \left(\frac{\partial}{\partial t} - \Lambda_s + \boldsymbol{Q}\frac{\partial}{\partial \boldsymbol{u}} + \boldsymbol{R}\frac{\partial}{\partial \boldsymbol{\varphi}}\right)^{-1} \boldsymbol{P}(\boldsymbol{u}, \boldsymbol{s}(\boldsymbol{u}, \boldsymbol{\varphi}, t), \boldsymbol{\varphi}, t) . \tag{7.3.6}$$

At this point we remember some of the results of Sect. 7.1. There we have seen that it is useful to express s as a series of inverse powers of β, to which Λ_s corresponds. As will become clear in a moment, this can be achieved by making the identifications

$$\left(\frac{\partial}{\partial t} - \Lambda_s\right) = A , \tag{7.3.7}$$

$$\boldsymbol{Q}\frac{\partial}{\partial \boldsymbol{u}} + \boldsymbol{R}\frac{\partial}{\partial \boldsymbol{\varphi}} = B \tag{7.3.8}$$

and using the formal power series expansion

$$\begin{aligned}(A+B)^{-1} &= A^{-1} - A^{-1}BA^{-1} + \dots \\ &= A^{-1}\sum_{\nu=0}^{\infty}(-BA^{-1})^{\nu} ,\end{aligned} \tag{7.3.9}$$

which is valid for operators (under suitable assumptions on A and B). In (7.3.9) we use the definition

$$(BA^{-1})^0 = 1 . \tag{7.3.10}$$

The proof of (7.3.9) will be left as an exercise at the end of this section.

Using (7.3.9) with the abbreviations (7.3.7, 8) we readily obtain

$$s = \left(\frac{\partial}{\partial t} - \Lambda_s\right)^{-1} \sum_{\nu=0}^{\infty}\left[-\left(\boldsymbol{Q}\frac{\partial}{\partial \boldsymbol{u}} + \boldsymbol{R}\frac{\partial}{\partial \boldsymbol{\varphi}}\right)\left(\frac{\partial}{\partial t} - \Lambda_s\right)^{-1}\right]^{\nu} \boldsymbol{P} . \tag{7.3.11}$$

In practical applications we will not extend the series in (7.3.11) up to infinity but rather to a finite number. Also from a more fundamental point of view the convergence of the formal series expansion is not secured. On the other hand, for practical applications we give an estimate of the rest term and we shall derive here some useful formulas. To define the rest term we decompose s according to (7.3.9)

$$s = \left[A^{-1} \sum_{\nu=0}^{m} (-BA^{-1})^{\nu} + A^{-1} \sum_{\nu=m+1}^{\infty} (-BA^{-1})^{\nu} \right] P . \quad (7.3.12)$$

One then readily finds the following decomposition

$$s = A^{-1} \sum_{\nu=0}^{m} (-BA^{-1})^{\nu} P + r , \quad (7.3.13)$$

where the rest term r is given by

$$r = (A+B)^{-1} (-BA^{-1})^{m+1} P . \quad (7.3.14)$$

(For the proof consult the exercises at the end of this section.)

For practical applications we note that

$$Q \frac{\partial}{\partial u} + R \frac{\partial}{\partial \varphi} \quad (7.3.15)$$

commutes with Λ_s. Further, Λ_s transforms the components of P, whereas (7.3.15) acts on each component in the same way. To make closer connection with the results of Sect. 7.1 dealing with an example and also to show what the inverse of the operator (7.3.7) means explicitly, we deal with s in still another way. To this end we start from (7.2.20) which possesses the formal solution

$$s = \int_{-\infty}^{t} \exp[\Lambda_s (t-\tau)] P(u, s, \varphi, \tau) d\tau , \quad \text{where} \quad (7.3.16)$$

$$s = s(u, \varphi, \tau) . \quad (7.3.17)$$

In analogy to (7.1.7) we perform a partial integration. To this end we assume that P is represented in the form (7.2.11). When performing the partial integration we make use of the fact that each term in (7.2.11) has the form

$$g(t) h(u, s, \varphi) . \quad (7.3.18)$$

We treat $g(t)$ as v and h as w in the well-known relation

$$\int v' w \, dt = v w - \int v w' \, dt . \quad (7.3.19)$$

Expressing the temporal derivatives $\dot{\boldsymbol{u}}$, $\dot{\boldsymbol{\varphi}}$ by the corresponding rhs of (7.2.19, 21), we may write (7.3.16) in the new form

$$\begin{aligned} s &= \exp(\Lambda_s t) \int_{-\infty}^{t} \exp(-\Lambda_s \tau) \boldsymbol{P}(\boldsymbol{u}(t), s(\boldsymbol{u}(t), \boldsymbol{\varphi}(t), t), \boldsymbol{\varphi}(t), \tau) d\tau \\ &- \exp(\Lambda_s t) \int_{-\infty}^{t} \left(\boldsymbol{Q} \frac{\partial}{\partial \boldsymbol{u}} + \boldsymbol{R} \frac{\partial}{\partial \boldsymbol{\varphi}} \right)_{\tau'} d\tau' \int_{-\infty}^{\tau'} \exp(-\Lambda_s \tau) \boldsymbol{P}(\boldsymbol{u}(\tau'), \dots, \tau) d\tau . \end{aligned} \tag{7.3.20}$$

Because (7.3.15) commutes with Λ_s we may insert

$$1 \equiv \exp(\Lambda_s \tau' - \Lambda_s \tau') \tag{7.3.21}$$

in the second expression of (7.3.20), eventually finding

$$\begin{aligned} s &= \int_{-\infty}^{t} \exp[\Lambda_s(t-\tau)] \boldsymbol{P}(\boldsymbol{u}(t), \dots, \tau) d\tau \\ &- \int_{-\infty}^{t} \exp[\Lambda_s(t-\tau')] \left(\boldsymbol{Q} \frac{\partial}{\partial \boldsymbol{u}} + \boldsymbol{R} \frac{\partial}{\partial \boldsymbol{\varphi}} \right)_{\tau'} d\tau' \int_{-\infty}^{\tau'} e^{\Lambda_s(\tau' - \tau)} \boldsymbol{P}(\boldsymbol{u}(\tau'), \dots, \tau) d\tau . \end{aligned} \tag{7.3.22}$$

The first part of (7.3.22) can be considered as the formal solution of the equation

$$\frac{\partial}{\partial t} s = \Lambda_s s + \boldsymbol{P}(\boldsymbol{u}, s, \boldsymbol{\varphi}, t) , \tag{7.3.23}$$

where $\boldsymbol{u}$, s, $\boldsymbol{\varphi}$ on the rhs are considered as given parameters. Under this assumption the formal solution of (7.3.23) can also be written in the form

$$s = \left(\frac{\partial}{\partial t} - \Lambda_s \right)^{-1} \boldsymbol{P} . \tag{7.3.24}$$

The first term of (7.3.22) therefore provides us with the interpretation of the inverse operator A^{-1}. In this way we may write (7.3.22) in the form

$$\begin{aligned} s &= \left(\frac{\partial}{\partial t} - \Lambda_s \right)^{-1} \boldsymbol{P}(\boldsymbol{u}(t), \dots, t) \\ &- \int_{-\infty}^{t} \exp[\Lambda_s(t-\tau')] \underbrace{\left(\boldsymbol{Q} \frac{\partial}{\partial \boldsymbol{u}} + \boldsymbol{R} \frac{\partial}{\partial \boldsymbol{\varphi}} \right)_{\tau'} \left(\frac{\partial}{\partial \tau'} - \Lambda_s \right)^{-1} \boldsymbol{P}(\boldsymbol{u}(\tau'), \dots, \tau') d\tau'}_{\boldsymbol{P}^{(1)}(\boldsymbol{u}(\tau'), \dots, \tau')} . \end{aligned} \tag{7.3.25}$$

Continuing the procedure of partial integrations, we again use the interpretation of the inverse operator A^{-1}. These subsequent partial integrations may then be written as

$$s = \left(\frac{\partial}{\partial t} - \Lambda_s\right)^{-1} P - \left(\frac{\partial}{\partial t} - \Lambda_s\right)^{-1} \left(Q\frac{\partial}{\partial u} + R\frac{\partial}{\partial \varphi}\right)\left(\frac{\partial}{\partial t} - \Lambda_s\right)^{-1} P + \dots \tag{7.3.26}$$

which evidently is the same result as before, namely (7.3.11).

Exercises. 1a) Prove the relation (7.3.9). *Hint:* multiply both sides of (7.3.9) by $A+B$ from the left and rearrange the infinite sum. 1b) Prove in the same way the following relation for the remainder [cf. (7.3.12, 14)]

$$A^{-1} \sum_{\nu=m+1}^{\infty} (-BA^{-1})^{\nu} = (A+B)^{-1} \cdot (-BA^{-1})^{m+1} . \tag{7.3.27}$$

2) Prove the relation (7.3.9) by iteration of the identity

$$\begin{aligned}(A+B)^{-1} &= (A+B)^{-1}(A+B-B)A^{-1} \\ &= A^{-1} + (A+B)^{-1}(-BA^{-1}) .\end{aligned} \tag{7.3.28}$$

7.4 The Iteration Procedure

So far we have assumed that s may be expressed as a function of u, φ and t [compare (7.3.1)] and have then derived a formal relation. In this section we want to show how we may actually construct s by an iteration procedure. To this end we use the formerly introduced small parameter δ. We introduce the approximation of s up to nth order by

$$s^{(n)}(u, \varphi, t) = \sum_{m=2}^{n} C^{(m)}(u, \varphi, t) , \tag{7.4.1}$$

where the individual terms $C^{(m)}$ are precisely of the order δ^m.

We further define the quantities

$$P^{(l)}(u, \varphi, t) \equiv P^{(l)}(u, \{s^{(k)}\}, \varphi, t): \quad \text{order } \delta^l \text{ precisely} \tag{7.4.2}$$

$$Q^{(l)}(u, \varphi, t) \equiv Q^{(l)}(u, \{s^{(k)}\}, \varphi, t): \quad \text{order } \delta^l \text{ precisely} \tag{7.4.3}$$

$$R^{(l)}(u, \varphi, t) \equiv R^{(l)}(u, \{s^{(k)}\}, \varphi, t): \quad \text{order } \delta^l \text{ precisely} \tag{7.4.4}$$

assuming that we may express s up to a certain order k as indicated in (7.4.2–4), so that just the correct final order δ^l of the corresponding expressions results.

As shown below, it is possible to construct the expressions $s^{(n)}$ consecutively. To find the correct expressions for $\boldsymbol{C}^{(n)}$ we consult the relation (7.3.11). From its rhs we select such terms which are precisely of order δ^n by decomposing $\boldsymbol{P}$ into those terms of the order of $n-m$, and those terms of the operator in front of $\boldsymbol{P}$ of order m ($m \leqslant n-2$). In this way we obtain

$$\boldsymbol{C}^{(n)}(\boldsymbol{u}, \varphi, t) = \sum_{m=0}^{n-2} \left(\frac{d}{dt} - \Lambda_s\right)^{-1}_{(m)} \boldsymbol{P}^{(n-m)}(\boldsymbol{u}, \varphi, t) \,. \tag{7.4.5}$$

A comparison with (7.3.11) shows further that the operator in front of $\boldsymbol{P}^{(n-m)}$ can be defined as

$$\left(\frac{d}{dt} - \Lambda_s\right)^{-1}_{(m)} = \left(\frac{d}{dt} - \Lambda_s\right)^{-1}_{(0)} \sum_{\substack{\text{all} \\ \text{products}}} \prod_{\substack{i \geqslant 1 \\ \sum i = m}} \left[\left(-\frac{d}{dt}\right)_{(i)} \left(\frac{d}{dt} - \Lambda_s\right)^{-1}_{(0)}\right], \tag{7.4.6}$$

where we have used the abbreviation

$$\left(\frac{d}{dt} - \Lambda_s\right)^{-1}_{(0)} \equiv \left(\frac{\partial}{\partial t} - \Lambda_s\right)^{-1}, \tag{7.4.7}$$

the relation

$$\left(-\frac{d}{dt}\right)_{(i)} = -\left(\frac{d}{dt}\right)_{(i)}, \tag{7.4.8}$$

and the definition

$$\left(\frac{d}{dt}\right)_{(i)} f(\boldsymbol{u}, \varphi, t) = \left(\boldsymbol{Q}^{(i+1)} \frac{\partial}{\partial \boldsymbol{u}} + \boldsymbol{R}^{(i)} \frac{\partial}{\partial \varphi}\right) f(\boldsymbol{u}, \varphi, t) \,. \tag{7.4.9}$$

The transition from (7.3.11) to (7.4.1) with the corresponding definitions can be easily done if we exhibit the order of each term explicitly. We have, of course, assumed that we may express $\boldsymbol{Q}^{(i+1)}$ and $\boldsymbol{R}^{(i)}$ by $\boldsymbol{u}$, φ and t. According to our previous assumptions in Sect. 7.2, we also have the relations

$$\boldsymbol{Q}^{(0)} \equiv \boldsymbol{Q}^{(1)} \equiv \boldsymbol{0} \quad \text{and} \tag{7.4.10}$$

$$\boldsymbol{R}^{(0)} \equiv \boldsymbol{R}^{(1)} \equiv \boldsymbol{0} \,. \tag{7.4.11}$$

A formula which is most useful for practical purposes, because it allows us to express subsequent $\boldsymbol{C}^{(n)}$ by the previous ones, is

$$C^{(n)} = \left(\frac{d}{dt} - \Lambda_s\right)^{-1}_{(0)} \left[P^{(n)} - \sum_{m=1}^{n-2} \left(\frac{d}{dt}\right)_{(m)} C^{(n-m)} \right]. \tag{7.4.12}$$

Its proof is of a more technical nature, so that we include it only for sake of completeness. In order to prove (7.4.12), we insert (7.4.5) on both sides of (7.4.12) which yields

$$\begin{aligned} &\sum_{m=0}^{n-2} \left(\frac{d}{dt} - \Lambda_s\right)^{-1}_{(m)} P^{(n-m)} \\ &\quad = \left(\frac{d}{dt} - \Lambda_s\right)^{-1}_{(0)} P^{(n)} - \left(\frac{d}{dt} - \Lambda_s\right)^{-1}_{(0)} \sum_{m=1}^{n-2} \left(\frac{d}{dt}\right)_{(m)} \\ &\qquad \cdot \sum_{l=0}^{n-2-m} \left(\frac{d}{dt} - \Lambda_s\right)^{-1}_{(l)} P^{(n-m-l)} . \end{aligned} \tag{7.4.13}$$

Obviously, the term with $m = 0$ cancels so that it remains to prove the relation

$$\begin{aligned} &\sum_{m=1}^{n-2} \left(\frac{d}{dt} - \Lambda_s\right)^{-1}_{(m)} P^{(n-m)} \\ &\quad = - \sum_{m'=1}^{n-2} \sum_{l=0}^{n-2-m'} \left(\frac{d}{dt} - \Lambda_s\right)^{-1}_{(0)} \left(\frac{d}{dt}\right)_{(m')} \left(\frac{d}{dt} - \Lambda_s\right)^{-1}_{(l)} P^{(n-m'-l)} . \end{aligned} \tag{7.4.14}$$

We now wish to change the sequence of summation in (7.4.14). To this end we check the domain in which the indices run according to the scheme

$$\left.\begin{array}{l} m' + l = m \\ 0 \leqslant l \leqslant n-2-m' \\ 1 \leqslant m' \leqslant n-2 \end{array}\right\} \quad m' \leqslant m \leqslant n-2 . \tag{7.4.15}$$

We then readily obtain for the rhs of (7.4.14)

$$- \sum_{m=1}^{n-2} \sum_{m'=1}^{m} \left(\frac{d}{dt} - \Lambda_s\right)^{-1}_{(0)} \left(\frac{d}{dt}\right)_{(m')} \left(\frac{d}{dt} - \Lambda_s\right)^{-1}_{(m-m')} P^{(n-m)} . \tag{7.4.16}$$

We now compare each term for fixed m on the lhs of (7.4.13) with that on the rhs and thus are led to check the validity of the relation

$$\left(\frac{d}{dt} - \Lambda_s\right)^{-1}_{(m)} = \sum_{m'=1}^{m} \left(\frac{d}{dt} - \Lambda_s\right)^{-1}_{(0)} \left(-\frac{d}{dt}\right)_{(m')} \left(\frac{d}{dt} - \Lambda_s\right)^{-1}_{(m-m')} . \tag{7.4.17}$$

For the rhs of (7.4.17) we obtain by use of definition (7.4.6)

$$\left(\frac{d}{dt}-\Lambda_s\right)^{-1}_{(0)}\sum_{m'=1}^{m}\left(-\frac{d}{dt}\right)_{(m')}\left(\frac{d}{dt}-\Lambda_s\right)^{-1}_{(0)}$$

$$\cdot\sum\prod_{\Sigma i=m-m'}\left(-\frac{d}{dt}\right)_{(i)}\left(\frac{d}{dt}-\Lambda_s\right)^{-1}_{(0)}. \tag{7.4.18}$$

This can be rewritten in the form

$$\left(\frac{d}{dt}-\Lambda_s\right)^{-1}_{(0)}\sum\prod_{\Sigma i=m}\left[\left(-\frac{d}{dt}\right)_{(i)}\left(\frac{d}{dt}-\Lambda_s\right)^{-1}_{(0)}\right] \tag{7.4.19}$$

which coincides with (7.4.6). This completes the proof of (7.4.12).

7.5 An Estimate of the Rest Term. The Question of Differentiability

As mentioned in Sect. 7.1.2, in practical applications it will be sufficient to retain only a few terms of the series expansion (7.3.11). Of course, it will then be important to know the size of the rest term. Instead of entering a full discussion of the size of this rest term, we rather present an example.

The rest term can be defined by

$$\int_{-\infty}^{t} e^{\Lambda_s(t-t')}\left[\left(\frac{d}{dt}\right)^{m}_{\infty}\boldsymbol{P}\right] \tag{7.5.1}$$

[compare (7.1.26)]. Let us consider the case in which in $\boldsymbol{P}$ only a variable u and not s occurs, and let us assume that

$$\boldsymbol{P}(u)\sim \boldsymbol{g}(t)u^k\,, \qquad \text{where} \tag{7.5.2}$$

$$|\boldsymbol{g}|<M\,, \quad u>0 \tag{7.5.3}$$

holds. Clearly

$$\left|\left(\frac{d}{dt}-\Lambda_s\right)^{-1}_{(0)}\boldsymbol{g}(t)\right|<|\gamma_i|^{-1}_{\min}M \tag{7.5.4}$$

holds provided Λ_s is diagonal and $|\gamma_i|_{\min}$ is the smallest modulus of the negative diagonal elements of Λ_s. We further assume $\boldsymbol{Q}$ in the form

$$Q \sim h(t) u^k \quad \text{with} \tag{7.5.5}$$

$$|h| < M. \tag{7.5.6}$$

We readily obtain

$$\left| \left(\frac{d}{dt} \right)_\infty P \right| < \frac{M}{|\gamma_i|_{\min}} |h(t)| k u^k u^{k-1}. \tag{7.5.7}$$

Similarly

$$\left| \left(\frac{d}{dt} \right)_\infty^2 P \right| \sim |h(t)| \left(\frac{M}{|\gamma_i|_{\min}} \right)^2 k u^k (2k-1) u^{2k-2} \tag{7.5.8}$$

and after applying $(d/dt)_\infty$ m times we obtain

$$\left(\frac{d}{dt} \right)_\infty^m P \sim |h(t)| \left(\frac{M}{|\gamma_i|_{\min}} \right)^m k(2k-1)(3k-2)\dots(mk-(m-1)) u^{(m+1)k-m}. \tag{7.5.9}$$

Since we are interested in large values of m, we may approximate rhs by

$$|h(t)| \left(\frac{M}{|\gamma_i|_{\min}} \right)^m k^m m!\, u^{(m+1)k-m}. \tag{7.5.10}$$

Similar estimates also hold for more complicated expressions for $\boldsymbol{P}$ and $\boldsymbol{Q}$ (and also for those of $\boldsymbol{R}$). The following conclusion can be drawn from these considerations. The rest term goes with some power of u where the exponent contains the factor m. Also all other factors go with powers of m. Thus, if u is chosen sufficiently small these contributions would decrease more and more. On the other hand, the occurrence of the factorial $m!$ leads to an increase of the rest term for sufficiently high m. In a strict mathematical sense we are therefore dealing with a semiconvergent series. As is well known from many practical applications in physics and other fields, such expansions can be most useful.

In this context it means that the approximation remains a very good one provided u is sufficiently small. In such a case the first terms will give an excellent approximation but when m becomes too big, taking higher-order rest terms into account will rather deteriorate the result.

In conclusion, we briefly discuss the differentiability properties of $\boldsymbol{s}(\boldsymbol{u}, \varphi, t)$ with respect to the variables $\boldsymbol{u}$ and φ. As indicated above, we can approximate $\boldsymbol{s}$ by a power series in $\boldsymbol{u}$ with φ- (and t-) dependent coefficients to any desired degree of accuracy, provided we choose δ in (7.4.1) small enough. If the rhs of (7.2.19 – 21) are polynomials (and analytic) in $\boldsymbol{u}$ and are analytic in φ in a certain domain, any of the finite approximations $\boldsymbol{s}^{(n)}$ (7.4.1) to $\boldsymbol{s}$ have the same (analyticity) property. Because of the properties of the rest term discussed above, with an increasing number of differentiations $\partial/\partial\varphi_j$, $\partial/\partial u_k$, the (differentiated) rest term may become bigger and bigger, letting the domain of $\boldsymbol{u}$, in which the approximation is valid, shrink, perhaps even to an empty set. This difficulty can be circumvented by "smoothing". Namely, we consider $\boldsymbol{s}^{(n)}$ (omitting the rest term) as a "smoothed" approximation to the manifold $\boldsymbol{s}(\boldsymbol{u}, \varphi, t)$. Then $\boldsymbol{s}^{(n)}$ possesses the just-mentioned differentiability (or analyticity) properties. If not otherwise mentioned, we shall use $\boldsymbol{s}^{(n)}$ later in this sense.

7.6 Slaving Principle for Discrete Noisy Maps*

The introduction showed that we may deal with complex systems not only by means of differential equations but also by means of maps using a discrete time sequence. The study of such maps has become a modern and most lively branch of mathematics and theoretical physics. In this section we want to show how the slaving principle can be extended to this case. Incidentally, this will allow us to derive the slaving principle for stochastic differential equations by taking an appropriate limit (Sect. 7.9).

Consider a dynamic system described by a state vector $\boldsymbol{q}_l$ which is defined at a discrete and equidistant time sequence l. The evolution of the system from one discrete time l to the next $l+1$ is described by an equation of the general form

$$\boldsymbol{q}_{l+1} = \boldsymbol{f}(\boldsymbol{q}_l, l) + \boldsymbol{G}(\boldsymbol{q}_l, l)\,\boldsymbol{\eta}_l\,, \tag{7.6.1}$$

in which $\boldsymbol{f}$ and $\boldsymbol{G}$ are nonlinear functions of $\boldsymbol{q}_l$ and which may depend explicitly on the index l. Here $\boldsymbol{\eta}_l$ is a random vector, whose probability distribution may but need not depend on the index l. In many physical cases $\boldsymbol{f}_l$ is of the order $O(\boldsymbol{q}_l)$

$$|\boldsymbol{f}(\boldsymbol{q}_l, l)| = O(\boldsymbol{q}_l)\,. \tag{7.6.2}$$

As shown by the case of differential equations at instability points where new ordered structures occur, it is possible to transform the variables to new collective modes which can be classified as so-called undamped or unstable modes $\boldsymbol{u}$, and damped or slaved modes $\boldsymbol{s}$.

We assume that we can make an analogous decomposition for discrete maps. To bring out the most important features we leave the phase variables φ aside. Thus we write the equations which are analogous to (7.2.1, 2) in the form (using the notation dl for the finite time interval)

$$\boldsymbol{u}_{l+1} - \boldsymbol{u}_l = \Lambda_u \boldsymbol{u}_l dl + d\boldsymbol{Q}(\boldsymbol{u}, \boldsymbol{s}, l) \tag{7.6.3}$$

and

$$\boldsymbol{s}_{l+1} - \boldsymbol{s}_l = \Lambda_s \boldsymbol{s}_l dl + d\boldsymbol{P}(\boldsymbol{u}, \boldsymbol{s}, l)\,. \tag{7.6.4}$$

We assume Λ_s in Jordan's normal form with negative diagonal elements. For simplicity, we also assume that Λ_u is diagonal and of smallness δ, but the procedure can easily be extended to a more general Λ_u.

Note that $d\boldsymbol{Q}$, $d\boldsymbol{P}$ may contain $\boldsymbol{u}$ and $\boldsymbol{s}$ also at retarded times $\boldsymbol{u}$, so that they are functions of $\boldsymbol{u}_l, \boldsymbol{u}_{l-1}, \dots$ and $\boldsymbol{s}_l, \boldsymbol{s}_{l-1}, \dots$ From our result it will transpire that we can also allow for a dependence of $d\boldsymbol{Q}$ and $d\boldsymbol{P}$ on future times, i.e., for instance, on $\boldsymbol{u}_{l+1}$. Our purpose is to devise a procedure by which we can express $\boldsymbol{s}$ in a unique and well-defined fashion by $\boldsymbol{u}$ and l alone. Let us therefore assume that such a replacement can be made. This allows us to establish a number of formal relations used later on.

7.7 Formal Relations*

Let us assume that $\boldsymbol{s}$ can be expressed by a function of $\boldsymbol{u}$ and l alone. This allows us to consider $d\boldsymbol{P}$ in (7.6.4) as a function of $\boldsymbol{u}$ and l alone

$$\boldsymbol{s}_{l+1} - \boldsymbol{s}_l = \Lambda_s \boldsymbol{s}_l dl + d\boldsymbol{P}(\boldsymbol{u}_l, \boldsymbol{u}_{l-1}, \dots, l)\,. \tag{7.7.1}$$

Equation (7.7.1) can be solved by

$$\boldsymbol{s}_{l+1} = \sum_{m=-\infty}^{l} (1 + \Lambda_s dm)^{l-m} d\boldsymbol{P}(\boldsymbol{u}_m, \boldsymbol{u}_{m-1}, \dots, m)\,. \tag{7.7.2}$$

We first show that (7.7.2) fulfills (7.7.1). Inserting (7.7.2) into (7.7.1) we readily obtain for the lhs

$$\begin{aligned}\boldsymbol{s}_{l+1} - \boldsymbol{s}_l = &\sum_{m=-\infty}^{l} (1 + \Lambda_s dm)^{l-m} d\boldsymbol{P}(\boldsymbol{u}_m, \boldsymbol{u}_{m-1}, \dots, m) \\ &- \sum_{m=-\infty}^{l-1} (1 + \Lambda_s dm)^{l-1-m} d\boldsymbol{P}(\boldsymbol{u}_m, \boldsymbol{u}_{m-1}, \dots, m)\,.\end{aligned} \tag{7.7.3}$$

Taking from the first sum the member with $m = l$ separately we obtain

$$\begin{aligned}\boldsymbol{s}_{l+1} - \boldsymbol{s}_l = &\, d\boldsymbol{P}(\boldsymbol{u}_l, \boldsymbol{u}_{l-1}, \dots, l) \\ &+ (1 + \Lambda_s dl) \sum_{m=-\infty}^{l-1} (1 + \Lambda_s dm)^{l-1-m} d\boldsymbol{P}(\boldsymbol{u}_m, \boldsymbol{u}_{m-1}, \dots, m) \\ &- \sum_{m=-\infty}^{l-1} (1 + \Lambda_s dm)^{l-1-m} d\boldsymbol{P}(\boldsymbol{u}_m, \boldsymbol{u}_{m-1}, \dots, m)\,.\end{aligned} \tag{7.7.4}$$

Taking the difference of the remaining sums gives

$$\begin{aligned}\boldsymbol{s}_{l+1} - \boldsymbol{s}_l = &\, d\boldsymbol{P}(\boldsymbol{u}_l, \boldsymbol{u}_{l-1}, \dots, l) \\ &+ \Lambda_s dl \sum_{m=-\infty}^{l-1} (1 + \Lambda_s dm)^{l-1-m} d\boldsymbol{P}(\boldsymbol{u}_m, \boldsymbol{u}_{m-1}, \dots, m)\,.\end{aligned} \tag{7.7.5}$$

According to its definition, the sum in the last term is [compare (7.7.2)] identical with s_l, so that rhs of (7.7.5) agrees with rhs of (7.7.1). To study the convergence of the sum over m we assume that the factor of

$$(1 + \Lambda_s dl)^{l-m} \tag{7.7.6}$$

is bounded. Because we have assumed that Λ_s is in Jordan's normal form, it suffices to study the behavior of (7.7.6) within a subspace of Jordan's decomposition. This allows us to consider

$$(1 + \lambda' + M_k dl)^l \tag{7.7.7}$$

instead of (7.7.6), where 1 is a unity matrix, $\lambda' = \lambda\, dl$ is a diagonal matrix and M_k is a $k \times k$ matrix given by

$$M_k = \begin{pmatrix} 0 & 1 & & & \\ & 0 & 1 & & 0 \\ & & 0 & 1 & \\ & 0 & & \ddots & 1 \\ & & & & 0 \end{pmatrix}. \tag{7.7.8}$$

In its first diagonal parallel to the main diagonal, M_k is equal to 1, whereas all other elements are 0. We treat the case that $l > 0$ (and is, of course, an integer). According to the binomial law

$$(1 + \lambda' + M_k dl)^l = \sum_{\nu=0}^{l} \binom{l}{\nu} (1 + \lambda')^{l-\nu} M_k^\nu (dl)^\nu . \tag{7.7.9}$$

As one can readily verify

$$M_k^\nu = 0 \quad \text{for} \quad \nu \geqslant k . \tag{7.7.10}$$

This allows us to replace (7.7.9) by

$$\sum_{\nu=0}^{k-1} \binom{l}{\nu} (1 + \lambda')^{l-\nu} M_k^\nu (dl)^\nu . \tag{7.7.11}$$

Because $\nu \leqslant k$ is bounded, the size of the individual elements of (7.7.11) can be estimated by treating $(1 + \lambda')^{l-\nu}$, or equivalently, $(1 + \lambda')^l$. The absolute value of this quantity tends to 0, for l to infinity, i.e.,

$$|(1 + \lambda')|^l \to 0 \quad \text{for} \quad l \to \infty , \tag{7.7.12}$$

provided

$$|(1 + \lambda')| < 1 . \tag{7.7.13}$$

Decomposing λ' into its real and imaginary parts λ_r and λ_i, respectively, we arrive at the necessary and sufficient condition

$$(1+\lambda_r')^2+\lambda_i'^2<1\,. \tag{7.7.14}$$

Equation (7.7.14) can be fulfilled only if

$$\lambda_r<0 \tag{7.7.15}$$

holds. Throughout this chapter we assume that conditions (7.7.14, 15) are fulfilled.

In order to establish formal relations we introduce the abbreviation

$$\Delta_- s_{l+1}=s_{l+1}-s_l \tag{7.7.16}$$

by which we may express s_l by means of s_{l+1}

$$s_l=(1-\Delta_-)s_{l+1}\,. \tag{7.7.17}$$

This allows us to write (7.7.1) in the form

$$(\Delta_- - \Lambda_s dl(1-\Delta_-))s_{l+1}=d\boldsymbol{P}(\boldsymbol{u}_l,l)\,. \tag{7.7.18}$$

The reader is remined that $d\boldsymbol{P}$ may be a function of $\boldsymbol{u}$ at various time indices l, $l-1, \ldots$

$$d\boldsymbol{P}(\boldsymbol{u}_l,l)=d\boldsymbol{P}(\boldsymbol{u}_l,\boldsymbol{u}_{l-1},\ldots,l)\,. \tag{7.7.19}$$

The formal solution of (7.7.18) reads

$$s_{l+1}=(\Delta_-(1+\Lambda_s dl)-\Lambda_s dl)^{-1}d\boldsymbol{P}(\boldsymbol{u}_l,l)\,. \tag{7.7.20}$$

A comparison of this solution with the former solution (7.7.2) yields

$$(\Delta_-(1+\Lambda_s dl)-\Lambda_s dl)^{-1}d\boldsymbol{P}(\boldsymbol{u}_l,l)=\sum_{m=-\infty}^{l}(1+\Lambda_s dm)^{l-m}d\boldsymbol{P}(\boldsymbol{u}_m,m)\,, \tag{7.7.21}$$

which may serve as a definition of the operator on lhs of (7.7.21). We now introduce the decomposition

$$\Delta_-=\Delta_-^{(l)}+\Delta_-^{(u)}T_-^{(l)} \tag{7.7.22}$$

where $\Delta_-^{(l)}$, $\Delta_-^{(u)}$, $T_-^{(l)}$ operate as follows:

$$\Delta_-^{(l)}f(\boldsymbol{u}_l,l)=f(\boldsymbol{u}_l,l)-f(\boldsymbol{u}_l,l-1)\,, \tag{7.7.23}$$

$$\Delta_-^{(u)} f(\boldsymbol{u}_l, l) = f(\boldsymbol{u}_l, l) - f(\boldsymbol{u}_{l-1}, l) , \tag{7.7.24}$$

$$T_-^{(l)} f(\boldsymbol{u}_l, l) \; = f(\boldsymbol{u}_l, l-1) . \tag{7.7.25}$$

Using operator techniques given in (7.3.9), one readily establishes the following identity

$$\begin{aligned} \{\Delta_-(1+\Lambda_s dl) - \Lambda_s dl\}^{-1} &= \{\Delta_-^{(l)}(1+\Lambda_s dl) - \Lambda_s dl\}^{-1} \\ &\quad - \{\Delta_-(1+\Lambda_s dl) - \Lambda_s dl\}^{-1} \cdot [\cdots] , \end{aligned} \tag{7.7.26}$$

where the square bracket represents an abbreviation defined by

$$[\cdots] = (1+\Lambda_s dl)\Delta_-^{(u)} T_-^{(l)} \cdot \{\Delta_-^{(l)}(1+\Lambda_s dl) - \Lambda_s dl\}^{-1} . \tag{7.7.27}$$

Because of (7.7.21), we may expect that the curly brackets in (7.7.27) possess an analog to the explicit form of rhs of (7.7.21). This is indeed the case.

Let us assume for the moment that we may decompose $d\boldsymbol{P}$ into a sum of expressions of the form [compare (7.3.18)]

$$h(\boldsymbol{u}_m, \boldsymbol{u}_{m-1}, \ldots) d\boldsymbol{g}(m) , \tag{7.7.28}$$

where h is a function of the variables $\boldsymbol{u}$ alone, but does not depend explicitly on m, whereas $d\boldsymbol{g}$ is a vector which depends on m alone. It is then simple to verify the following relation

$$\begin{aligned} &\sum_{m=-\infty}^{l} (1+\Lambda_s dm)^{l-m} h(\boldsymbol{u}_m, \boldsymbol{u}_{m-1}, \ldots) d\boldsymbol{g}(m) \\ &\quad = h(\boldsymbol{u}_l, \boldsymbol{u}_{l-1}, \ldots) \sum_{m=-\infty}^{l} (1+\Lambda_s dm)^{l-m} d\boldsymbol{g}(m) \\ &\qquad - \sum_{m=-\infty}^{l} (1+\Lambda_s dm)^{l+1-m} \sum_{m'=-\infty}^{m-1} (1+\Lambda_s dm)^{m-1-m'} \{\cdots\}_{m,m'} \end{aligned} \tag{7.7.29}$$

where the brace is defined by

$$\{\cdots\}_{m,m'} - \Delta_- h(\boldsymbol{u}_m, \boldsymbol{u}_{m-1}, \ldots) d\boldsymbol{g}(m') . \tag{7.7.30}$$

The essence of (7.7.29) can be expressed as follows. While lhs contains $\boldsymbol{u}$ at all previous times, the first term on rhs contains $\boldsymbol{u}$ only at those earlier times which are initially present in $h(\boldsymbol{u}_l, \boldsymbol{u}_{l-1}, \ldots)$. Thus, instead of an infinite regression we have only a finite regression. From a comparison of (7.7.26) with (7.7.29) we may derive the following definition

$$\begin{aligned} &[\Delta_-^{(l)}(1+\Lambda_s dl) - \Lambda_s dl]^{-1} d\boldsymbol{P}(\boldsymbol{u}_l, \boldsymbol{u}_{l-1}, \ldots, l) \\ &\quad = \sum_{m=-\infty}^{l} (1+\Lambda_s dm)^{l-m} d\boldsymbol{P}(\boldsymbol{u}_l, \boldsymbol{u}_{l-1}, \ldots, m) , \end{aligned} \tag{7.7.31}$$

where on rhs $\boldsymbol{u}_l$, $\boldsymbol{u}_{l-1}$, etc., are kept fixed and the summation runs only over m. Equation (7.7.26) with (7.7.27) may be iterated, giving

$$\{\Delta_-(1+\Lambda_s dl)-\Lambda_s dl\}^{-1}=\{\Delta_-^{(l)}(1+\Lambda_s dl)-\Lambda_s dl\}^{-1}\cdot(\cdots) \tag{7.7.32}$$

where the parentheses on rhs are given by

$$(\cdots)=\sum_{\nu=0}^{\infty}[\cdots]^{\nu} \tag{7.7.33}$$

and these square brackets are defined in (7.7.27).

As already noted, in practical cases one does not extend the summation to infinity because it is known from a number of applications that few terms suffice. Furthermore, the convergence of the whole series (7.7.33) need not hold so that we are dealing here with semiconvergent series. Therefore it is important to give an estimate for the size of the rest term, provided we take only a finite sum. The rest term for rhs of (7.7.32) can be derived from the formula

$$\begin{aligned}\{\cdots\}^{-1}\sum_{\nu=n+1}^{\infty}[\cdots]^{\nu}&=\{\cdots\}^{-1}\sum_{\nu=0}^{\infty}[\cdots]^{\nu}[\cdots]^{n+1}\\&=\{\cdots\}^{-1}(1-[\cdots])^{-1}[\cdots]^{n+1}.\end{aligned} \tag{7.7.34}$$

Since a more explicit discussion of the rest term entails many mathematical details going far beyond the scope of this book, we shall skip the details here.

While the operator containing $\Delta_-^{(l)}$ is defined by (7.7.31), we have still to explain in more detail how to evaluate the operator $\Delta_-^{(u)}$. It has the following properties:

1) It acts only on $\boldsymbol{u}_l, \boldsymbol{u}_{l-1}, \ldots$.

2) For any function f of $\boldsymbol{u}_l$ it is defined by (7.7.24)

$$\Delta_-^{(u)}f(\boldsymbol{u}_l)=f(\boldsymbol{u}_l)-f(\boldsymbol{u}_{l-1}). \tag{7.7.35}$$

If f contains variables at several "times" we may write

$$\Delta_-^{(u)}f(\{\boldsymbol{u}_l\})=f(\{\boldsymbol{u}_l\})-f(\{\boldsymbol{u}_{l-1}\}),\qquad\text{where} \tag{7.7.36}$$

$$\{\boldsymbol{u}_l\}=(\boldsymbol{u}_l,\boldsymbol{u}_{l-1},\ldots) \tag{7.7.37}$$

denotes the set of variables containing different indices l.

3) For the product of the functions v and w we readily obtain

$$\Delta_-^{(u)}(v(\{\boldsymbol{u}_l\})\,w(\{\boldsymbol{u}_l\}))=w(\{\boldsymbol{u}_l\})\,\Delta_-^{(u)}v(\{\boldsymbol{u}_l\})+v(\{\boldsymbol{u}_{l-1}\})\,\Delta_-^{(u)}w(\{\boldsymbol{u}_l\}). \tag{7.7.38}$$

In the following we shall assume that the functions f can be considered as polynomials.

4) In order to treat (7.7.36) still more explicitly we use (7.7.38) and the following rules

$$\Delta_-^{(u)} u_l^n = u_l^n - u_{l-1}^n = \left(\sum_{\nu_1+\nu_2=n-1} u_l^{\nu_1} u_{l-1}^{\nu_2} \right) \Delta_- u_l . \tag{7.7.39}$$

The rhs may be written in the form

$$\left(\mathcal{S} \frac{\partial u_l^n}{\partial u_l} \right) \cdot \Delta_- u_l , \tag{7.7.40}$$

where the factor and parentheses can be considered as a symmetrized derivative, which is identical with the parentheses in (7.7.39).

5) From relations (7.7.39, 40) we readily derive for an arbitrary function of $f(u)$ alone

$$\Delta_-^{(u)} f(u_l) = \mathcal{S} \frac{\partial f(u_l)}{\partial u_l} \cdot \Delta_-^{(u)} u_l , \tag{7.7.41}$$

or, more generally,

$$\Delta_-^{(u)} f(\boldsymbol{u}_l, \ldots, \boldsymbol{u}_{l'}) = \sum_{\nu=0}^{l-l'} g_\nu(\boldsymbol{u}_l, \ldots, \boldsymbol{u}_{l'-1}) \Delta_- \boldsymbol{u}_{l'+\nu} . \tag{7.7.42}$$

We may now replace $\Delta_- \boldsymbol{u}_{l'+\nu}$ according to (7.6.3) by

$$\Delta_-^{(u)} \boldsymbol{u}_{l'+\nu} = \Lambda_u dl \boldsymbol{u}_{l'+\nu} + d\boldsymbol{Q}(\ldots, l' + \nu - 1) , \tag{7.7.43}$$

where we abbreviate the rhs by

$$d\bar{\boldsymbol{Q}}(\ldots, l' + \nu) . \tag{7.7.44}$$

This gives (7.7.42) in its final form

$$\Delta_-^{(u)} f(\boldsymbol{u}_l, \ldots, \boldsymbol{u}_{l'}) = \sum_{\nu=0}^{l-l'} g_\nu(\boldsymbol{u}_l, \ldots, \boldsymbol{u}_{l'-1}) d\bar{\boldsymbol{Q}}(\ldots, l' + \nu) . \tag{7.7.45}$$

For sake of completeness we derive a formula for

$$\Delta_-^{(u)} \sum_{\nu_0+\ldots+\nu_\varkappa=n} u_{m+\varkappa+1}^{\nu_\varkappa} \ldots u_{m+1}^{\nu_0} , \tag{7.7.46}$$

which, according to the definition (7.7.35), can be written as

$$\sum_{\nu_0+\ldots+\nu_\varkappa=n} (u_{m+\varkappa+1}^{\nu_\varkappa} \ldots u_{m+1}^{\nu_0} - u_{m+\varkappa}^{\nu_\varkappa} \ldots u_m^{\nu_0}) . \tag{7.7.47}$$

It can be rearranged to

$$\sum u_{m+\varkappa}^{\nu_\varkappa} u_{m+\varkappa-1}^{\nu_{\varkappa-1}} \dots u_m^{\nu_1} (u_{m+\varkappa+1}^{\nu_0} - u_m^{\nu_0}) \,, \tag{7.7.48}$$

where the parentheses can be written in the form

$$\sum u_{m+\varkappa+1}^{\nu_0-1-\lambda} u_m^{\lambda} (u_{m+\varkappa+1} - u_m) \,. \tag{7.7.49}$$

We therefore find for (7.7.46)

$$\sum_{\nu_0+\dots+\nu_{\varkappa+1}=n-1} u_{m+\varkappa+1}^{\nu_{\varkappa+1}} \dots u_m^{\nu_0} (u_{m+\varkappa+1} - u_m) \,, \tag{7.7.50}$$

where we may write

$$\begin{aligned} (u_{m+\varkappa+1} - u_m) &= u_{m+\varkappa+1} - u_{m+\varkappa} + u_{m+\varkappa} - \dots + u_{m+1} - u_m \\ &= \sum_{\nu=0}^{\varkappa} \Delta_- u_{m+1+\nu} \,. \end{aligned} \tag{7.7.51}$$

The sum in (7.7.50) in front of the parenthesis can again be considered as the symmetrized derivative, while the parentheses in (7.7.50) can be expressed by individual differences $u_{k+1} - u_k$.

7.8 The Iteration Procedure for the Discrete Case*

Taking all the above formulas together we obtain again a well-defined procedure for calculating s as a function of u and l, provided dP is prescribed as a function of u and l alone. In practice, however, dP depends on s. Therefore we must devise a procedure by which we may express s by u and l stepwise by an inductive process.

To this end we introduce a smallness parameter δ. In general, dQ and dP contain a nonstochastic and a stochastic part according to

$$dQ = Q_0(u,s,l)\,dl + dF_u(u,s,l) \,, \tag{7.8.1}$$

$$dP = P_0(u,s,l)\,dl + dF_s(u,s,l) \,, \tag{7.8.2}$$

where for the stochastic part the following decomposition is assumed

$$dF_i(u,s,l) = M_i(u_l, u_{l-1}, \dots, s_l, s_{l-1}, \dots)\,dF_0(l) \,, \quad i = u,s \,. \tag{7.8.3}$$

We assume that Λ_u in (7.6.3) is of the order δ and that the functions occurring in (7.8.1, 2) can be expressed as polynomials of u and s with l-dependent coefficients. The coefficients are either continuous functions (contained in Q_0 and P_0) or quantities describing a Wiener process [in $dF_0(l)$] which is defined by

(4.2.2, 3). We assume that the constant terms which are independent of $\boldsymbol{u}$ and $\boldsymbol{s}$ are of smallness δ^2, the coefficients of the linear terms of smallness δ, while $\boldsymbol{u}$ and $\boldsymbol{s}$ are of order δ, δ^2, respectively. In order to devise an iteration procedure we represent $\boldsymbol{s}$ in the form

$$\boldsymbol{s}(\boldsymbol{u}, l) = \sum_{k=2}^{\infty} \boldsymbol{C}^{(k)}(\boldsymbol{u}, l)\,, \tag{7.8.4}$$

where $\boldsymbol{C}^{(k)}$ is a term which contains expressions of precisely order k

$$\boldsymbol{C}^{(k)} \propto \delta^k\,. \tag{7.8.5}$$

Similarly, we introduce the decompositions

$$d\boldsymbol{Q} = \sum_{k=2}^{\infty} d\boldsymbol{Q}^{(k)} \qquad \text{and} \tag{7.8.6}$$

$$d\boldsymbol{P} = \sum_{k=2}^{\infty} d\boldsymbol{P}^{(k)}\,. \tag{7.8.7}$$

We now proceed in two steps. We apply (7.7.32) with (7.7.33 and 27) on $d\boldsymbol{P}$, but on both sides we take only the term of the order δ^k.

Starting from (7.7.45) we define $g_\nu^{(k-k')}$ as a function which is precisely of order $\delta^{k-k'}$ and $d\bar{\boldsymbol{Q}}^{(k')}$ as a function being precisely of order k'. We put

$$\underline{\varDelta}^{(u)(k)}\boldsymbol{f} = \sum_{k'=0}^{k} \sum_{\nu=0}^{l-l'} g_\nu^{(k-k')}(\boldsymbol{u}_l, \ldots, \boldsymbol{u}_{l'-1})\, d\bar{\boldsymbol{Q}}^{(k')}(\ldots, l'+\nu)\,, \tag{7.8.8}$$

which is obviously of order δ^k. After these preparations the final formula is given by

$$\boldsymbol{C}^{(k)} = \{\underline{\varDelta}^{(l)}(1 + \Lambda_s dl) - \Lambda_s dl\}^{-1} \sum_{k_1+\ldots+k_r=k'} \prod_{i=1}^{\nu} [\cdots]_{k_i} d\boldsymbol{P}^{(k-k')} \tag{7.8.9}$$

together with

$$[\cdots]_{k_i} = -(1 + \Lambda_s dm)\underline{\varDelta}^{(u)(k_i)} \underline{T}^{(l)} \cdot \{\underline{\varDelta}^{(l)}(1 + \Lambda_s dl) - \Lambda_s dl\}^{-1}\,. \tag{7.8.10}$$

This formula allows us to express $\boldsymbol{s}$ by $\boldsymbol{u}$ and l, where $\boldsymbol{u}$ is taken at l and a finite number of retarded time indices. In order to elucidate the above procedure it is worthwhile to consider special cases.

1) If $d\boldsymbol{Q}$ and $d\boldsymbol{P}$ do not explicitly depend on time and there are no fluctuating forces, (7.8.9) reduces to

$$\boldsymbol{C}^{(k)} = \{-\Lambda_s dl\} \sum_{k_1+\ldots+k_r=k'} \prod_{i=1}^{\nu} [\cdots]_{k_i} \boldsymbol{P}_0^{(k-k')} dl\,, \tag{7.8.11}$$

where

$$[\cdots]_{k_i} = -(1+\Lambda_s dm)\Delta_-^{(u)(k_i)} T_-^{(l)} \{-\Lambda_s dl\}^{-1} . \tag{7.8.12}$$

2) If l becomes a continuous variable and there are no fluctuating forces we refer to the results of Sect. 7.4. For convenience we put $dl = \tau$. Using the notation of that section we immediately find

$$\{\Delta_-(1+\Lambda_s dl) - \Lambda_s dl\}^{-1} \rightarrow \frac{1}{\tau}\left(\frac{d}{dt} - \Lambda_s + O(\tau)\right)^{-1}_{(0)} \quad \text{and} \tag{7.8.13}$$

$$(1+\Lambda_s dl)\Delta_-^{(u)(k_i)} T_-^{(l)} \rightarrow \tau\left(\dot{\boldsymbol{u}}^{k_i+1}\frac{\partial}{\partial \boldsymbol{u}}\right) = \tau\left(\frac{d}{dt}\right)_{(k)_i} . \tag{7.8.14}$$

In the limit $\tau \rightarrow 0$ we finally obtain

$$\boldsymbol{C}^{(k)} = \left(\frac{d}{dt} - \Lambda_s\right)^{-1}_{(0)} \sum_{k_1+\ldots+k_r=k'} \prod_{i=1}^{\nu} \left[-\left(\frac{d}{dt}\right)_{(k_i)} \left(\frac{d}{dt} - \Lambda_s\right)^{-1}_{(0)}\right] \boldsymbol{P}_0^{(k-k')} . \tag{7.8.15}$$

7.9 Slaving Principle for Stochastic Differential Equations*

We are now in a position to derive the corresponding formulas for stochastic differential equations. To introduce the latter, we consider

$$d\boldsymbol{u} = \Lambda_u \boldsymbol{u}\, dt + d\boldsymbol{Q}(\boldsymbol{u},\boldsymbol{s},t) , \tag{7.9.1}$$

$$d\boldsymbol{s} = \Lambda_s \boldsymbol{s}\, dt + d\boldsymbol{P}(\boldsymbol{u},\boldsymbol{s},t) . \tag{7.9.2}$$

Although it is not difficult to include the equations for $\boldsymbol{\varphi}$ [cf. Sects. 7.2, 4] and a $\boldsymbol{\varphi}$-dependent $d\boldsymbol{Q}$ and $d\boldsymbol{P}$ in our analysis, to simplify our presentation they will be omitted.

Because of the Wiener process we must be careful when taking the limit to a continuous time sequence. Because of the jumps which take place at continuous times we must carefully distinguish between precisely equal times and time differences which are infinitesimally small. The fundamental regression relation to be used stems from (7.7.29) and can be written in the form

$$\begin{aligned}
&\int_{-\infty}^{t} \exp[\Lambda_s(t-\tau)]\, h(\boldsymbol{u}_\tau, \boldsymbol{u}_{\tau-dt}, \boldsymbol{u}_{\tau-2dt,\ldots})\, d\boldsymbol{g}(\tau) \\
&\quad = \int_{-\infty}^{t} \exp[\Lambda_s(t-\tau)]\, d\boldsymbol{g}(\tau)\, h(\boldsymbol{u}_t, \boldsymbol{u}_{t-dt}, \ldots) \\
&\qquad - \int_{-\infty}^{t} \exp[\Lambda_s(t-\tau')]\, d_-h(\boldsymbol{u}_\tau, \boldsymbol{u}_{\tau-dt}, \ldots) \int_{-\infty}^{\tau-dt} \exp[\Lambda_s(\tau-\tau')\, d\boldsymbol{g}(\tau')
\end{aligned} \tag{7.9.3}$$

with

$$d_- h(\boldsymbol{u}_\tau, \boldsymbol{u}_{\tau-dt}, \dots) = h(\boldsymbol{u}_\tau, \boldsymbol{u}_{\tau-dt}, \dots) - h(\boldsymbol{u}_{\tau-dt}, \boldsymbol{u}_{\tau-2dt}, \dots) . \tag{7.9.4}$$

We shall first discuss (7.9.3, 4) in detail, allowing us eventually to translate relations (7.7.26 – 33) to the present case.

For the first integral on the rhs of (7.9.3), we consider whether we may replace $\boldsymbol{u}_{t-dt}$, $\boldsymbol{u}_{t-2dt}$, etc., by $\boldsymbol{u}_t$.

To this end we replace firstly $\boldsymbol{u}_{t-dt}$ by $\boldsymbol{u}_t$ and compensate for this change by adding the corresponding two other terms. We thus obtain

$$\begin{aligned} &\int_{-\infty}^{t} \exp[\Lambda_s(t-\tau)]\, dg(\tau)\, h(\boldsymbol{u}_t, \boldsymbol{u}_{t-dt}, \boldsymbol{u}_{t-2dt}, \dots) \\ &\quad = \int_{-\infty}^{t} \exp[\Lambda_s(t-\tau)]\, dg(\tau)\, h(\boldsymbol{u}_t, \boldsymbol{u}_t, \boldsymbol{u}_{t-2dt}, \dots) \\ &\qquad - \int_{-\infty}^{t} \exp[\Lambda_s(t-\tau)]\, dg(\tau)\, [h(\boldsymbol{u}_t, \boldsymbol{u}_t, \boldsymbol{u}_{t-2dt}, \dots) \\ &\qquad - h(\boldsymbol{u}_t, \boldsymbol{u}_{t-dt}, \boldsymbol{u}_{t-2dt}, \dots)] . \end{aligned} \tag{7.9.5}$$

The difference on rhs of (7.9.5) can be formally written as

$$d_{\boldsymbol{u}_{t-dt}} h(\boldsymbol{u}_t, \boldsymbol{u}_{t-dt}, \dots) . \tag{7.9.6}$$

We assume the fluctuating forces in the form

$$d\boldsymbol{F}_u(\boldsymbol{u}, \boldsymbol{s}, t) = \boldsymbol{F}_{u,i}(\boldsymbol{u}, \boldsymbol{s}, t)\, dw_i(t) , \tag{7.9.7}$$

where we shall assume the Îto-calculus, the convention to sum over indices which occur twice, and that w_i describes a Wiener process.

It is not difficult to include in our approach the case where $\boldsymbol{F}_{u,i}$ depends on $\boldsymbol{u}$ and $\boldsymbol{s}$ at previous times and on time integrals over dw_j over previous times (see below). We shall further assume that $dw_i\, dw_k$ converges in probability according to

$$dw_i(t)\, dw_k(t) = dt\, \delta_{ik} . \tag{7.9.8}$$

We use the Îto transformation rule for a function $\varphi(\boldsymbol{u})$ according to which

$$d\varphi(\boldsymbol{u}) = dt \left[\frac{\partial \varphi}{\partial u_k} Q_{0k}(\boldsymbol{u}, \boldsymbol{s}, t) + \frac{1}{2} \frac{\partial^2 \varphi}{\partial u_i \partial u_k} F_{u_i,p} F_{u_k,p} \right] + \frac{\partial \varphi}{\partial u_k} F_{u_k,m}\, dw_m . \tag{7.9.9}$$

With help of this relation, (7.9.6) acquires the form

$$d_{u_{t-dt}} h(u_t, u_{t-dt}, \dots) = \left\{ \frac{\partial h}{\partial u_{i,t-dt}} Q_{0k}(u_{t-dt}, s_{t-dt}, t-dt) + \frac{1}{2} \frac{\partial^2 h}{\partial u_{i,t-dt} u_{k,t-dt}} F_{u_i,p} F_{u_k,p} \right\} dt + \frac{\partial h}{\partial u_{k,t-dt}} F_{u_k,m}(t-dt) \, dw_m(t-dt) \, . \quad (7.9.10)$$

In the next step the expression on rhs of (7.9.5) will be inserted for the stable mode into the equation for the unstable mode. We now have to distinguish between two cases:

a) s occurs in

$$Q_0(u, s, t) \, dt \, . \quad (7.9.11)$$

Because (7.9.10) gives rise to terms of order $O(dt)$, $O(dw)$, we obtain terms of the form dt^2 or $dt \, dw$, which can both be neglected compared to dt.

b) s occurs in

$$F_{u,i}(u(t), s(t), t) \, dw_i(t) \, , \quad (7.9.12)$$

leading to terms $dt \, dw$ which tend to 0 more strongly than dt, and to

$$dw_i(t) \, dw_m(t-dt) \to 0 \, . \quad (7.9.13)$$

These considerations show that

$$h(u_t, u_{t-dt}, \dots) \to h(u_t) \, . \quad (7.9.14)$$

We may continue in this way which proves that we may replace everywhere u_{t-kdt}, $k = 1, 2, \dots$, by u_t.

We now consider the second term on rhs of (7.9.3), i.e.,

$$\int_{-\infty}^{t} \exp[\Lambda_s(t-\tau)] \, d_- h(u_\tau, u_{\tau-1}, \dots) \int_{-\infty}^{\tau-dt} \exp[\Lambda_s(\tau-\tau')] \, dg(\tau') \, . \quad (7.9.15)$$

When we use the definition of d_- of (7.9.4), we may write (7.9.15) in the form

$$\int_{-\infty}^{t} \exp[\Lambda_s(t-\tau)] [h(u_\tau, u_{\tau-dt}, \dots) - h(u_{\tau-dt}, u_{\tau-2dt}, \dots)] \cdot \int_{-\infty}^{\tau-dt} \exp[\Lambda_s(\tau-\tau')] \, dg(\tau') \, . \quad (7.9.16)$$

Using

$$du_i = Q_{0i}(\boldsymbol{u}_t, \boldsymbol{u}_{t-dt}, \ldots, t)\,dt + \underbrace{F_{u_i,p}(\boldsymbol{u}_t, u_{t-dt}, t)\,dw_p}_{dF_{u_i}} \tag{7.9.17}$$

and the Îto rule, we cast (7.9.16) into the form

$$\begin{aligned}
&\int_{-\infty}^{t} \exp[\Lambda_s(t-\tau)] \left\{ \sum_{n=1}^{k} \frac{\partial h(\boldsymbol{u}_{\tau-dt}, \boldsymbol{u}_{\tau-2d\tau}, \ldots)}{\partial u_{i,\tau-ndt}} [Q_{0i}(\boldsymbol{u}_{\tau-ndt}, \ldots, \tau-ndt)\,d\tau \right.\\
&\quad + F_{u_i,p}(\boldsymbol{u}_{\tau-ndt}, \ldots, \tau-ndt)\,dw_p(\tau-ndt)]\\
&\quad + \frac{1}{2}\sum_{n=1}^{k} \frac{\partial^2 h(\boldsymbol{u}_{\tau-dt}, \boldsymbol{u}_{\tau-2dt}, \ldots)}{\partial u_{i,\tau-ndt}\,\partial u_{k,\tau-ndt}} F_{u_i,p}(\boldsymbol{u}_{\tau-ndt}, \ldots)\,F_{u_k,p}(\boldsymbol{u}_{\tau-ndt}, \ldots)\,d\tau\\
&\quad \times \int_{-\infty}^{\tau-dt} \exp[\Lambda_s(\tau-\tau')]\,d\boldsymbol{g}(\tau')\,.
\end{aligned} \tag{7.9.18}$$

In the first term of (7.9.18), when we replace $\boldsymbol{u}_{\tau-nd\tau}$, $n = 2, 3, \ldots,$ by $\boldsymbol{u}_{\tau-dt}$ we obtain correction terms which can be rigorously neglected because all expressions must still be multiplied by $d\tau$. This allows us to replace h which depends on $\boldsymbol{u}$'s at different times by a new function which depends only on $\boldsymbol{u}$ at time $\tau-d\tau$. Denoting this function by $h(\{\boldsymbol{u}\})$, the first term of (7.9.18) acquires the form

$$\int_{-\infty}^{t} \exp[\Lambda_s(t-\tau)] \frac{\partial h(\{\boldsymbol{u}\})}{\partial u_i} Q_{0i}(\{\boldsymbol{u}\})\,d\tau \int_{-\infty}^{\tau} \exp[\Lambda_s(\tau-\tau')\,d\boldsymbol{g}(\tau')\,. \tag{7.9.19}$$

Because we shall use the last sum in the curly brackets in the same form as it occurs in (7.9.19), we have to deal only with the second term containing $\boldsymbol{F}$ linearly.

We now study this term

$$\begin{aligned}
&\int_{-\infty}^{t} \exp[\Lambda_s(t-\tau)] \sum_{n=1}^{k} \frac{\partial h(\boldsymbol{u}_{\tau-dt}, \boldsymbol{u}_{\tau-2dt}, \ldots)}{\partial u_{i,\tau-ndt}} F_{u_i,p}(\boldsymbol{u}_{\tau-ndt}, \ldots)\,dw_p(\tau-ndt)\\
&\quad \times \int_{-\infty}^{\tau-dt} \exp[\Lambda_s(\tau-\tau')]\,d\boldsymbol{g}(\tau')\,.
\end{aligned} \tag{7.9.20}$$

Our goal is to replace the $\boldsymbol{u}$'s occurring in h with different time arguments by $\boldsymbol{u}$'s which depend on a single time argument only. All $\boldsymbol{u}$'s standing in front of $\boldsymbol{u}_{\tau-nt}$ can be replaced by $\boldsymbol{u}_{\tau-ndt}$, and similarly all $\boldsymbol{u}$'s following $\boldsymbol{u}_{\tau-ndt}$ can be replaced by $\boldsymbol{u}_{\tau-(n+1)dt}$. Similarly the arguments in $\boldsymbol{F}$ can be replaced by $\boldsymbol{u}_{\tau-ndt}$. These substitutions give contributions of higher order and can therefore be rigorously neglected in the limit $dt \to 0$. In this way (7.9.20) is transformed into

$$\int_{-\infty}^{t} \exp[\Lambda_s(t-\tau)] \sum_{n=1}^{k} \frac{\partial h(\{\boldsymbol{u}_{\tau-(n-1)dt}\}, \boldsymbol{u}_{\tau-ndt}, \{\boldsymbol{u}_{\tau-(n+1)dt}\})}{\partial u_{i,\tau-ndt}} F_{u_i,p}(\{\boldsymbol{u}_{\tau-ndt}\})$$

$$\cdot\, dw_p(\tau-ndt) \int_{-\infty}^{\tau-dt} \exp[\Lambda_s(\tau-\tau')]\, dg(\tau') + O(d\tau) + O(dw)\,. \qquad (7.9.21)$$

The next step is to replace the argument $\boldsymbol{u}_{\tau-(n-1)dt}$ by $\boldsymbol{u}_{\tau-(n+1)dt}$. In order to correct for this substitution we encounter contributions containing dw_p at time $\tau-ndt$. This dw_p taken together with another dw_p occurring in (7.9.20) will give rise to terms proportional to $d\tau$. These terms must be carried along our calculation if they occur under an integral. Carrying out these calculations explicitly, we may cast (7.9.20) into the form

$$\int_{-\infty}^{t} \exp[\Lambda_s(t-\tau)] \sum_{n=1}^{k} \left[\frac{\partial h(\boldsymbol{u}_{\tau-ndt}, \{\boldsymbol{u}_{\tau-(n+1)dt}\})}{\partial u_{i,\tau-ndt}} F_{u_i,p}(\{\boldsymbol{u}_{\tau-ndt}\})\, dw_p(\tau-ndt) \right.$$

$$\left. + \sum_{m=1}^{n-1} \frac{\partial^2 h(\boldsymbol{u}_{\tau-dt}, \boldsymbol{u}_{\tau-2dt} \dots)}{\partial u_{k,\tau-mdt}\, \partial u_{i,\tau-ndt}} F_{u_i,p}(\{\boldsymbol{u}_{\tau-ndt}\}) F_{u_k,p}(\{\boldsymbol{u}_{\tau-mdt}\})\, d\tau \right]$$

$$\times \int_{-\infty}^{\tau-dt} \exp[\Lambda_s(\tau-\tau')]\, dg(\tau')\,. \qquad (7.9.22)$$

We are now in a position to replace the arguments of h by a single argument $\boldsymbol{u}_{\tau-dt}$. The second term in (7.9.22) remains the same so that we obtain

$$\int_{-\infty}^{t} \exp[\Lambda_s(t-\tau)] \left\{ \frac{\partial h(\{\boldsymbol{u}_{\tau-dt}\})}{\partial u_i} F_{u_i,p}(\{\boldsymbol{u}_{\tau-dt}\})\, dw_p(\tau-dt) \right.$$

$$\left. + \sum_{n=1}^{k} \sum_{m=1}^{n-1} \frac{\partial^2 h(\boldsymbol{u}_{\tau-dt}, \boldsymbol{u}_{\tau-2dt}, \dots)}{\partial u_{k,\tau-mdt}\, \partial u_{i,\tau-ndt}} F_{u_i,p}(\{\boldsymbol{u}_{\tau-ndt}\})\, F_{u_k,p}(\{\boldsymbol{u}_{\tau-mdt}\})\, d\tau \right\}$$

$$\times \int_{-\infty}^{\tau-dt} \exp[\Lambda_s(\tau-\tau')]\, dg(\tau')\,. \qquad (7.9.23)$$

We now put all terms occurring in (7.9.18) together. These transformations require a study of the explicit form of the integral over dg which in fact is a multiple integral over dw's and functions of time.

Since this study is somewhat lengthy without adding new results we give rather the final formula only. According to it, the operator d_-, which was defined in (7.9.4), is given by

$$d_- = \left\{ Q_{0i}(\{\boldsymbol{u}_{t-dt}\}) \frac{\partial}{\partial u_i} + \frac{1}{2} F_{u_i,p}(\{\boldsymbol{u}_{t-dt}\}) F_{u_k,p}(\{\boldsymbol{u}_{t-dt}\}) \frac{\partial^2}{\partial u_i \partial u_k} \right\} dt$$

$$+ F_{u_i,p}(\{\boldsymbol{u}_{t-dt}\})\, dw_p(t-dt) \frac{\partial}{\partial u_i}\,. \qquad (7.9.24)$$

We can now proceed in complete analogy to the discrete time case by devising an iteration procedure in which we collect terms of the same order of magnitude, i.e., the same powers of $\boldsymbol{u}$. To this end we define the operator

$$d_-^{(k_r)} = \left\{ \left(\frac{d}{dt} \right)_{(k_r)(t-dt)} + \frac{1}{2} \sum_r F_{u_i,p}^{(r+1)}(\{\boldsymbol{u}_{t-dt}\}) F_{u_k,p}^{(k_r-r+1)}(\{\boldsymbol{u}_{t-dt}\}) \frac{\partial^2}{\partial u_i \partial u_k} \right\} dt$$

$$+ F_{u_i,p}^{(k_r+1)}(\{\boldsymbol{u}_{t-dt}\}) dw_p(t-dt) \frac{\partial}{\partial u_i} \tag{7.9.25}$$

in which the first operator is defined by

$$\left(\frac{d}{dt} \right)_{(m)} = \left(\dot{\boldsymbol{u}}^{(m+1)} \frac{\partial}{\partial \boldsymbol{u}} \right) \tag{7.9.26}$$

and $\dot{\boldsymbol{u}}^{(m+1)}$ means that we have to take from (7.9.17) only those terms which contain the $m+1$ power of $\boldsymbol{u}$ (after $\boldsymbol{s}$ has been replaced by $\boldsymbol{u}$). After these steps we are in a position to write the final result. Accordingly, $\boldsymbol{s}(t)$ can be written as

$$\boldsymbol{s}(t) = \sum_{k=2}^{\infty} \boldsymbol{C}^{(k)}, \quad \text{where} \tag{7.9.27}$$

$$\boldsymbol{C}^{(k)} = \left(\frac{d}{dt} - \Lambda_s \right)_{(0)}^{-1} \sum_{k_1+\ldots+k_\nu=k'} \prod_{i=1}^{\nu} [\cdots]_{k_i} d\boldsymbol{P}^{(k-k')} \quad \text{with} \tag{7.9.28}$$

$$[\cdots]_{k_i} = d_-^{(k_i)} \left(\frac{d}{dt} - \Lambda_s \right)_{(0)}^{-1} . \tag{7.9.29}$$

The inverse operator occurring in (7.9.29) denotes the performance of integration of the time variable which occurs explicitly in all functions except for the function $\boldsymbol{u}(t)$.

It is simple to study stochastic differential equations along similar lines using the Stratonovich calculus. The only difference of the final result consists in the definition of d_- (or $d_-^{(k_i)}$) in the usual way.

8. Nonlinear Equations. Qualitative Macroscopic Changes

In this and the subsequent chapter we deal with a problem central to synergetics, namely qualitative macroscopic changes of complex systems. Though it is possible to treat the various instabilities under the impact of noise by means of a single approach, for pedagogical reasons we shall deal with the special cases individually. For the same reasons we first start with equations which do not contain fluctuations (noise) and shall treat the corresponding problems only later. The general philosophy of our approach was outlined in Sect. 1.14.

8.1 Bifurcations from a Node or Focus. Basic Transformations

We start with probably the simplest case, namely the bifurcation of solutions from a node. Our starting point is a set of nonlinear differential equations for the state vector $\boldsymbol{q}(t)$, namely

$$\dot{\boldsymbol{q}}(t) = \boldsymbol{N}(\boldsymbol{q}(t), \alpha) . \tag{8.1.1}$$

We assume that the nonlinear function $\boldsymbol{N}(\boldsymbol{q}, \alpha)$ depends on a control parameter α. We make the following assumptions.

1) For a value $\alpha = \alpha_0$ there exists a time-independent solution of (8.1.1)

$$\boldsymbol{q}_0(\alpha_0) . \tag{8.1.2}$$

2) When we change α_0 continuously to other, say bigger, values α, $\boldsymbol{q}_0$ can be extended to this new parameter value, so that

$$\boldsymbol{N}(\boldsymbol{q}_0(\alpha), \alpha) = \boldsymbol{0} . \tag{8.1.3}$$

We now consider the stability of $\boldsymbol{q}_0$ at this new control parameter α, focusing attention on a situation where the solution (8.1.2) becomes unstable. We study the stability by linear analysis, making the hypothesis

$$\boldsymbol{q}(t) = \boldsymbol{q}_0(\alpha) + \boldsymbol{w}(t) . \tag{8.1.4}$$

Inserting (8.1.4) into (8.1.1), we obtain the still exact relation

$$\dot{\boldsymbol{w}}(t) = \boldsymbol{N}(\boldsymbol{q}_0(\alpha) + \boldsymbol{w}(t), \alpha) . \tag{8.1.5}$$

We now assume that $\boldsymbol{w}$ is a small quantity which allows us to linearize (8.1.5). That means we expand the rhs into a power series in the components w_l of $\boldsymbol{w}$. Because of (8.1.3), the first term vanishes and the second leads to the result

$$\dot{\boldsymbol{w}}(t) = \sum_l \frac{\partial \boldsymbol{N}(\boldsymbol{q}_0(\alpha), \alpha)}{\partial q_l} w_l(t), \tag{8.1.6}$$

whereas, as stated before, we neglected all terms of higher order. Equation (8.1.6) can be cast into the general form

$$\dot{\boldsymbol{w}}(t) = L(\alpha)\boldsymbol{w}(t), \tag{8.1.7}$$

where L depends on α but does not depend on time t. As we know from Sect. 2.6, the solutions of (8.1.7) have the general form

$$\boldsymbol{w}^{(k)}(t) = \exp(\lambda_k t)\boldsymbol{v}^{(k)}. \tag{8.1.8}$$

If the eigenvalues $\lambda_k(\alpha)$ of the matrix $L(\alpha)$ are nondegenerate, the vectors $\boldsymbol{v}$ are time independent. In case of degeneracy, $\boldsymbol{v}^{(k)}$ may depend on powers of t.

In order not to overload our presentation we shall assume in the following that the vectors $\boldsymbol{v}^{(k)}$ are time independent which is secured if, e.g., the eigenvalues λ_k are all different from each other, i.e., nondegenerate. Since the vectors $\boldsymbol{v}^{(k)}$ form a complete set in the vector space of $\boldsymbol{q}$, we may represent the desired solution $\boldsymbol{q}$ as a superposition of the vectors $\boldsymbol{v}^{(k)}$. Therefore, the most general solution may be written in the form

$$\boldsymbol{q}(t, \alpha) = \boldsymbol{q}_0(\alpha) + \underbrace{\sum_k \xi_k(t)\boldsymbol{v}^{(k)}}_{\boldsymbol{W}(t)}, \tag{8.1.9}$$

where the coefficients $\xi_k(t)$ are still unknown time-dependent variables. It will be our first task to determine ξ_k. To this end we insert (8.1.9) into (8.1.1)

$$\sum_k \dot{\xi}_k(t)\boldsymbol{v}^{(k)} = \boldsymbol{N}(\boldsymbol{q}_0(\alpha) + \underbrace{\sum_k \xi_k(t)\boldsymbol{v}^{(k)}}_{\boldsymbol{W}(t)}, \alpha). \tag{8.1.10}$$

To be as explicit as possible we assume that we may expand $\boldsymbol{N}$ into a power series of $\boldsymbol{W}$ and that it is sufficient to keep only its first terms. Our procedure will make it clear, however, how to proceed in the general case in which the polynomial extends to higher orders. We thus write $\boldsymbol{N}$ in the form

$$\boldsymbol{N}(\boldsymbol{q}_0(\alpha) + \boldsymbol{W}, \alpha) = \underbrace{\boldsymbol{N}(\boldsymbol{q}_0(\alpha), \alpha)}_{1} + \underbrace{L\boldsymbol{W}}_{2} + \underbrace{\boldsymbol{N}^{(2)}:\boldsymbol{W}:\boldsymbol{W}}_{3} + \underbrace{\boldsymbol{N}^{(3)}:\boldsymbol{W}:\boldsymbol{W}:\boldsymbol{W}}_{4} + \dots, \tag{8.1.11}$$

where according to (8.1.3) the first term vanishes

$$N(q_0(\alpha),\alpha) = \mathbf{0}\,. \tag{8.1.12}$$

According to (8.1.6 or 7) and using (8.1.8), the second term can be cast into the form

$$LW = \sum_k \xi_k(t) L v^{(k)} = \sum_k \xi_k(t)\lambda_k v^{(k)}\,. \tag{8.1.13}$$

The third term is a shorthand notation which is defined as follows

$$N^{(2)}\colon W\colon W = \frac{1}{2}\sum_{k'k''} \xi_{k'}(t)\,\xi_{k''}(t) \underbrace{\sum_{j'j''} (\partial^2 N/\partial q_{j''}\,\partial q_{j'})\, v_{j'}^{(k')} v_{j''}^{(k'')}}_{2N^{(2)}_{k'k''}}\,. \tag{8.1.14}$$

The higher-order terms are defined in a similar fashion. Since we want to derive equations for ξ_k, we should get rid of the vectors $v^{(k)}$. To this end we use the dual eigenvectors introduced in Sect. 2.5. These eigenvectors,

$$\bar{v}^{(k)}, \tag{8.1.15}$$

which are in the present case as v time independent, have the property

$$\langle \bar{v}^{(k)} v^{(k')}\rangle = \delta_{kk'}\,. \tag{8.1.16}$$

Taking the scalar product of (8.1.10) with (8.1.15) and making use of the decomposition (8.1.11), we find the following equations

$$\dot{\xi}_k(t) = \lambda_k \xi_k(t) + \sum_{k'k''} \xi_{k'}(t)\,\xi_{k''}(t) \underbrace{\langle \bar{v}_k N^{(2)}_{k'k''}\rangle}_{A^{(2)}_{kk'k''}} + \dots, \tag{8.1.17}$$

where we have introduced the abbreviation $A^{(2)}$. The higher-order terms are, of course, defined in an analogous fashion, giving the final result

$$\dot{\xi}_k = \lambda_k \xi_k + \sum_{k'k''} A^{(2)}_{kk'k''}\xi_{k'}\xi_{k''} + \sum_{k'k''k'''} A^{(3)}_{kk'k''k'''}\xi_{k'}\xi_{k''}\xi_{k'''} + \dots\,. \tag{8.1.18}$$

8.2 A Simple Real Eigenvalue Becomes Positive

We numerate the eigenvalues in such a way that the one with the biggest real part carries the index 1. We assume that this eigenvalue is real and nondegenerate and that the control parameters are such that $\lambda_1(\alpha)$ changes its sign from a negative to a positive value. We shall focus our attention on a situation in which

$$\lambda_1(\alpha) \geqslant 0 \tag{8.2.1}$$

holds and in which the real parts of all other eigenvalues are still negative. Since (8.2.1) in the linear stability analysis indicates that the corresponding solution is unstable, whereas

$$\mathrm{Re}\{\lambda_2\}, \ldots \leqslant C_0 < 0 \tag{8.2.2}$$

indicates that all other solutions are still stable, we shall express this fact by a change of notation. To this end we put for

$$\lambda_1(\alpha): \quad \xi_1(t) = u(t), \lambda_1 = \lambda_u , \tag{8.2.3}$$

where u refers to "unstable" and for

$$\lambda_{k'}, k' \geqslant 2: \quad \xi_{k'}(t) = s_{\underbrace{k'-1}_{k}}, \lambda_{k'-1} \Rightarrow \lambda_k , \tag{8.2.4}$$

where s refers to "stable". We must stress here that our notation may lead to some misunderstanding which must be avoided at any rate. Note that "unstable" and "stable" refer to the *linear stability analysis only.* We shall show that in many cases, due to the nonlinearities, the mode which was formerly unstable in the linear analysis becomes stabilized, and it will be our main purpose to explore the new stable regions of u.

So the reader is well advised to use u and s as abbreviations distinguishing those modes which in the linear stability analysis have eigenvalues characterized by the properties (8.2.1, 2). After these introductory remarks we split (8.1.18) according to the indices 1 and 2, ..., respectively, into the following equations

$$\dot{u} = \lambda_u u + A_u^{(2)} u^2 + A_u^{(3)} u^3 + \ldots + \sum_k A_{uk}^{(2)} u \cdot s_k + \ldots , \tag{8.2.5}$$

$$\dot{s}_k = \lambda_k s_k + A_s^{(2)} u^2 + A_s^{(3)} u^3 + \ldots + \sum_k A_{sk}^{(2)} u \cdot s_k + \ldots . \tag{8.2.6}$$

In this section and Sect. 8.3 we shall assume u real. Provided u is small enough and (8.2.2) holds, we may apply the slaving principle. This allows us to express s_k as a function of u which may be approximated by a polynomial

$$s_k = s_k(u) = a_k u^2 + b_k u^3 + \ldots . \tag{8.2.7}$$

Inserting (8.2.7) in (8.2.5) we readily obtain the order parameter equation

$$\dot{u} = \lambda_u u + A_u^{(2)} u^2 + (A_u^{(3)} + C) u^3 + \text{rest} , \tag{8.2.8}$$

where C is given by

$$C = \sum_k A_{uk}^{(2)} a_k . \tag{8.2.9}$$

The rest term contains higher powers of u. Of course in practical applications it will be necessary to check the size of the rest term. Here we shall assume that it is sufficiently small so that we can neglect it. To exhibit the main features of (8.2.8) we start with a few examples. We write (8.2.8) in the form

$$\dot{u} = \lambda_u u + f(u) \tag{8.2.10}$$

and first treat the example

$$f(u) = -\beta u^3, \quad \beta \text{ real} \tag{8.2.11}$$

so that (8.2.10) reads

$$\dot{u} = \lambda_u u - \beta u^3 . \tag{8.2.12}$$

For those readers who are acquainted with mechanics we note that (8.2.12) can be visualized as the overdamped motion of a particle in a force field which has a potential V according to the equation

$$\dot{u} = -\frac{\partial V}{\partial u} . \tag{8.2.13}$$

In the present case V is given by

$$V = -\lambda_u \frac{u^2}{2} + \beta \frac{u^4}{4} , \tag{8.2.14}$$

which easily allows us to visualize the behavior of the solution u of the nonlinear equation (8.2.12). An inspection of the potential curve V reveals immediately that depending on the sign of λ_u two entirely different situations occur. For $\lambda_u < 0$ the value $u = 0$ is a stationary stable solution. This solution persists for $\lambda_u > 0$ but becomes unstable. The new stable solutions are given by

$$u = \pm \sqrt{\lambda_u/\beta} . \tag{8.2.15}$$

Note that this solution becomes imaginary for negative λ_u and must therefore be excluded for that branch of λ_u.

Taking the roots (8.2.15) and

$$u = 0 \tag{8.2.16}$$

together we may decompose the rhs of (8.2.12), which is a polynomial, into a product of its roots according to

$$\dot{u} = -\beta \cdot u(u - \sqrt{\lambda_u/\beta})(u + \sqrt{\lambda_u/\beta}) . \tag{8.2.17}$$

Introducing the abbreviations

$$u_0 = 0\,, \quad u_1 = \sqrt{\lambda_u/\beta}\,, \quad u_2 = -\sqrt{\lambda_u/\beta} \tag{8.2.18}$$

for the individual roots, we may write (8.2.17) in the more condensed form

$$\dot{u} = -\beta(u-u_0)(u-u_1)(u-u_2)\,, \quad u_k = u_k(\alpha)\,. \tag{8.2.19}$$

Because λ_u depends on the control parameter α or may even be identified with it in a number of cases, the roots u_k are functions of this control parameter α. These considerations can be generalized in a straightforward manner to cases in which $f(u)$ is a general polynomial

$$\dot{u} = \lambda_u u + f(u) \equiv P(u)\,. \tag{8.2.20}$$

Again it is helpful to interpret (8.2.20) as the overdamped motion of a particle in the potential field V from which the rhs of (8.2.20) may be derived

$$\dot{u} = -\frac{\partial V}{\partial u}\,. \tag{8.2.21}$$

In analogy to (8.2.19) we may decompose the polynomial (8.2.20) into a product combining the individual roots

$$\dot{u} = C[u-u_1(\alpha)]\,[u-u_2(\alpha)]\cdots[u-u_\varkappa(\alpha)]\,. \tag{8.2.22}$$

Knowing the dependence of u_k on the control parameter α we may find a number of branches of the possible stationary solutions when α changes. It is a nice game to study the emergence, coalescence, or disappearance of such branches, some examples of which are given in the exercises.

To study the stability of these individual branches of solutions we may expand $P(u)$ around the respective root u_k. In the immediate vicinity of u_k we put

$$u = u_k + \delta u\,, \tag{8.2.23}$$

linearize (8.2.20) around u_k, and study the stability by means of the linearized equations

$$\delta\dot{u} = \left.\frac{\partial P(u)}{\partial u}\right|_{u=u_k} \delta u\,. \tag{8.2.24}$$

Equation (8.2.24) is of the general form

$$\delta\dot{u} = \gamma \cdot \delta u\,, \tag{8.2.25}$$

where the sign of γ decides upon stability or instability,

$$\gamma = \left.\frac{\partial P(u)}{\partial u}\right|_{u=u_k} \lessgtr 0\,. \tag{8.2.26}$$

The temporal evolution in the nonlinear domain can be studied also because (8.2.20) is a differential equation whose variables can be separated so that we immediately find

$$\int_{u_0}^{u} \frac{du'}{\lambda_u u' + f(u')} = t - t_0\,. \tag{8.2.27}$$

Using (8.2.22) the integrand can be decomposed into partial fractions so that the integral can be explicitly evaluated.

Exercise. Evaluate the integral in (8.2.27) in the case (8.2.19) and determine $u(t)$.

8.3 Multiple Real Eigenvalues Become Positive

We assume that a number M of the eigenvalues with the greatest real parts coincide and that they are real

$$\lambda_1 = \lambda_2 = \ldots = \lambda_M = \lambda_u(\alpha) \tag{8.3.1}$$

and are zero or positive while all other eigenvalues have negative real parts. In extension of the previous section we denote the mode amplitudes which belong to (8.3.1) by

$$u_1, \ldots, u_M\,. \tag{8.3.2}$$

Splitting the general system (8.1.18) into the mode amplitudes (8.3.2) and the rest and applying the slaving principle we find the order parameter equations

$$\left.\begin{aligned} \dot{u}_1 &= \lambda_u u_1 + f_1(u_1, \ldots, u_M, \lambda_u, \alpha) \\ &\vdots \\ \dot{u}_M &= \lambda_u u_M + f_M(u_1, \ldots, u_M, \lambda_u, \alpha) \end{aligned}\right\} \tag{8.3.3}$$

or, in shorter notation (not denoting explicit α-dependence of the P's),

$$\left.\begin{aligned} \dot{u}_1 &= P_1(\boldsymbol{u}, \lambda_u(\alpha)) \\ &\vdots \\ \dot{u}_M &= P_M(\boldsymbol{u}, \lambda_u(\alpha)) \end{aligned}\right\}. \tag{8.3.4}$$

In principle, these equations are of precisely the same form as the original equations (8.1.1). However, the following two points should be observed. In contrast to the general equations (8.1.1), the number of variables of the reduced equations (8.3.3) is in very many practical cases much smaller. Especially in complex systems, where we deal with very many degrees of freedom, an enormous reduction of the number of degrees of freedom is achieved by going from (8.1.1) to (8.3.3). Furthermore, when eigenvalues are degenerate, symmetry properties can often be used. Here we indicate how to cope with (8.3.3) in the special case in which stationary solutions

$$\dot{u}_j = 0: \quad P_j(\boldsymbol{u}, \lambda_u(\alpha)) = 0 \tag{8.3.5}$$

exist. Note, however, that (8.3.3) may also have solutions other than (8.3.5), for instance oscillatory solutions. We shall come back to such equations later. Here we make a few comments for the case when (8.3.5) holds.

A rather simple case occurs when the rhs of (8.3.4) can be written as a gradient of a potential. A class of such problems is treated by catastrophe theory ([Ref. 1, Chap. 5] and additional reading material given at the end of the present book). We shall not dwell on that problem here further. Another class of equations of the form (8.3.4) is provided by

$$P_j = u_j[\Sigma a_{lm}^{(j)} u_l u_m + \lambda_u(\alpha)] \,. \tag{8.3.6}$$

The expressions (8.3.6) vanish if u_j or the bracket connected with it vanish. If some of the u_j vanish, say those with the indices $k+1$, $k+2, \ldots$, we need only require that for $1, \ldots, k$ the brackets vanish

$$\Sigma a_{lm}^{(j)} u_l u_m + \lambda_u(\alpha) = 0 \,. \tag{8.3.7}$$

Equations (8.3.7) describe ellipses, straight lines or hyperbola in a plane, or their corresponding counter parts, namely certain hypersurfaces in n-dimensional space. The possible stationary solutions (8.3.5) are then defined as the cross sections of such hypersurfaces.

Let us illustrate this by a most simple example in which we deal with two variables, and one of the equations (8.3.7) reads

$$u_1^2 + u_2^2 = \lambda_u(\alpha) \tag{8.3.8}$$

and the other

$$\frac{u_1^2}{a^2} + \frac{u_2^2}{b^2} = \lambda_u(\alpha) \,. \tag{8.3.9}$$

Depending on the constants a and b we may have 0, 2, or 4 cross sections. Evidently with increasing $\lambda_u(\alpha)$ in the case of possible cross sections the variables u_1 and u_2 increase. Each of the cross sections represents one of the newly evolving possible solutions.

Exercise. Show that at least in general (8.3.8) and (8.3.9) do not possess a potential.

8.4 A Simple Complex Eigenvalue Crosses the Imaginary Axis. Hopf Bifurcation

Our starting point is the set of differential equations (8.1.1). Again we assume that for a certain control parameter $\alpha = \alpha_0$ a stable time-independent solution $\boldsymbol{q}_0$ exists. We assume that when we change the control parameter, this time-independent solution can be extended and we again perform a stability analysis. We first consider

$$\dot{\boldsymbol{w}} = L\boldsymbol{w} \tag{8.4.1}$$

which is identical with (8.1.7). Inserting the usual hypothesis

$$\boldsymbol{w} = \mathrm{e}^{\lambda t}\boldsymbol{v} \tag{8.4.2}$$

into (8.4.1), we are led to the eigenvalue problem

$$L\boldsymbol{v} = \lambda\boldsymbol{v} \,. \tag{8.4.3}$$

It is assumed that the eigenvalue with the biggest real part, which we denote by λ_1, is nondegenerate and complex,

$$\lambda_1 = \lambda_1' + \mathrm{i}\lambda_1'' \,, \qquad \text{where} \tag{8.4.4}$$

$$\lambda_1' \geqslant 0 \,. \tag{8.4.5}$$

We shall use the notation λ_u instead of λ_1 again

$$\lambda_1 = \lambda_u \,. \tag{8.4.6}$$

Furthermore, we assume that

$$\mathrm{Re}\{\lambda_j\} \leqslant C_0 < 0 \,, \quad j = 2, \dots \tag{8.4.7}$$

which allows us to use the slaving principle for small enough u (which is a new notation of ξ_1).

Because (8.4.4) is a complex quantity, it is plausible that u is complex. Exhibiting only the terms up to 3rd order in u, the order parameter equation for u reads

$$\begin{aligned}\dot{u} = {} & (\lambda_u' + \mathrm{i}\lambda_u'')u + A^{(2)}_{u,uu}u^2 + 2A^{(2)}_{u,uu^*}uu^* + A^{(2)}_{u,u^*u^*}u^{*2} \\ & + B_{uuuu}u^3 + 3B_{uuuu^*}u \cdot uu^* + 3B_{uuu^*u^*}uu^{*2} \\ & + B_{uu^*u^*u^*}u^{*3} + \text{rest} \,.\end{aligned} \tag{8.4.8}$$

In the following we shall focus our attention on important special cases for (8.4.8) which occur repeatedly in many applications.

We start with the case (λ_u', λ_u'' real)

$$\dot{u} = (\lambda_u' + \mathrm{i}\lambda_u'')u - bu|u|^2, \tag{8.4.9}$$

where b is assumed real and

$$b \equiv -B_{uuuu^*} > 0. \tag{8.4.10}$$

Equation (8.4.9) can be readily solved by means of the hypothesis

$$u(t) = r(t)\exp[\mathrm{i}\psi(t) + \mathrm{i}\lambda_u'' t], \quad r \geqslant 0, \quad \psi \text{ real}. \tag{8.4.11}$$

Inserting it into (8.4.9) we obtain

$$\dot{r} + (\mathrm{i}\dot{\psi} + \mathrm{i}\lambda_u'')r = (\lambda_u' + \mathrm{i}\lambda_u'')r - br^3 \tag{8.4.12}$$

or, after separation of the real and imaginary parts,

$$\dot{\psi} = 0, \tag{8.4.13}$$

$$\dot{r} = \lambda_u' r - br^3. \tag{8.4.14}$$

Equation (8.4.14) is identical with (8.2.12), with the only difference being that r must be a positive quantity. Therefore the stationary solutions read

$$r_0 = 0 \quad \text{for} \quad \lambda_u' < 0, \tag{8.4.15}$$

$$r_0 = 0, \quad r_0 = +\sqrt{\lambda_u'/b} \quad \text{for} \quad \lambda_u' \geqslant 0. \tag{8.4.16}$$

Because (8.4.13) holds for all times, the transient solution can be readily found by solving (8.4.14) [1].

The stationary solution has the form

$$u = r_0 \exp(\mathrm{i}\psi_0 + \mathrm{i}\lambda_u'' t). \tag{8.4.17}$$

It shows that (for $r_0 > 0$) the total system undergoes an harmonic oscillation or, when we plot u in the phase plane consisting of its real and imaginary parts as coordinates, we immediately find a limit cycle. The transient solution tells us that all points of the plane converge from either the outside or the inside towards this limit cycle. An interesting modification arises if the constant b occurring in (8.4.9) is complex

$$b = b' + \mathrm{i}b'', \quad b', b'' \text{ real}. \tag{8.4.18}$$

We assume

$$b' > 0, \text{ real} \tag{8.4.19}$$

and the equation we have to study reads

$$\dot{u} = (\lambda_u' + \mathrm{i}\lambda_u'')u - (b' + \mathrm{i}b'')u|u|^2 . \tag{8.4.20}$$

For its solutions we make the hypothesis

$$u = r(t)\,\mathrm{e}^{\mathrm{i}\varphi(t)}, \quad r \geqslant 0 , \quad \varphi \text{ real} . \tag{8.4.21}$$

Inserting (8.4.21) into (8.4.20) and separating real and imaginary parts we find

$$\dot{r} = \lambda_u' r - b' r^3 \qquad \text{and} \tag{8.4.22}$$

$$\dot{\varphi} = \lambda_u'' - b'' r^2 . \tag{8.4.23}$$

These are two coupled differential equations, but the type of coupling is very simple. We may first solve (8.4.22) for $r(t)$ in the stationary or the transient state exactly. Then we may insert the result for r^2 into (8.4.23) which can now be solved immediately. We leave the discussion of the transient case as an exercise for the reader and write down the solution for the steady state

$$\varphi = (\lambda_u'' - b'' r_0^2)t + \varphi_0 . \tag{8.4.24}$$

This is an important result because it shows that the frequency of the oscillation depends on the amplitude. Thus, in the case of Hopf bifurcation we shall not only find new values for r, i.e., a nonzero radius of the limit cycle, but also shifted frequencies. The present case can be generalized in a very nice way to the case in which (8.4.20) is replaced by the more general equation

$$\dot{u} = (\lambda_u' + \mathrm{i}\lambda_u'')u - u[f(|u|^2) + \mathrm{i}g(|u|^2)] . \tag{8.4.25}$$

Let us assume that f and g are real functions of their arguments and are of the form of polynomials. Then again the hypothesis (8.4.21) can be made. Inserting it into (8.4.25) and splitting the resulting equation into its real and imaginary parts, we obtain

$$\dot{r} = \lambda_u' r - r f(r^2) , \tag{8.4.26}$$

$$\dot{\varphi} = \lambda_u'' - g(r^2) . \tag{8.4.27}$$

Equation (8.4.26) has been discussed before [compare (8.2.20)]. The additional equation (8.4.27) can then be solved by a simple quadrature.

8.5 Hopf Bifurcation, Continued

Let us now consider a more general case than (8.4.25) which one frequently meets in practical applications. In it the order parameter equation has the following form

$$\dot{u} = (\lambda_u' + \mathrm{i}\lambda_u'')u - (b' + \mathrm{i}b'')u|u|^2 + f(u, u^*) , \tag{8.5.1}$$

where λ_u', λ_u'', b' and b'' are real constants.

We shall assume that f is of order $|u|^3$ but no longer contains the term $u|u|^2$ which is exhibited explicitly in (8.5.1). The specialization of the general equation (8.4.8) therefore consists in the assumption that the rhs of (8.5.1) does not contain quadratic or bilinear terms in u, u^*. By making the hypothesis

$$u(t) = r(t)\,\mathrm{e}^{\mathrm{i}\varphi(t)} , \quad r \geqslant 0 , \quad \varphi \text{ real} \tag{8.5.2}$$

we may transform (8.5.1) into the equations

$$\dot{r} = \lambda_u' r - b' r^3 + g(r, \mathrm{e}^{\mathrm{i}\varphi}, \mathrm{e}^{-\mathrm{i}\varphi}) \qquad \text{and} \tag{8.5.3}$$

$$\dot{\varphi} = \underbrace{\lambda_u'' - b'' r^2}_{\omega} + h(r, \mathrm{e}^{\mathrm{i}\varphi}, \mathrm{e}^{-\mathrm{i}\varphi}) , \tag{8.5.4}$$

with the abbreviations

$$g = \mathrm{Re}\{\mathrm{e}^{-\mathrm{i}\varphi} f\} \qquad \text{and} \tag{8.5.5}$$

$$h = \mathrm{Im}\{\mathrm{e}^{-\mathrm{i}\varphi} f\}/r . \tag{8.5.6}$$

In accordance with our specification of f we require that g is of order r^3 and h of order r^2. We may put

$$g = g_1 + g_2(\varphi) , \tag{8.5.7}$$

where g_1 is φ independent and of order r^4 and g_2 fulfills the condition

$$\int_0^{2\pi} g_2 d\varphi = 0 . \tag{8.5.8}$$

Furthermore, we may put

$$h = h_1 + h_2(\varphi) , \tag{8.5.9}$$

where h_1 is of order r^3 and h_2 fulfills the condition

$$\int_0^{2\pi} h_2 d\varphi = 0 . \tag{8.5.10}$$

We assume $b' > 0$. While for $\lambda_u' < 0$ the solution $r = 0$ is stable, it becomes unstable for $\lambda_u' > 0$, so that we make the hypothesis

$$r = r_0 + \xi \tag{8.5.11}$$

and require

$$\lambda_u' r_0 - b' r_0^3 = 0 . \tag{8.5.12}$$

By this we may transform (8.5.3) into

$$\dot{\xi} = -2\lambda_u' \xi - 3b' r_0 \xi^2 - b' \xi^3 + g(r_0 + \xi, e^{i\varphi}, e^{-i\varphi}) . \tag{8.5.13}$$

Since we want to study the behavior of the solutions of (8.5.1) close to the transition point $\lambda_u' = 0$, the smallness of λ_u' is of value. To exhibit this smallness we introduce a smallness parameter ε,

$$\lambda_u' = \varepsilon^2 \hat{\lambda} . \tag{8.5.14}$$

Equation (8.5.12) suggests that we may put

$$r_0 = \varepsilon \tilde{r}_0 . \tag{8.5.15}$$

To find appropriate terms of the correct order in ε on both sides of (8.5.13), we rescale the amplitude by

$$\xi = \varepsilon \eta \tag{8.5.16}$$

and time by

$$t = \tau / \varepsilon^2 . \tag{8.5.17}$$

Because of the dependence of g on r it follows that

$$g = \varepsilon^3 \tilde{g} . \tag{8.5.18}$$

Introducing the scaling transformations (8.5.14 – 18) into (8.5.13) we readily find

$$\frac{d\eta}{d\tau} = -2\hat{\lambda}\eta - 3b'\eta^2 - b'\eta^3 + \tilde{g}(\tilde{r}_0 + \eta, \varepsilon, e^{i\varphi}, e^{-i\varphi}) \tag{8.5.19}$$

after having divided (8.5.13) by ε^3. Making the corresponding scaling transformations in (8.5.4) and putting

$$h = \varepsilon^2 \tilde{h} \tag{8.5.20}$$

we transform (8.5.4) into

$$\frac{d\varphi}{d\tau} = \underbrace{(\lambda_u'' - b''\varepsilon^2\tilde{r}^2)/\varepsilon^2}_{\tilde{\omega}} + \tilde{h}(\tilde{r}, \mathrm{e}^{\mathrm{i}\varphi}, \mathrm{e}^{-\mathrm{i}\varphi}) \,. \tag{8.5.21}$$

This equation shows that instead of the original frequency ω occurring in (8.5.4), we now have to deal with the renormalized frequency $\tilde{\omega}$ where

$$\tilde{\omega} = \omega/\varepsilon^2 \,. \tag{8.5.22}$$

Because ε is a small quantity we recognize that on the new time scale τ under consideration, $\tilde{\omega}$ is a very high frequency. We want to show that therefore $\tilde{g}$ and $\tilde{h}$ in (8.5.19) and (8.5.21), respectively, can be considered as small quantities. From (8.5.21) we may assume that in lowest approximation $\exp(\mathrm{i}\varphi) \approx \exp(\mathrm{i}\tilde{\omega}\tau)$ is a very rapidly oscillating quantity.

Now consider a function of such a quantity, e.g., $\tilde{g}_2$ or $\tilde{h}_2$. Since $\tilde{g}_2$ and $\tilde{h}_2$ are functions of $\exp(\mathrm{i}\varphi)$, they can both be presented in the form

$$\sum_{m \neq 0} c_m \mathrm{e}^{\mathrm{i}m\omega\tau/\varepsilon^2} \,. \tag{8.5.23}$$

Let us assume that $\tilde{g}$ and $\tilde{h}$ are small quantities. Then it is possible to make a perturbation expansion in just the same way as in the quasiperiodic case in Chap. 6.

As we have seen we then have to evaluate an indefinite integral over (8.5.23) which acquires the form

$$\frac{\varepsilon^2}{\mathrm{i}\omega} \sum_{m \neq 0} \frac{c_m}{m} \mathrm{e}^{\mathrm{i}m\omega\tau/\varepsilon^2} \,. \tag{8.5.24}$$

If (8.5.23) converges absolutely so does (8.5.24) because of

$$\left| \frac{c_m}{m} \right| \leqq |c_m| \,. \tag{8.5.25}$$

However, the factor ε^2 in front of the sum in (8.5.24) can be arbitrarily small. This means $\tilde{g}_2$ and $\tilde{h}_2$ can be treated as small quantities. In addition, according to our assumptions above, the φ-independent terms $\tilde{g}_1$ and $\tilde{h}_1$ are still of higher order and can again be considered as small quantities so that $\tilde{g}$ and $\tilde{h}$ act as small perturbations.

In order to bring out the smallness of the terms $\tilde{g}$ and $\tilde{h}$ explicitly we introduce a further scaling, namely by

$$\tilde{g} = \varepsilon'^2 \hat{g} \,, \tag{8.5.26}$$

$$\tilde{h} = \varepsilon' \hat{h} \tag{8.5.27}$$

jointly with

$$\eta = \varepsilon' \hat{\eta} . \tag{8.5.28}$$

This brings us from (8.5.19, 21) to

$$\frac{d\hat{\eta}}{d\tau} = -2\hat{\lambda}\hat{\eta} - 3b'\varepsilon'\hat{\eta}^2 - b'\varepsilon'^2\hat{\eta}^3 + \varepsilon'\hat{g}(\tilde{r}_0 + \varepsilon'\hat{\eta}, \varepsilon, e^{i\varphi}, e^{-i\varphi}) , \tag{8.5.29}$$

$$\frac{d\varphi}{d\tau} = \frac{1}{\varepsilon^2}\lambda_u'' - b''(\tilde{r}_0 + \varepsilon'\hat{\eta})^2 + \varepsilon'\hat{h}(\tilde{r}_0 + \varepsilon'\hat{\eta}, \varepsilon, e^{i\varphi}, e^{-i\varphi}) , \tag{8.5.30}$$

respectively. Lumping all quantities which contain the smallness parameter ε' together, we may write (8.5.29, 30) in the final forms

$$\frac{d\hat{\eta}}{d\tau} = -2\hat{\lambda}\hat{\eta} + \varepsilon'\bar{g}(\hat{\eta}, \varepsilon', e^{i\varphi}, e^{-i\varphi}) \quad \text{and} \tag{8.5.31}$$

$$\frac{d\varphi}{d\tau} = \omega_u + \varepsilon'\bar{h}(\hat{\eta}, \varepsilon', e^{i\varphi}, e^{-i\varphi}) , \tag{8.5.32}$$

where ω_u is given by

$$\omega_u = \lambda_u''/\varepsilon^2 - b''\tilde{r}_0^2 . \tag{8.5.33}$$

For reasons which will transpire immediately, we rewrite (8.5.31 and 32) in the form

$$\dot{\eta} = -\hat{\lambda}_u \eta + \varepsilon g(\eta, \varphi, \varepsilon) , \tag{8.5.34}$$

$$\dot{\varphi} = \omega_u + \varepsilon h(\eta, \varphi, \varepsilon) , \tag{8.5.35}$$

where g and h are 2π-periodic in φ [and must not be confused with g and h in (8.5.3, 4)].

Equations (8.5.34, 35) are now in a form which allows us to make contact with results in Sect. 6.2, where we presented a theorem of Moser that will now enable one to cope with such equations [compare Eqs. (6.1.9, 10)].

It should be noted that the full power of that theorem is not needed here because it deals with *quasiperiodic* motion whereas here we deal with periodic motion only, which excludes the problem of small divisors. (We may treat (8.5.34, 35) more directly by perturbation expansions also.) But nevertheless we shall use similar reasoning later on in the more complicated case of quasiperiodic motion and therefore we present our arguments in the simpler case here.

Let us compare (8.5.34, 35) with (6.1.9, 10). There the aim was to introduce counterterms D and Δ (and perhaps d) such that the quantities Λ and ω were not changed by the iteration procedure.

Here our iteration is somewhat different because (8.5.34, 35) do not contain these counterterms. Rather, we have to encounter the fact that due to the perturbations g and h, Λ and ω will be changed. It is rather simple to make contact with the results of Sect. 6.1 nevertheless. To this end we put $-\hat{\lambda}_u = \lambda_u$ [where this λ_u must not be confused with λ_u in (8.4.6)] and write

$$\lambda_u = \underbrace{\lambda_u - D}_{\lambda_r} + D \tag{8.5.36}$$

and similarly

$$\omega_u = \underbrace{\omega_u - \Delta}_{\omega_r} + \Delta\,. \tag{8.5.37}$$

Introducing (8.5.36, 37) into (8.5.34, 35), respectively, we cast these equations into precisely the form required by Moser's theorem. This procedure is well known from physics, namely from quantum field theory where the quantities with the index u, e.g., ω_u, are referred to as "unrenormalized" quantities whereas the quantities ω_r, λ_r are referred to as "renormalized". According to Moser's theorem we may calculate D and Δ explicitly from (8.5.34, 35) provided we use the decomposition (8.5.36, 37). Then D and Δ become functions of λ_r, ω_r and ε so that we have the relations

$$\lambda_r = \lambda_u - D(\lambda_r, \omega_r, \varepsilon)\,, \tag{8.5.38}$$

$$\omega_r = \omega_u - \Delta(\lambda_r, \omega_r, \varepsilon)\,. \tag{8.5.39}$$

Here D and Δ are analytic in ε and by an inspection of the iteration process described in Sect. 6.3 we may readily convince ourselves that D and Δ are continuous and even differentiable functions of λ_r, ω_r. At least for small enough ε this allows us to resolve equations (8.5.38, 39) for λ_r and ω_r

$$\lambda_r = F(\lambda_u, \omega_u, \varepsilon)\,, \tag{8.5.40}$$

$$\omega_r = G(\lambda_u, \omega_u, \varepsilon)\,. \tag{8.5.41}$$

It is even possible to give a simple algorithm in order to do this inversion explicitly. Namely, we just have to substitute the quantities λ_r and ω_r in the arguments of D and Λ, which occur on rhs of (8.5.38, 39), by the corresponding rhs in lowest order of (8.5.38, 39) consecutively. Thus in lowest approximation we obtain

$$\lambda_r = \lambda_u - D(\lambda_u, \omega_u, \varepsilon)\,. \tag{8.5.42}$$

For sake of completeness we note that in the case of $\lambda_r = 0$ a constant counterterm d may occur. In this case

$$\dot{\eta} = d - \lambda_u \eta + \varepsilon g(\eta, \varphi, \varepsilon), \tag{8.5.43}$$

$$\lambda_u r_0 - b' r_0^3 = d \tag{8.5.44}$$

hold. We merely note that d can be generated by choosing (8.5.44) instead of (8.5.12).

After having described a procedure which allows us to determine the shifted, i.e., renormalized, frequency (8.5.39), and the renormalized relaxation constant (8.5.38), we now turn so study the form of solutions η and φ. Instead of (8.5.34, 35), we studied a general class of equations of which (8.5.34, 35) are a special case. Those equations referred to the phase angles φ and real quantities ξ. In Sect. 6.3 we showed that the corresponding equations for φ and ξ can be cast into a simple form by making the transformations

$$\varphi = \psi + \varepsilon \boldsymbol{u}(\psi, \varepsilon) \quad \text{and} \tag{8.5.45}$$

$$\xi = \chi + \varepsilon \boldsymbol{v}(\psi, \varepsilon) + \varepsilon V(\psi, \varepsilon)\chi. \tag{8.5.46}$$

In Sect. 6.3 we also described an iteration procedure by which $\boldsymbol{u}$, $\boldsymbol{v}$ and V could be constructed explicitly. In particular these functions were 2π-periodic in ψ. The resulting equations could be written in the form

$$\dot{\psi} = \omega + O(\chi) \quad \text{and} \tag{8.5.47}$$

$$\dot{\chi} = \Lambda \chi + O(\chi^2) \tag{8.5.48}$$

which allows for solutions of the form

$$\psi = \omega t \quad \text{and} \tag{8.5.49}$$

$$\dot{\chi} = \mathrm{e}^{\Lambda t} \chi_0 \tag{8.5.50}$$

when we retained only the lowest orders in (8.5.47, 48). Inserting (8.5.49, 50) in (8.5.45, 46) yields the explicit time dependence of φ and ξ in the form

$$\varphi = \omega t + \varepsilon \boldsymbol{u}(\omega t, \varepsilon) \quad \text{and} \tag{8.5.51}$$

$$\xi = \mathrm{e}^{\Lambda t} \chi_0 + \varepsilon \boldsymbol{v}(\omega t, \varepsilon) + \varepsilon V(\omega t, \varepsilon) \mathrm{e}^{\Lambda t} \chi_0. \tag{8.5.52}$$

The only step still to be performed is to specialize these results to the case of our equations (8.5.34, 35). Making the identifications

$$\xi \to \eta, \tag{8.5.53}$$

$$\Lambda \to \lambda_r, \tag{8.5.54}$$

$$\varphi \to \varphi, \tag{8.5.55}$$

$$\omega \to \omega_r , \tag{8.5.56}$$

our final result reads

$$\varphi = \omega_r t + \varepsilon u(\omega_r t, \varepsilon) \tag{8.5.57}$$

$$\eta = \chi_0 \exp(\lambda_r t) + \varepsilon v(\omega_r t, \varepsilon) + \varepsilon V(\omega_r t, \varepsilon) \chi_0 \exp(\lambda_r t), \quad \lambda_r < 0 . \tag{8.5.58}$$

It gives us the form of the solution close to a bifurcation point. From (8.5.58) we may readily conclude that the solution is stable and relaxes towards a steady state given by

$$\eta = \varepsilon v(\omega_r t, \varepsilon) . \tag{8.5.59}$$

In particular it follows that an oscillation takes place at the renormalized frequency ω_r. This approach goes considerably beyond conventional approaches because it not only determines the bifurcation solution in the steady state, but also allows us to discuss its behavior close to the steady state, i.e., its relaxation properties. Such a treatment is particularly necessary when we wish to study the stability, and beyond that, to take fluctuations into account (as we will show later).

8.6 Frequency Locking Between Two Oscillators

As an example we shall consider the bifurcation starting from a focus in a case where *two* complex eigenvalues acquire positive real parts. Here we have to deal with two equations for the two complex order parameters u_1 and u_2. To illustrate the basic ideas we write these equations in the form

$$\dot{u}_1 = (\lambda_1' + \mathrm{i}\omega_1) u_1 - b_1 u_1 |u_1|^2 + c_1 u_2 |u_2|^2 , \tag{8.6.1}$$

$$\dot{u}_2 = (\lambda_2' + \mathrm{i}\omega_2) u_2 - b_2 u_2 |u_2|^2 + c_2 u_1 |u_1|^2 . \tag{8.6.2}$$

For their solution we make the hypothesis

$$u_j = r_j \exp[\mathrm{i}\varphi_j(t)] , \quad j = 1, 2 . \tag{8.6.3}$$

In the following we shall use the decomposition

$$\varphi_j = \omega_j' t + \psi_j(t) , \quad j = 1, 2 , \tag{8.6.4}$$

where the constant frequencies ω_j' are such that $\psi_j, j = 1, 2$ remains bounded for all times. We shall call ω_j' renormalized frequency. Inserting (8.6.3), e.g., in (8.6.1) leads us to

$$\dot{r}_1 + \mathrm{i}\dot{\varphi}_1 r_1 = (\lambda' + \mathrm{i}\omega_1) r_1 - b_1 r_1^3 + c_1 r_2^3 \exp[\mathrm{i}(\varphi_2 - \varphi_1)] , \tag{8.6.5}$$

which can be split into its real and imaginary parts, respectively,

$$\dot{r}_1 = \lambda_1' r_1 - b_1 r_1^3 + c_1 r_2^3 \operatorname{Re}\{\exp[\mathrm{i}(\varphi_2 - \varphi_1)]\}\,, \tag{8.6.6}$$

$$\dot{\varphi}_1 = \omega_1 + c_1 (r_2^3/r_1) \operatorname{Im}\{\exp[\mathrm{i}(\varphi_2 - \varphi_1)]\}\,. \tag{8.6.7}$$

A similar equation arises for r_2, φ_2. Let us assume that c_1 is a small quantity so that in a first approximation the corresponding term can be neglected in (8.6.6). This allows us to find in a good approximation the stationary solution $r_{1,0}$ by

$$r_{1,0} = \sqrt{\lambda_1'/b_1}\,. \tag{8.6.8}$$

Similarly we assume that a corresponding time-independent solution $r_{2,0}$ can be found.

Let us now study (8.6.7) and its corresponding equation for φ_2 under the assumption (8.6.8) and its corresponding one. In order to discuss the corresponding equations

$$\dot{\varphi}_1 = \omega_1 + c_1 (r_2^3/r_1)_0 \sin(\varphi_2 - \varphi_1) \qquad \text{and} \tag{8.6.9}$$

$$\dot{\varphi}_2 = \omega_2 - c_2 (r_1^3/r_2)_0 \sin(\varphi_2 - \varphi_1) \tag{8.6.10}$$

we introduce a new variable

$$\Psi = \varphi_2 - \varphi_1 \tag{8.6.11}$$

and the frequency difference

$$\Omega = \omega_2 - \omega_1\,. \tag{8.6.12}$$

Subtracting (8.6.9) from (8.6.10) we obtain

$$\dot{\Psi} = \Omega - \alpha \sin \Psi \qquad \text{with} \tag{8.6.13}$$

$$\alpha = c_2 (r_1^3/r_2)_0 + c_1 (r_2^3/r_1)_0\,. \tag{8.6.14}$$

Equation (8.6.13) can be readily solved by separation of the variables, giving

$$t = \int_{\Psi_0}^{\Psi} \frac{d\Psi'}{\Omega - \alpha \sin \Psi'}\,. \tag{8.6.15}$$

In it Ψ_0 denotes the initial value of Ψ at time $t = 0$. The behavior of the integral on the rhs of (8.6.15) is entirely different depending on the size of α and Ω. If

$$|\alpha| < |\Omega|\,, \tag{8.6.16}$$

the integrand in (8.6.15) never diverges. In particular, we may expand the denominator into a power series with respect to α. In that case the integral can be represented in the form

$$\int \dots = \frac{1}{\Omega}(\Psi - \Psi_0) + \text{small pulsations} . \tag{8.6.17}$$

Therefore, aside from small pulsations we find for (8.6.15) the relation

$$\Psi - \Psi_0 = t\Omega = t(\omega_2 - \omega_1) . \tag{8.6.18}$$

In view of (8.6.11, 12) and the decomposition (8.6.4), (8.6.18) implies

$$(\omega_2' - \omega_1')t + \psi_2 - \psi_1 = t(\omega_2 - \omega_1) + \Psi_0 \qquad \text{or} \tag{8.6.19}$$

$$\omega_2' - \omega_1' = \omega_2 - \omega_1 . \tag{8.6.20}$$

This means that the renormalized frequencies ω_j' retain the same distance as the frequencies ω_j without nonlinear interaction. Quite different behavior occurs, however, if the condition

$$|\alpha| > |\Omega| \tag{8.6.21}$$

is fulfilled. Then the integral diverges at a finite Ψ or, in other words, $t = \infty$ is reached for that finite Ψ. Perhaps aside from transients, the solution of the differential equation (8.6.13) reads

$$\Psi = \arcsin(\Omega/\alpha) = \text{const} , \tag{8.6.22}$$

which in view of (8.6.11 and 4) implies

$$\varphi_2 - \varphi_1 = (\omega_2' - \omega_1')t + \psi_2 - \psi_1 = \text{const} . \tag{8.6.23}$$

This relation can be fulfilled only if the renormalized frequencies ω_j' coincide

$$\omega_2' = \omega_1' , \tag{8.6.24}$$

i.e., if the two frequencies acquire the same size or, in other words, if they are locked together. Also the phases φ_2 and φ_1 are locked together by means of (8.6.22).

It should be noted that such frequency locking effects can occur in far more general classes of equations for which (8.6.1, 2) are only a very simple example.

Exercise. Treat three coupled oscillators and show that frequency locking may occur between the frequency differences $\omega_2' - \omega_1'$ and $\omega_3' - \omega_2'$ provided

$$(\omega_2 - \omega_1) - (\omega_3 - \omega_2) \tag{8.6.25}$$

is small enough.

Take as an example the equations

$$\left.\begin{aligned}\dot{u}_1 &= (\lambda' + \mathrm{i}\,\omega_1)u_1 - b_1 u_1 |u_1|^2 + c_1 u_2^2 u_3^* \\ \dot{u}_2 &= (\lambda' + \mathrm{i}\,\omega_2)u_2 - b_2 u_2 |u_2|^2 + c_2 u_2^* u_1 u_3 \\ \dot{u}_3 &= (\lambda' + \mathrm{i}\,\omega_3)u_3 - b_3 u_3 |u_3|^2 + c_3 u_2^2 u_1^*\end{aligned}\right\} . \tag{8.6.26}$$

Hint: use the same decomposition as (8.6.3, 4) and introduce as a new variable

$$\varphi_1 + \varphi_3 - 2\varphi_2 = \Psi . \tag{8.6.27}$$

Form a suitable linear combination of the equations for φ_j which results in an equation for (8.6.27).

8.7 Bifurcation from a Limit Cycle

In the preceding chapter we learned that a system which was originally quiescent can start a coherent oscillation or, in other words, its trajectory in the corresponding multi-dimensional space of the system lies on a limit cycle. When we change the external control parameter further it may happen that the motion on such a limit cycle becomes unstable and we wish to study some of the new kinds of motion which may evolve.

We first perform a general analysis allowing us to separate the order parameters from the slaved mode amplitudes and then we shall discuss a number of typical equations for the resulting order parameters. We shall confine our analysis to autonomous systems described by nonlinear evolution equations of the form

$$\dot{\boldsymbol{q}} = \boldsymbol{N}(\boldsymbol{q}, \alpha) , \tag{8.7.1}$$

where $\boldsymbol{q}(t)$ is, in general, a multi-dimensional vector describing the state of the system. We assume that we have found the limit cycle solution of (8.7.1) for a certain range of the control parameter

$$\boldsymbol{q}_0(t, \alpha) . \tag{8.7.2}$$

We note that farther away from the solution (8.7.2) other possible solutions of (8.7.1) for the same control parameter may exist, but these shall not be considered.

We now change the value of the control parameter α and assume that $\boldsymbol{q}_0$ can be continued into this new region. Because of the assumed periodicity we require

$$\boldsymbol{q}_0(t + T, \alpha) = \boldsymbol{q}_0(t, \alpha) , \tag{8.7.3}$$

where the period T may depend on α,

$$T = T(\alpha) . \tag{8.7.4}$$

In order to study the stability of (8.7.2) we make the usual hypothesis

$$\boldsymbol{q}(t) = \boldsymbol{q}_0(t) + \boldsymbol{w}(t) \tag{8.7.5}$$

and insert it into (8.7.1). This leaves us with

$$\dot{\boldsymbol{q}}_0(t) + \dot{\boldsymbol{w}}(t) = \boldsymbol{N}(\boldsymbol{q}_0(t) + \boldsymbol{w}(t)) . \tag{8.7.6}$$

Assuming that $\boldsymbol{w}$ is a small quantity we may linearize (8.7.6) with respect to $\boldsymbol{w}$ to obtain

$$\dot{\boldsymbol{w}}(t) = L\boldsymbol{w}(t) . \tag{8.7.7}$$

In it the matrix

$$L = (L_{ij}) \tag{8.7.8}$$

has the elements

$$L_{ij} = \left. \frac{\partial N_i(\boldsymbol{q}, \alpha)}{\partial q_j} \right|_{\boldsymbol{q} = \boldsymbol{q}_0(t)} . \tag{8.7.9}$$

Because $\boldsymbol{N}$ contains the periodic function $\boldsymbol{q}_0$ and depends on the control parameter α, L depends on time t and control parameter α

$$L = L(t, \alpha) \tag{8.7.10}$$

and has the same periodicity property as $\boldsymbol{q}_0$, i.e.,

$$L(t + T, \alpha) = L(t, \alpha) . \tag{8.7.11}$$

According to Sect. 2.7 we know that the solutions of (8.7.7) can be written in the form

$$\boldsymbol{w}_k(t) = \mathrm{e}^{\lambda_k t} \boldsymbol{v}_k(t) . \tag{8.7.12}$$

If the λ_k's are nondegenerate, $\boldsymbol{v}_k$ is periodic with period T. Otherwise $\boldsymbol{v}_k$ may contain a finite number of powers of t with periodic coefficients. In order not to overload our presentation we shall assume that all $\boldsymbol{v}_k$ are periodic in time though it is not difficult to extend the following elimination scheme to the more general case. When we use (8.7.1) for $\boldsymbol{q}_0$,

$$\dot{\boldsymbol{q}}_0 = \boldsymbol{N}(\boldsymbol{q}_0) , \tag{8.7.13}$$

and differentiate it with respect to time on both sides, we readily obtain

$$\ddot{q}_0 = L\dot{q}_0 , \tag{8.7.14}$$

which by comparison with (8.7.7) reveals that one of the solutions of (8.7.7) is identical with $\dot{q}_0$

$$w_1 \equiv \dot{q}_0(t) \rightarrow \lambda_1 = 0 . \tag{8.7.15}$$

Evidently w_1 always points along the tangent of the trajectory at each point q_0 at time t. Because the solutions of (8.7.7) span a complete vector space we find that at each time t this space can be spanned by a vector tangential to the trajectory and $n-1$ additional vectors which are transversal to the trajectory. Since we wish to study new trajectories evolving smoothly from the present limit cycle, we shall construct a coordinate system based on the coordinates with respect to the old limit cycle. We expect that due to the nonlinearities contained in (8.7.6) the new trajectory will shift its phase with respect to the original limit cycle, and that it will acquire a certain distance in changing directions away from the limit cycle. Therefore we shall construct the new solution $q(t)$ for the evolving trajectory using the phase angle $\Phi(t)$ and deviations $\xi(t)$ away from the limit cycle. Thus, our hypothesis reads

$$q(t) = q_0(t+\Phi(t)) + \underbrace{\sum_k{}' \xi_k(t)\, v_k(t+\Phi(t))}_{W} . \tag{8.7.16}$$

The prime at the sum means that we shall exclude the solution (8.7.15), i.e.,

$$k \neq 1 , \tag{8.7.17}$$

because this direction is taken care of by the phase angle $\Phi(t)$. In (8.7.16) W is a function of $\xi_k(t)$, t, and $\Phi(t)$, i.e.,

$$W = W(\xi, t+\Phi(t)) . \tag{8.7.18}$$

Inserting hypothesis (8.7.16) into (8.7.1) we readily obtain [omitting the argument $t+\Phi(t)$ in q_0, $\dot{q}_0$, v_k, $\dot{v}_k$] [1]

$$\dot{q}_0 + \dot{q}_0\dot{\Phi} + \sum_k{}' \dot{\xi}_k v_k + \sum_k{}' \xi_k \dot{v}_k (1+\dot{\Phi}) = N(q_0 + W) , \tag{8.7.19}$$

where rhs can be expanded into a power series of the vector W

$$N(q_0 + W) = N(q_0) + LW + H(W) . \tag{8.7.20}$$

[1] The dot ($\dot{}$) means derivative with respect to the total argument, $t + \Phi$, of q, v, or t of Φ, respectively.

In it N and L have been defined before, whereas $\boldsymbol{H}(\boldsymbol{W})$ can be imagined as a power series which starts with second-order terms

$$\boldsymbol{H}(\boldsymbol{W}) = N^{(2)}:\boldsymbol{W}:\boldsymbol{W} + \dots . \tag{8.7.21}$$

Because of (8.7.18), (8.7.21) can be considered as a function of ξ and $t + \Phi$, i.e.,

$$\boldsymbol{H}(\boldsymbol{W}) = \boldsymbol{H}(\xi, t+\Phi) = \sum_{kk'}{}' \xi_k \xi_{k'} N^{(2)}:\boldsymbol{v}_k:\boldsymbol{v}_{k'} + \dots . \tag{8.7.22}$$

We now use the fact that (8.7.2) fulfills (8.7.1), and also use (8.7.7) and the decomposition (8.7.12) to obtain

$$\dot{\boldsymbol{q}}_0 \dot{\Phi} + \sum_k{}' \dot{\xi}_k \boldsymbol{v}_k + \sum_k{}' \xi_k \dot{\boldsymbol{v}}_k \dot{\Phi} = \sum_k{}' \xi_k \lambda_k \boldsymbol{v}_k + \boldsymbol{H}(\xi, t+\Phi) . \tag{8.7.23}$$

In Sect. 2.5 it was shown that we may construct a bi-orthogonal set of functions $\bar{\boldsymbol{v}}_k$ $(k = 1, 2, \dots)$, [compare (2.5.16)]. Multiplying (8.7.23) with $\bar{\boldsymbol{v}}_1$ and using the orthogonality property we readily obtain

$$\dot{\Phi} + \sum_k{}' \xi_k \langle \bar{\boldsymbol{v}}_1 \dot{\boldsymbol{v}}_k \rangle \dot{\Phi} = \langle \bar{\boldsymbol{v}}_1 \boldsymbol{H} \rangle . \tag{8.7.24}$$

We may write explicitly

$$\langle \bar{\boldsymbol{v}}_1 \boldsymbol{H} \rangle \equiv \hat{H}_1 = \sum_{kk'}{}' \xi_k \xi_{k'} \underbrace{\langle \bar{\boldsymbol{v}}_1 N^{(2)}:\boldsymbol{v}_k:\boldsymbol{v}_{k'} \rangle}_{A_{1kk'}(t+\Phi(t))} + \dots . \tag{8.7.25}$$

Similarly, using $\bar{\boldsymbol{v}}_k$ $(k = 2, 3, \dots)$ we readily obtain from (8.7.23)

$$\dot{\xi}_k = \lambda_k \xi_k - \sum{}' \xi_{k'} \underbrace{\langle \bar{\boldsymbol{v}}_k \dot{\boldsymbol{v}}_{k'} \rangle}_{a_{kk'}(t+\Phi)} \dot{\Phi} + \underbrace{\langle \bar{\boldsymbol{v}}_k \boldsymbol{H} \rangle}_{\hat{H}_k} , \quad k, k' \neq 1 \tag{8.7.26}$$

where

$$\hat{H}_k = \sum_{k'k''}{}' A_{kk'k''}(t+\Phi) \xi_{k'} \xi_{k''} + \dots . \tag{8.7.27}$$

We may readily solve (8.7.24) for $\dot{\Phi}$ which yields

$$\dot{\Phi} = \left[1 + \sum_k{}' \xi_k(t) a_{1k}(t+\Phi)\right]^{-1} \hat{H}_1(\xi, t+\Phi) \equiv \hat{f}(\xi, t+\Phi) . \tag{8.7.28}$$

In this way we may also eliminate $\dot{\Phi}$ from (8.7.26):

$$\dot{\xi}_k = \lambda_k \xi_k - \sum_{k'}{}' \xi_{k'} a_{kk'}(t+\Phi) \left[1 + \sum_{k''}{}' \xi_{k''}(t) a_{1k''}(t+\Phi)\right]^{-1} H_1(\xi, t+\Phi)$$
$$+ \hat{H}_k \equiv \lambda_k \xi_k + \hat{G}_k(\xi, t+\Phi) . \tag{8.7.29}$$

We remind the reader that $\hat{H}_k$, $k = 1, 2, \ldots$, depends on $t + \Phi$ via $\boldsymbol{q}_0$, $\dot{\boldsymbol{q}}_0 \equiv \boldsymbol{v}_1$, and $\boldsymbol{v}_k$. Let us study this behavior in more detail. Because $\boldsymbol{q}_0$ is a smooth and periodic function we may expand it into a Fourier series

$$\boldsymbol{q}_0(t) = \sum_m \boldsymbol{c}_m \mathrm{e}^{\mathrm{i}m\omega t}, \quad \text{where} \tag{8.7.30}$$

$$\omega = 2\pi/T. \tag{8.7.31}$$

Similarly we may write

$$\boldsymbol{v}_k(t) = \sum_m \boldsymbol{d}_m \mathrm{e}^{\mathrm{i}m\omega t}. \tag{8.7.32}$$

Because of the assumed form of $\hat{H}_k$ we may put

$$\hat{H}_k(t+\Phi) = \sum_m C_m^{(k)} \mathrm{e}^{\mathrm{i}m\omega(t+\Phi)}. \tag{8.7.33}$$

In particular, in order to make contact with theorems presented earlier, especially in Sect. 6.2, we introduce a new variable $\varphi(t)$ by

$$\varphi(t) = \omega[t + \Phi(t)]. \tag{8.7.34}$$

Differentiating (8.7.34) with respect to time yields

$$\dot{\varphi} = \omega + \omega\dot{\Phi} \tag{8.7.35}$$

into which we insert rhs of (8.7.28). This yields

$$\dot{\varphi} = \omega + f(\xi, \varphi), \tag{8.7.36}$$

where we have used the abbreviation

$$f(\xi, \varphi) = \omega\hat{f}(\xi, \varphi\omega^{-1}). \tag{8.7.37}$$

Similarly, we find from (8.7.29)

$$\dot{\xi}_k(t) = \lambda_k \xi_k(t) + g_k(\xi, \varphi), \tag{8.7.38}$$

where we have used the abbreviation

$$g_k(\xi, \varphi) = \hat{G}_k(\xi, \varphi\omega^{-1}). \tag{8.7.39}$$

We note that f and g_k are 2π-periodic in φ. Equation (8.7.38) can be cast into a vector notation in the form

$$\dot{\xi} = \Lambda\xi + \boldsymbol{G}(\xi, \varphi). \tag{8.7.40}$$

Equations (8.7.36, 40) are still exact and represent the final result of this section. We now study special cases.

8.8 Bifurcation from a Limit Cycle: Special Cases

8.8.1 Bifurcation into Two Limit Cycles

Remember that in our present notation $\lambda_1 = 0$. We assume that we have labeled the other eigenvalues in such a way that the one with the biggest real part carries the index 2. We assume that λ_2 is real and nondegenerate

$$\lambda_2 \text{ real}, \quad \geqq 0, \tag{8.8.1}$$

and that

$$\mathrm{Re}\{\lambda_k\} \leqq C < 0, \quad k = 3, 4, \ldots . \tag{8.8.2}$$

For small enough ξ we may invoke the slaving principle of Chap. 7 which allows us to do away with all the slaved variables with $k = 3, 4, \ldots$ and to derive equations for the order parameters alone,

$$\left.\begin{array}{ll} \xi(t) \equiv u(t), & \text{real} \\ \varphi(t) & \end{array}\right\} \begin{array}{l} \text{order} \\ \text{parameters} . \end{array} \tag{8.8.3}$$

According to the slaving principle the original set of (8.7.36, 40) reduces to the equations

$$\dot{u} = \lambda_u u + g(u, \varphi), \tag{8.8.4}$$

$$\dot{\varphi} = \omega + f(u, \varphi) . \tag{8.8.5}$$

In it g and f can be constructed explicitly as power series (of u) whose coefficients are 2π-periodic functions of φ. We may decompose g according to

$$g = g_2 u^2 + g_3 u^3 + \ldots . \tag{8.8.6}$$

Each coefficient g_j may be decomposed again into a constant term and another one which is 2π-periodic in φ

$$g_j = g_j(\varphi) = g_{j,1} + g_{j,2}(\varphi) \tag{8.8.7}$$

so that

$$\int_0^{2\pi} d\varphi\, g_{j,2}(\varphi) = 0 \tag{8.8.8}$$

holds. A similar decomposition can be used for f

$$f = f_1 u + f_2 u^2 + \dots, \tag{8.8.9}$$

$$f_j = f_j(\varphi) = f_{j,1} + f_{j,2}(\varphi) \tag{8.8.10}$$

with the property

$$\int_0^{2\pi} d\varphi \, f_{j,1}(\varphi) = 0 . \tag{8.8.11}$$

We shall first focus our attention on the special case

$$g_2 = f_1 = 0 , \quad g_3 \equiv -b < 0 . \tag{8.8.12}$$

From Sect. 8.4 it will be clear that the resulting equations (8.8.4, 5) acquire a form treated there, with one difference, however, which we shall explain now. When we make the hypothesis

$$u = u_0 + \eta \tag{8.8.13}$$

and require

$$\lambda_u u_0 - b u_0^3 = 0 \tag{8.8.14}$$

we find the two solutions

$$u_0 = \pm \sqrt{\lambda_u / b} \tag{8.8.15}$$

which are distinguished by the sign in front of the square root. We note that in the case of Hopf bifurcations we must choose the positive sign because r_0 has to be positive. Here, however, we take the two signs. Either choice of u_0 takes (8.8.4, 5) [under the assumption of (8.8.12)] to equations which we know from Sect. 8.4 so that we need not treat this case here again explicitly. We can rather discuss the final result.

To illustrate the resulting solution we consider the case in which the original limit cycle lies in a plane. Then the two solutions (8.8.15) mean that two limit cycles evolve to the outer and inner side of the limit cycle. According to the analysis of Sect. 8.7, the phase motion may suffer a shift of frequency with respect to the original limit cycle. Furthermore, the phase may have superimposed on its steady motion oscillations with that renormalized frequency. Also deviations from the new distance u_0 may occur which are in the form of oscillations at the renormalized frequency ω_r. If the limit cycles lie in a higher-dimensional space, the newly evolving two trajectories need not lie in a plane but may be deformed curves.

8.8.2 Period Doubling

Let us assume a situation similar to that in the preceding section, with the only difference that λ_2 is complex

$$\lambda_2 = \lambda_u' + \mathrm{i}\,\omega_2 \,. \tag{8.8.16}$$

We shall admit that also $\omega_2 = 0$ is included provided that λ_u' remains twofold degenerate. After having applied the slaving principle we again find an order parameter equation which we assume to be in the specific form

$$\dot{u} = (\lambda_u' + \mathrm{i}\,\omega_2)u - bu^3 \mathrm{e}^{-\mathrm{i}\omega t} \,. \tag{8.8.17}$$

For simplicity we assume b real though this is not essential. It is assumed that there may be additional terms to (8.8.17) but that these terms can be treated as a perturbation. It is assumed with (8.8.17) that the equation for φ simply reads

$$\dot{\varphi} = \omega \,. \tag{8.8.18}$$

We shall show that (8.8.17) allows for a solution which oscillates at half the frequency or, in other words, whose period has doubled with respect to the original limit cycle. To this end we make the hypothesis

$$u = \mathrm{e}^{\mathrm{i}\omega t/2} y(t) \,, \tag{8.8.19}$$

where y is a complex variable. Inserting (8.8.19) into (8.8.17) yields

$$\dot{y} = (\lambda_u' + \mathrm{i}\,\omega_2 - \mathrm{i}\,\omega/2)y - by^3 \,. \tag{8.8.20}$$

It allows for a time-independent stable solution $y_0 = 0$ for $\lambda_u' < 0$ and for solutions (besides $y_0 = 0$)

$$\left.\begin{aligned} &y_0^2 = [\lambda_u' + \mathrm{i}(\omega_2 - \omega/2)]/b = \lambda_0\,\mathrm{e}^{\mathrm{i}\psi_0}, \quad \text{i.e.,} \\ &y_0 = \pm\sqrt{\lambda_0}\,\mathrm{e}^{\mathrm{i}\psi_0/2} \end{aligned}\right\} \text{ for } \lambda_u' > 0 \,. \tag{8.8.21}$$

One may readily show that the solutions (8.8.21) are stable (provided $\lambda_u' > 0$). Thus we have shown that a stable solution with double period exists. Of course, in the general case one now has to study higher order terms and show that they can be treated as perturbations.

8.8.3 Subharmonics

We now want to show that period doubling is just a special case of the generation of subharmonics in which the frequency ω is just the nth part of the original frequency of the limit cycle. To substantiate our claim we assume that the order parameter equation has the form

$$\dot{u} = \lambda_u - bu^{n+1}\mathrm{e}^{-\mathrm{i}\omega t}. \tag{8.8.22}$$

Making the hypothesis

$$u = y\mathrm{e}^{\mathrm{i}\omega t/n}, \tag{8.8.23}$$

where y is assumed to be constant, we readily obtain

$$y(\lambda - \mathrm{i}\omega/n) - by^{n+1} = 0, \quad \lambda \equiv \lambda_u, \tag{8.8.24}$$

which allows for the solutions

$$y_0 = 0 \qquad \text{or} \tag{8.8.25}$$

$$y_0^n = (\lambda - \mathrm{i}\omega/n)/b = \lambda_0\,\mathrm{e}^{\mathrm{i}\psi_0}, \tag{8.8.26}$$

or, after taking the nth roots

$$y = \exp(2\pi\mathrm{i}m/n)\sqrt[n]{\lambda_0}\,\mathrm{e}^{\mathrm{i}\psi_0/n}, \quad m \text{ integer}. \tag{8.8.27}$$

By linear stability analysis we can show that (8.8.23) with (8.8.27) is a stable solution provided $\lambda_u > 0$. While here we have assumed that λ is real and positive, one may readily extend this analysis to a complex λ and complex b provided the real part of $\lambda > 0$.

Let us briefly discuss examples in which the other remaining terms omitted in our above analysis can be treated as perturbation. Let us consider an equation of the form

$$\dot{u} = \lambda_u u - bu^3\mathrm{e}^{-\mathrm{i}\omega t} + cu^5 + du|u|^4 + \dots, \tag{8.8.28}$$

where the perturbative terms are of order u^5 or higher. Making the hypothesis

$$u = \mathrm{e}^{\mathrm{i}\omega t/2}y(t) \tag{8.8.29}$$

we find

$$\dot{y} = (\lambda_u - \mathrm{i}\omega/2)y - by^3 + cy^5\mathrm{e}^{\mathrm{i}\omega t(5/2)} + d\mathrm{e}^{\mathrm{i}\omega t/2}y|y|^4\dots. \tag{8.8.30}$$

From a comparison between this equation and (8.5.1), it transpires that further analysis is similar to that in Sect. 8.5. Thus we must make the transformation

$$y = y_0 + \eta \tag{8.8.31}$$

with y_0

$$(\lambda_u - \mathrm{i}\omega/2) - by_0^2 = 0. \tag{8.8.32}$$

We leave the further discussion to the reader.

We now present an example indicating that there are cases in which it is not at all evident from the beginning which terms can be considered as perturbations. Some kind of bifurcation may occur in which, if one solution is singled out, one class of terms can be considered as small, but if another class of solutions is singled out, another class of terms can be considered as small. Let us consider an example,

$$\dot{u} = \lambda_u u - b u^3 (e^{-i\omega t} + e^{i\omega t}), \tag{8.8.33}$$

where λ_u and b are assumed real. Then in lowest approximation we may choose

$$u = e^{i(\omega/2)t} y \quad \text{or} \tag{8.8.34}$$

$$u = e^{-i(\omega/2)t} y \tag{8.8.35}$$

as a solution. Depending on the choice (8.8.34 or 35), (8.8.33) is cast into the form

$$\dot{y} = (\lambda_u \mp i\omega/2) y - b y^3 (1 + e^{\pm i\varphi}) \tag{8.8.36}$$

with the additional equation

$$\dot{\varphi} = 2\omega . \tag{8.8.37}$$

These equations are the same type as (8.5.3, 4), already discussed in detail (here r is complex). There, by suitable scaling, we have put in evidence that the terms containing $\exp(\pm i\varphi)$ can be considered as small perturbations. Thus (8.8.33) allows for two equivalent solutions, (8.8.34, 35) in which y is a constant on which small amplitude oscillations at frequencies $m \cdot \omega/2$ are superimposed. In physics the neglect of the term $\exp(\pm i\varphi)$ compared to 1 (where $\varphi = 2\omega t$) is known as "rotating wave approximation".

8.8.4 Bifurcation to a Torus

In Sect. 8.8.1 we discussed the case in which λ_2, i.e., the eigenvalue with the biggest real part of all eigenvalues, is real. Here we assume that this λ_2 is complex

$$\lambda_2 = \lambda_u' + i\omega_2 , \quad \lambda_u' \geqslant 0 . \tag{8.8.38}$$

Again we assume that for all other eigenvalues

$$\mathrm{Re}\{\lambda_k\} \leqslant C < 0 , \quad k = 3, 4, \dots . \tag{8.8.39}$$

Invoking the slaving principle, we have to study an order parameter equation for u alone. In a first preparatory step we assume that this order parameter equation has the form

$$\dot{u} = (\lambda_u' + i\omega_2) u - b u |u|^2 , \tag{8.8.40}$$

where for simplicity we assume b real and independent of an additional phase angle φ. We are familiar with this type of equation from discussion of the Hopf bifurcation (Sect. 8.4). We know that the solution can be written in the form

$$u = r e^{i\omega_2 t}, \tag{8.8.41}$$

where r obeys the equation

$$\dot{r} = \lambda_u' r - b r^3, \tag{8.8.42}$$

which for $\lambda_u' > 0$ possesses a nonvanishing stable stationary solution $r = r_0$. To visualize the meaning of the corresponding solution we remind the reader that the solution of the total system $\boldsymbol{q}$ can be written in the form

$$\boldsymbol{q} = \boldsymbol{q}_0(t+\Phi) + u(t)\boldsymbol{v}_2(t+\Phi) + u^*(t)\boldsymbol{v}_2^*(t+\Phi) + \dots, \tag{8.8.43}$$

where the dots indicate the amplitudes of the slaved variables which are smaller than those which are explicitly exhibited. Here $\boldsymbol{v}_2$ and its conjugate complex are eigensolutions to (8.7.7) with (8.7.12). Inserting (8.8.41) into (8.8.43) we find

$$\boldsymbol{q} = \boldsymbol{q}_0 + \cos(\omega_2 t) \cdot \mathrm{Re}\{\boldsymbol{v}_2\} + \sin(\omega_2 t) \cdot \mathrm{Im}\{\boldsymbol{v}_2\}. \tag{8.8.44}$$

The real part of $\boldsymbol{v}_2$ and the imaginary part of $\boldsymbol{v}_2$ are linearly independent and present two vectors which move along the original limit cycle. The solution (8.8.44) indicates a rotating motion in the frame of these two basic vectors $\mathrm{Re}\{\boldsymbol{v}_2\}$ and $\mathrm{Im}\{\boldsymbol{v}_2\}$, i.e., the end point of the vector spirals around the limit cycle while the origin of the coordinate system moves along $\boldsymbol{q}_0$. If the two frequencies ω_2 and $\omega_1 \equiv \omega = \dot{\varphi} \equiv \dot{\varphi}_1$ are not commensurable the trajectory spans a whole torus. Thus in our present case we have a simple example of the bifurcation of a limit cycle to a two-dimensional torus. In most practical applications the order parameter equation for u will not have the simple form (8.8.40) but several complications may occur. In particular, we may have additional terms which introduce a dependence of the constant b on φ_1. A simple example is the equation

$$\dot{u} = (\lambda_u' + i\omega_2)u - bu|u|^2 - c(\underbrace{\omega_1 t}_{\varphi_1})u|u|^2, \tag{8.8.45}$$

where c is 2π-periodic in φ_1. Hypothesis (8.8.41) takes us from (8.8.45) to the equations

$$\dot{r} = \lambda_u' r - b r^3 - r^3 c(\varphi_1) \quad \text{and} \tag{8.8.46}$$

$$\dot{\varphi}_1 = \omega_1 t, \tag{8.8.47}$$

which are of precisely the same form encountered previously in the case of Hopf bifurcation. From this analysis we know that basically the bifurcation of the limit cycle of a torus is still present.

Things become considerably more complicated if the equation for the order parameter u is of the form

$$\dot{u} = (\lambda_u' + \mathrm{i}\omega_2)u - bu|u|^2 + g(u, u^*, \varphi_1) \tag{8.8.48}$$

and that for the phase angle $\varphi \equiv \varphi_1$ of the form

$$\dot{\varphi}_1 = \omega_1 + f_1(u, u^*, \varphi_1) \,. \tag{8.8.49}$$

The functions g and f_1 are 2π-periodic in φ_1. The hypothesis

$$u = r(t)\,\mathrm{e}^{\mathrm{i}\varphi_2(t)} \tag{8.8.50}$$

leads to the equations

$$\dot{r} = \lambda_u' r - br^3 + \hat{g}(r, \varphi_1, \varphi_2) \tag{8.8.51}$$

$$\dot{\varphi}_2 = \omega_2' + \hat{f}_2(r, \varphi_1, \varphi_2) \,, \tag{8.8.52}$$

to which we must add the equation

$$\dot{\varphi}_1 = \omega_1 + \hat{f}_1(r, \varphi_1, \varphi_2) \tag{8.8.53}$$

which stems from (8.8.49) and where $\hat{f}_1$ (like $\hat{g}$ and f_2) is 2π-periodic in φ_1, φ_2. The discussion of the solutions of these equations is somewhat subtle, at least in the general case, because a perturbation expansion with respect to $\hat{g}$ and $\hat{f}_1$, $\hat{f}_2$ requires the validity of a KAM condition on ω_1, ω_2. Since this whole discussion will turn out to be a special case of the discussion on bifurcations from a torus to other tori, we shall postpone it to the end of Sect. 8.10.2.

8.9 Bifurcation from a Torus (Quasiperiodic Motion)

As in our preceding chapters we start from an equation of the form

$$\dot{\boldsymbol{q}}(t) = \boldsymbol{N}(\boldsymbol{q}(t), \alpha) \tag{8.9.1}$$

with the by now familiar notation. We assume that (8.9.1), for a certain range of control parameters α, allows quasiperiodic solutions $\boldsymbol{q}_0$ which we may write in the form

$$\boldsymbol{q}_0(t, \alpha) = \sum_{\boldsymbol{m}} \boldsymbol{c}_{\boldsymbol{m}} \mathrm{e}^{\mathrm{i}\boldsymbol{m}\cdot\boldsymbol{\omega}t} \,, \tag{8.9.2}$$

where we have used the abbreviation

$$\boldsymbol{m}\cdot\boldsymbol{\omega} = m_1\omega_1 + m_2\omega_2 + \ldots + m_M\omega_M \tag{8.9.3}$$

and where the m's run over all integers. As discussed in Sects. 1.12 and 8.8, such a motion can be visualized as motion on a torus. We assume in the following that the trajectories lie dense on that torus. We may parametrize such a torus by phase angles Φ_j, $j = 1, \ldots, M$, so that by use of these variables the vector $\boldsymbol{r}$ whose end point lies on the torus can be written in the form

$$\boldsymbol{r}(\Phi_1, \Phi_2, \ldots, \Phi_M) . \tag{8.9.4}$$

Because the trajectories lie dense on the torus[1] we may subject them to the initial condition

$$q_0(0, \alpha) = \boldsymbol{r} . \tag{8.9.5}$$

This allows us to construct the solutions (8.9.2) which fulfill the initial condition (8.9.5) explicitly by putting

$$q_0(t, \alpha; \boldsymbol{\Phi}) = \sum_m c_m \exp[\mathrm{i}\boldsymbol{m}(\omega t + \hat{\boldsymbol{\Phi}})] , \tag{8.9.6}$$

where we have used the abbreviation

$$\hat{\boldsymbol{\Phi}} = (\omega_1 \Phi_1, \omega_2 \Phi_2, \ldots, \omega_m \Phi_M) ; \tag{8.9.7}$$

whereas the vector $\boldsymbol{\Phi}$ is defined by

$$\boldsymbol{\Phi} = (\Phi_1, \Phi_2, \ldots) . \tag{8.9.8}$$

The first steps of our subsequent procedure are now well known from the preceding chapters. We assume that the form of the solution (8.9.6) still holds when we increase the control parameter α further, in particular into a region where the old torus loses its linear stability.

To study the stability we perform the stability analysis, considering the equation

$$\dot{w}(t) = L(t) w(t) , \tag{8.9.9}$$

where the matrix

$$L = (L_{jk}) \tag{8.9.10}$$

---

[1] More precisely speaking, we assume that with (8.9.2) as solution of (8.9.1) also

$$q_0(t, \alpha, \boldsymbol{\Phi}) = \sum_m c_m \exp[\mathrm{i}\boldsymbol{m}(\omega t + \hat{\boldsymbol{\Phi}})] \tag{8.9.6}$$

is a solution to (8.9.1), where q_0 is assumed to be $C^k (k \geqslant 1)$ with respect to $\boldsymbol{\Phi}$ and k will be specified later. For certain cases, discussed later, we require that $q_0(t, \alpha, \boldsymbol{\Phi})$ is analytic with respect to $\boldsymbol{\Phi}$ in a certain domain.

is defined by

$$L_{jk} = \left. \frac{\partial N_j(\boldsymbol{q}, \alpha)}{\partial q_k} \right|_{\boldsymbol{q}_0 = \boldsymbol{q}_0(t, \alpha; \boldsymbol{\Phi})} . \tag{8.9.11}$$

Because $\boldsymbol{q}_0$ is given by (8.9.6), L can be cast into a similar form

$$L_{jk} = L_{jk}(t, \alpha, \boldsymbol{\Phi}) = \sum_{m} L_{jk;m} \exp\left[\mathrm{i}\boldsymbol{m}(\boldsymbol{\omega} t + \hat{\boldsymbol{\Phi}})\right] . \tag{8.9.12}$$

In the following we shall drop the index α. Clearly (8.9.12) is a quasiperiodic function. We now refer to the results of Chap. 3, especially Sect. 3.7. There we studied the general form of solutions of (8.9.9) with quasiperiodic coefficients, showing that under certain conditions on the generalized characteristic exponents and some more technical conditions the form of the solution of (8.9.9) can be explicitly constructed, namely as

$$\boldsymbol{w}_k(t) = \mathrm{e}^{\lambda_k t} \boldsymbol{v}_k(t) . \tag{8.9.13}$$

This form holds especially if the characteristic exponents (or generalized characteristic exponents) are all different from each other, if L is sufficiently often differentiable with respect to Φ_j, $j = 1, \ldots, M$, and if a suitable KAM condition is fulfilled by the ω's. In this case, the $\boldsymbol{v}_k$'s are quasiperiodic functions of time, whose dependence on $\boldsymbol{\Phi}$ or $\hat{\boldsymbol{\Phi}}$, respectively, is of the general form of rhs of (8.9.6). Here we need a version of that theorem in which all λ_k are different from each other except those for which $\lambda_k = 0$, whose $\boldsymbol{v}_k$ will be constructed now. To this end we start from the equation

$$\dot{\boldsymbol{q}}_0 = \boldsymbol{N}(\boldsymbol{q}_0, \alpha) , \tag{8.9.14}$$

which we differentiate on both sides with respect to Φ_l. On account of (8.9.11), one immediately verifies the relation

$$\frac{d}{dt} (\partial \boldsymbol{q}_0 / \partial \Phi_l) = L (\partial \boldsymbol{q}_0 / \partial \Phi_l) . \tag{8.9.15}$$

This relation tells us that we have found solutions to (8.9.9) given by

$$\partial \boldsymbol{q}_0 / \partial \Phi_l = \boldsymbol{w}_l , \quad l = 1, \ldots, M . \tag{8.9.16}$$

Since lhs of (8.9.16) is quasiperiodic, we have thus found solutions in the form (8.9.13) with $\lambda_l = 0$. We shall denote all other solutions to (8.9.9) by

$$\boldsymbol{w}_k , \quad k = M + 1, \ldots . \tag{8.9.17}$$

We now wish to find a construction for the solutions which bifurcate from the old torus which becomes unstable. We do this by generalizing our considerations on the bifurcation of a limit cycle. We assume self-consistently that the new torus or the new tori are not far away from the old torus, so that we may utilize small vectors pointing from the old torus to the new torus. On the other hand, we have to take into account the fact that when bifurcation takes place, the corresponding points on the old trajectory and the new one can be separated arbitrarily far from each other when time elapses.

To take into account these two features we introduce the following coordinate system. We utilize the phase angles as local coordinates on the old torus, and introduce vectors $\boldsymbol{v}$ which point from each local point on the old torus to the new point at the bifurcating new torus. The local coordinates which are transversal to the old torus are provided by (8.9.17). These arguments lead us to the following hypothesis for the bifurcating solutions (the prime on the sum means exclusion of $k = 1, 2, \dots, M$)

$$\boldsymbol{q}(t) = \boldsymbol{q}_0(t, \boldsymbol{\Phi}) + \underbrace{\sum{}' \xi_k \boldsymbol{v}_k(t, \boldsymbol{\Phi})}_{\boldsymbol{W}}, \tag{8.9.18}$$

where we assume that

$$\boldsymbol{\Phi} = \boldsymbol{\Phi}(t) \tag{8.9.19}$$

is a function of time still to be determined. Also $\boldsymbol{v}_k$ is a function of time

$$\boldsymbol{v}(t, \boldsymbol{\Phi}(t)) . \tag{8.9.20}$$

Inserting (8.9.18) into (8.9.1) and performing the differentiations we readily find

$$\begin{aligned} &\dot{\boldsymbol{q}}_0 + \sum_l \frac{\partial \boldsymbol{q}_0}{\partial \Phi_l} \dot{\Phi}_l + \sum_k{}' \dot{\xi}_k \boldsymbol{v}_k + \sum_{k,l}{}' \xi_k \frac{\partial \boldsymbol{v}_k}{\partial \Phi_l} \dot{\Phi}_l + \sum_k{}' \xi_k \frac{\partial \boldsymbol{v}_k}{\partial t} \\ &\quad = \boldsymbol{N}(\boldsymbol{q}_0) + \boldsymbol{L} \sum_k{}' \xi_k \boldsymbol{v}_k + \boldsymbol{H}(\boldsymbol{W}), \end{aligned} \tag{8.9.21}$$

where we have expanded the rhs into powers of $\boldsymbol{W}$. In this way, $\boldsymbol{H}$ is given by an expression of the form

$$\boldsymbol{H}(\boldsymbol{W}) = \boldsymbol{N}^{(2)} : \sum_k \xi_k \boldsymbol{v}_k : \sum_{k'} \xi_{k'} \boldsymbol{v}_{k'} + \dots, \tag{8.9.22}$$

where we have indicated higher-order terms by dots. It is, of course, a simple task to write down such higher-order terms explicitly in the way indicated by the second-order term.

We now make use of results of Sect. 2.5, where a set of vectors $\bar{\boldsymbol{v}}$ was constructed orthogonal on the solutions $\boldsymbol{v}$ of (8.9.13). Multiplying (8.9.21) by

$$\bar{\boldsymbol{v}}_l, \quad l = 1, \dots, M, \tag{8.9.23}$$

we readily obtain

$$\dot{\Phi}_l + \sum_j{}' \sum_{l'} \xi_j \underbrace{\langle \bar{v}_l \cdot \partial v_j / \partial \Phi_{l'} \rangle}_{a_{jll'}} \dot{\Phi}_{l'} = \underbrace{\langle \bar{v}_l H \rangle}_{\hat{H}_l} , \tag{8.9.24}$$

while by multiplication of (8.9.21) by

$$\bar{v}_k , \quad k = M+1, \dots , \tag{8.9.25}$$

we obtain

$$\dot{\xi}_k = \lambda_k \xi_k - \sum_j{}' \sum_{l'} \xi_j \underbrace{\langle \bar{v}_k \cdot \partial v_j / \partial \Phi_{l'} \rangle}_{a_{jkl'}} \dot{\Phi}_{l'} + \underbrace{\langle \bar{v}_k H \rangle}_{\hat{H}_k} . \tag{8.9.26}$$

Introducing the matrix K with matrix elements

$$K_{ll'} = \delta_{ll'} + \sum_j{}' \xi_j a_{jll'} \tag{8.9.27}$$

and taking $\boldsymbol{\Phi}$ and $\hat{\boldsymbol{H}}$ as column vectors, we may cast the set of equations (8.9.24) into the form

$$K \dot{\boldsymbol{\Phi}} = \hat{\boldsymbol{H}} . \tag{8.9.28}$$

Because we assume that ξ_j is a small quantity we may immediately solve (8.9.28) [cf. (8.9.24)] by means of

$$\dot{\boldsymbol{\Phi}} = K^{-1} \hat{\boldsymbol{H}} = \hat{\boldsymbol{f}}(\xi, t, \boldsymbol{\Phi}) , \tag{8.9.29}$$

where the last term of this equation is just an abbreviation. Expressing $\dot{\Phi}_{l'}$ which occurs in (8.9.26) by rhs of (8.9.29), we transform (8.9.26) into

$$\dot{\xi}_k = \lambda_k \xi_k + \hat{G}_k(\xi, t, \boldsymbol{\Phi}) . \tag{8.9.30}$$

We note that t and $\boldsymbol{\Phi}$, on which the right-hand sides of (8.9.29, 30) depend, occur only in the combination

$$\omega_j t + \omega_j \Phi_j = \varphi_j . \tag{8.9.31}$$

Therefore we introduce φ_j as a new variable which allows us to cast (8.9.29) into the form

$$\dot{\varphi}_j = \omega_j + \tilde{f}_j(\xi, \varphi) , \quad j = 1, \dots M , \tag{8.9.32}$$

where we used the abbreviation

$$\tilde{f}_j(\xi, \varphi) = \omega_j \hat{f}_j(\xi, 0, \{\varphi_j \omega_j^{-1}\}) . \tag{8.9.33}$$

In a similar fashion we may cast equations (8.9.30) into the form

$$\dot{\xi}_k = \lambda_k \xi_k + \tilde{g}_k(\xi, \varphi) \tag{8.9.34}$$

with the abbreviation

$$\tilde{g}_k(\xi, \varphi) = \hat{G}_k(\xi, 0, \{\varphi_j \omega_j^{-1}\}) . \tag{8.9.35}$$

We note that $\tilde{f}$ and $\tilde{g}$ are 2π-periodic in φ_j.

Within the limitation that we may use the form (8.9.13) of the solutions of the linearized equations (8.9.9), (8.9.32, 34) are still general. Of course, another approach would be to forget the way in which we arrive at (8.9.32, 34) from (8.9.1) and take (8.9.32, 34) just as given starting equations. These equations describe for $\tilde{f} = \tilde{g} = \xi = \mathbf{0}$ the motion on the old torus while for nonvanishing $\tilde{f}$ and $\tilde{g}$ we have to establish the form of the newly evolving trajectories.

8.10 Bifurcation from a Torus: Special Cases

8.10.1 A Simple Real Eigenvalue Becomes Positive

We now numerate the eigenvalues λ_k in such a way that λ_{M+1} has the biggest real part. We assume

$$\lambda_{M+1} \geqslant 0 , \quad \lambda_{M+1} \text{ real} \qquad \text{and} \tag{8.10.1}$$

$$\mathrm{Re}\{\lambda_k\} \leqslant C < 0 , \quad k = M+2, \dots \tag{8.10.2}$$

and put

$$\lambda_{M+1} = \lambda_u , \quad \xi_{M+1} = u . \tag{8.10.3}$$

Here u plays the role of the order parameter, while ξ_k, $k \geqslant M+2$ plays the role of the slaved variables. Applying the slaving principle we reduce the set of equations (8.9.34 and 32) to equations of the form

$$\dot{u} = \lambda_u u + g(u, \varphi) \tag{8.10.4}$$

$$\dot{\varphi} = \omega + f(u, \varphi) . \tag{8.10.5}$$

We shall focus our attention on the case in which g can be written in the form

$$g(u, \varphi) = -bu^3 + u^3 h(u, \varphi) , \quad b > 0 , \tag{8.10.6}$$

where we may split h further into the following two parts:

$$h = h_1(u) + h_2(u, \varphi) , \tag{8.10.7}$$

where h_1, h_2 are assumed to have the following properties

$$h_1(u) = O(u)\,, \tag{8.10.8}$$

$$h_2(u, \varphi) = O(1)\,, \tag{8.10.9}$$

$$\int_0^{2\pi} \dots \int_0^{2\pi} d\varphi_1 \dots d\varphi_M h_2 = 0\,. \tag{8.10.10}$$

In addition we assume

$$f(u, \varphi) = f_1(u) + f_2(u, \varphi)\,, \qquad \text{where} \tag{8.10.11}$$

$$f_1(u) = O(u^3)\,, \tag{8.10.12}$$

$$f_2(u) = O(u^2)\,, \tag{8.10.13}$$

$$\int_0^{2\pi} \dots \int_0^{2\pi} d\varphi_1 \dots d\varphi_M f_2 = \mathbf{0}\,. \tag{8.10.14}$$

(These conditions can be easily weakened, e.g., by requiring $f_1(u) = O(u^2)$ and $f_2(u) = O(u)$ [Sect. 8.8.1].)

We are familiar with similar properties from our treatment of the bifurcation of a limit cycle. To continue, we now consider a region somewhat above the transition point and assume

$$\lambda_u = \lambda_0 \varepsilon^2 \qquad \text{and} \tag{8.10.15}$$

$$u = \varepsilon(u_0 + \eta)\,. \tag{8.10.16}$$

We determine u_0 by

$$u_0^2 = \lambda_0 / b \tag{8.10.17}$$

so that

$$u_0 = \pm\sqrt{\lambda_0 / b} \tag{8.10.18}$$

holds. This relation indicates that two new tori split off the old torus and keep a mean distance u_0 from the old torus. Introducing in addition to (8.10.15, 16) the new scaling of time

$$t = \tau / \varepsilon^2\,, \tag{8.10.19}$$

by using relation (8.10.17), we may cast (8.10.4) into

$$d\eta/d\tau = -2\lambda_0\eta - 3bu_0\eta^2 - b\eta^3$$
$$+ (u_0+\eta)^3 - [\varepsilon\tilde{h}_1(u_0+\eta,\varepsilon) + \tilde{h}_2(\varepsilon, u_0+\eta,\varphi)] , \tag{8.10.20}$$

where

$$\varepsilon\tilde{h}_1(u_0+\eta,\varepsilon) = h_1(\varepsilon(u_0+\eta)) , \tag{8.10.21}$$

$$\tilde{h}_2(\varepsilon, u_0+\eta,\varphi) = h_2(\varepsilon(u_0+\eta),\varphi) . \tag{8.10.22}$$

On account of assumption (8.10.8), $\tilde{h}_1$ is $O(1)$ so that $\varepsilon\tilde{h}$ can be treated as a small perturbation. Because $\tilde{h}_2$ depends on rapid oscillations due to φ through scaling of time, $\tilde{h}_2$ can also be considered as a perturbation $\propto \varepsilon$ from a formal point of view. Note that in the present case of quasiperiodic motion such an argument holds only if the ω's fulfill a KAM condition. The same scaling brings us from (8.10.5) to

$$d\varphi/d\tau = \omega/\varepsilon^2 + \varepsilon\tilde{f}_1(u_0+\eta,\varepsilon) + \tilde{f}_2(u_0+\eta,\varepsilon,\varphi) , \quad \text{where} \tag{8.10.23}$$

$$\varepsilon\tilde{f}_1(u+\eta,\varepsilon) = \frac{1}{\varepsilon^2} f_1(\varepsilon(u+\eta)) \tag{8.10.24}$$

is of order $O(\varepsilon)$, and

$$\tilde{f}_2(u_0+\eta,\varepsilon,\varphi) = \frac{1}{\varepsilon^2} f_2(\varepsilon(u_0+\eta),\ \varphi) \tag{8.10.25}$$

is of order unity, but due to its dependence on φ acts like a perturbation $\propto \varepsilon$.

Superficially, (8.10.20, 23) are familiar from Sects. 8.7, 8.8. The reader should be warned, however, not to apply the previous discussion directly to the case of (8.10.20, 23) without further reflection. The basic difference between (8.10.20, 23) and the equations of Sects. 8.7, 8.8 is that here we are dealing with quasiperiodic motion, while Sects. 8.7, 8.8. dealt with periodic motion only. It is appropriate here to recall the intricacies of Moser's theorem. Before we discuss this question any further we shall treat the case in which λ_{M+1} is complex. We shall show that in such a case we are led to equations which again have the structure of equations (8.10.20, 23), so that we can discuss the solutions of these equations simultaneously.

8.10.2 A Complex Nondegenerate Eigenvalue Crosses the Imaginary Axis

The initial steps of our procedure are by now well known to the reader. We put

$$\lambda_{M+1} = \lambda_u' + \mathrm{i}\,\omega_{M+1} \quad \text{and} \tag{8.10.26}$$

$$\xi_{M+1} = u \tag{8.10.27}$$

and assume

$$\mathrm{Re}\{\lambda_k\} \leqslant C < 0\,, \quad k = M+2, \ldots\,. \tag{8.10.28}$$

Application of the slaving principle allows the reduction of the original set of equations (8.9.32, 34) to the two equations

$$\dot{u} = (\lambda_u' + \mathrm{i}\omega_{M+1})u + g(u, \varphi) \qquad \text{and} \tag{8.10.29}$$

$$\dot{\varphi} = \omega + f(u, \varphi) \tag{8.10.30}$$

for the order parameters u and φ. Let us consider as an example

$$g(u, \varphi) = -bu|u|^2 + m(u, \varphi)\,, \quad b = b' + \mathrm{i}b'', \quad b' > 0\,, \tag{8.10.31}$$

where we shall specify $m(u, \varphi)$ below.

The hypothesis

$$u = r(t)\exp[\mathrm{i}\varphi_{M+1}(t)] \tag{8.10.32}$$

allows us to transform (8.10.29) into

$$\dot{r} = \lambda_u' r - br^3 + \mathrm{Re}\{\exp(-\mathrm{i}\varphi_{M+1})m(r\exp(\mathrm{i}\varphi_{M+1}), \varphi)\}\,. \tag{8.10.33}$$

Introducing the new vector $\hat{\varphi}$ which contains the additional phase φ_{M+1},

$$\hat{\varphi} = \begin{pmatrix} \varphi_1 \\ \vdots \\ \varphi_{M+1} \end{pmatrix}\,. \tag{8.10.34}$$

we may cast (8.10.33) into the form

$$\dot{r} = \lambda_u' r - b' r^3 + \hat{g}(r, \hat{\varphi})\,. \tag{8.10.35}$$

In a similar fashion we obtain from (8.10.29)

$$\dot{\varphi}_{M+1} = \omega_{M+1} - b'' r^2 + \underbrace{\mathrm{Im}\{\exp(-\mathrm{i}\varphi_{M+1})m(r\exp(\mathrm{i}\varphi_{M+1}), \varphi)r^{-1}\}}_{f_{M+1}} \tag{8.10.36}$$

which can be cast into the form

$$d\hat{\varphi}/dt = \hat{\omega} + \hat{f} \qquad \text{with} \tag{8.10.37}$$

$$\hat{f} = \begin{pmatrix} f_1 \\ \vdots \\ f_{M+1} \end{pmatrix} \qquad \text{and} \tag{8.10.38}$$

$$\hat{\omega} = \begin{pmatrix} \omega_1 \\ \vdots \\ \omega_{M+1} \end{pmatrix}\,. \tag{8.10.39}$$

From a formal point of view (8.10.35, 37) have the same structure as (8.9.32), (8.9.34) which then can be transformed into (8.10.20, 23).

In precisely the same way as we discussed in Sect. 8.10.1, (8.10.35, 37) allow for the introduction of scaled variables. It is a simple matter to formulate conditions on $\hat{g}$ and $\hat{f}$ [and thus on $m(u, \varphi)$] which are analogs to the conditions (8.10.7 – 14). This allows us to exhibit explicitly the smallness of the perturbative terms by which we may write (8.10.35, 37) in the form

$$\dot{\eta} = -\hat{\lambda}_u \eta + \varepsilon g(\varepsilon\eta, \varphi), \tag{8.10.40}$$

$$\dot{\varphi} = \omega_u + \varepsilon \boldsymbol{h}(\varepsilon\eta, \varphi). \tag{8.10.41}$$

The functions g and $\boldsymbol{h}$ are 2π-periodic in φ.

We now apply the same reasoning as in Sect. 8.5 in order to apply Moser's theorem. We put $-\hat{\lambda}_u = \lambda_u$ [where this λ_u must not be confused with λ_u occurring in (8.10.4)] and split λ_u according to

$$\lambda_u = \underbrace{\lambda_u - D}_{\lambda_r} + D \tag{8.10.42}$$

and similarly

$$\omega_u = \underbrace{\omega_u - \Delta}_{\omega_r} + \Delta. \tag{8.10.43}$$

In this way, (8.10.40, 41) are cast into the form of the equations treated by Moser's theorem. Let us postpone the discussion whether the conditions of this theorem can be met, and assume that these conditions are fulfilled. Then we can apply the procedure we described after (8.5.36, 37) in Sect. 8.5. In particular we can, at least in principle, calculate D and Δ so that

$$\lambda_r = \lambda_u - D(\lambda_r, \omega_\lambda, \varepsilon) \quad \text{and} \tag{8.10.44}$$

$$\omega_r = \omega_u - \Delta(\lambda_r, \omega_r, \varepsilon). \tag{8.10.45}$$

In addition, we can construct the steady state *and* transient solutions of (8.10.40, 41) in complete analogy to the procedure following (8.5.45). Thus the transient solutions (close to the bifurcating torus) acquire the form

$$\varphi = \omega_r t + \varepsilon \boldsymbol{u}(\omega_r t, \varepsilon), \tag{8.10.46}$$

$$\eta = \chi_0 \exp(\lambda_r t) + \varepsilon v(\omega_r t, \varepsilon) + \varepsilon V(\omega_r t, \varepsilon)\chi_0 \exp(\lambda_r t), \quad \lambda_r < 0. \tag{8.10.47}$$

$\boldsymbol{u}$, v, and V are 2π-periodic in each component of $\omega_r t$. The steady state solutions are given by

$$\varphi = \omega_r t + \varepsilon \boldsymbol{u}(\omega_r t, \varepsilon), \tag{8.10.48}$$

$$\eta = \varepsilon v(\omega_r t, \varepsilon). \tag{8.10.49}$$

The rest of this section will be devoted to a discussion under which circumstances the conditions of Moser's theorem can be fulfilled, in particular under which conditions ω_r fulfills the KAM condition[1]. We first turn to the question under which hypothesis we derive (8.10.29, 30). If we use these equations or (8.10.40, 41) in the form of a model, no further specifications on ω_u must be made. In this case we find the bifurcation from a torus provided we can find for given λ_u and ω_u such λ_r and ω_r so that (8.10.44, 45) are fulfilled and ω_r obeys a KAM condition. If, on the other hand, (8.10.40, 41) have been derived starting from the original autonomous equations (8.9.1) then we must take into account the hypothesis we made when going from (8.9.1) to these later equations. The main hypothesis involved concerned the structure of the functions (8.9.13), to secure that $\boldsymbol{v}_k$ are quasiperiodic. This implied in particular that the frequencies $\omega_1, \ldots, \omega_M$ obeyed the KAM condition.

We now face a very profound problem, namely whether it is possible to check whether or not such frequencies as ω_r and ω_u obey a KAM condition. Maybe with the exception of a few special cases this problem implies that we know each number ω_j with absolute precision. This is, however, a task impossible to solve. Further discussion depends on our attitude, whether we take the point of view of a mathematician, or a physicist or chemist.

A mathematician would proceed as follows. In order to check whether a bifurcation from one torus to another can be expected to occur "in reality", we investigate how probable it is that out of a given set of ω's a specific set fulfills the KAM condition. As is shown in mathematics, there is a high probability for the fulfillment of such a condition provided the constant K occurring in (6.2.6) is small enough. It should be noted that this constant K occurs jointly with the scaling factor ε^2, as can be seen by the following argument.

We start from the KAM condition (6.2.6). The scaling of time transforms the frequencies ω_r and Λ_μ into ω_r/ε^2, $\Lambda_\mu/\varepsilon^2$, respectively. This transforms the KAM condition into

$$\left| \mathrm{i} \sum_{\nu=1}^{n} j_\nu \omega_\nu + \sum_{\mu=1}^{m} k_\mu \Lambda_\mu \right| \geqq \varepsilon^2 K(\|j\|^\tau + 1)^{-1} . \tag{8.10.50}$$

From the mathematical point of view we thus are led to the conclusion that the probability of finding suitable combinations of ω's is very high, or more precisely expressed, the bifurcation from one torus to another takes place with nonvanishing measure.

[1] From a rigorous mathematical point of view we require that g and $\boldsymbol{h}$ in (8.10.40, 41) are analytic in η and φ. It should be noted that the slaving principle introduced in Chap. 7 does not guarantee this property even if the initial equations (i.e., those before applying the slaving principle) only contained right-hand sides analytic in the variables. Thus we have either to invoke "smoothing" as explained in Sect. 7.5, or to introduce the equations (8.10.40, 41) by way of a model (which appeals to a physicist or engineer), or by making stronger assumptions on the original equations so that the slaving principle can be given a correspondingly sharper form (which appeals to a mathematician). A further possibility is provided by weakening the assumptions of Moser's theorem (which is also possible in certain cases, e.g., if certain symmetries are fulfilled).

From a physical point of view we can hardly think that nature discriminates in a precise way between ω's fulfilling a KAM condition and others. Rather, at a microscopic level all the time fluctuations take place which will have an impact on the actual frequencies. Therefore, the discussion on the bifurcation of a torus will make sense only if we take fluctuations into account. We shall indicate a possible treatment of such a problem in a subsequent chapter. Here we just want to mention that fluctuations, at least in general, will let the phase angles diffuse over the whole torus so that, at least in general, such delicate questions as to whether a KAM condition is fulfilled or not will not play a role. Rather, a system averages over different frequencies in one way or another.

Some other important cases must not be overlooked, however. One of them is frequency locking which has been dicussed in Sect. 8.6. In this case a torus collapses into a limit cycle. It is worth mentioning that fluctuations may play an important role in this case, too. Finally it may be noted that by a combination of the methods developed in Sects. 8.5 – 10, further phenomena, e.g. period doubling of the motion on a torus, can easily be treated.

8.11 Instability Hierarchies, Scenarios, and Routes to Turbulence

As we saw in the introduction, systems may pass through several instabilities when a control parameter is changed. There are different kinds of patterns which occur after each instability. If we consider only temporal patterns, we may find a time-independent state, periodic motion, quasiperiodic motion, chaos, and various transitions between these states at instability points leading, for example, to frequency locking or to period doubling (subharmonics). Which sequence of transitions is adopted by a specific system is, of course, an important question. Such a sequence is often called a route, in particular if a sequence leads to turbulence, or chaos. In such a case one speaks of routes to turbulence. The theoretical discussion of a route is often referred to as a "scenario", or "picture".

8.11.1 The Landau-Hopf Picture

For a number of systems, e.g., fluids, a typical route is as follows: a time-independent (spatially homogeneous) state bifurcates[1] into other time-independent (but spatially inhomogeneous) states. A new such state then bifurcates into an oscillating state, i.e., a limit cycle occurs (Hopf bifurcation). Then two basic frequencies occur; i.e., the limit cycle bifurcates into a torus. Landau conjectured that these kinds of transitions are continued in such a way that systems exhibit subsequent bifurcations to tori of higher and higher dimensions. At each step a new frequency ω_j is added to the set of basic frequencies at which the systems oscillate (quasiperiodic motion). Thus it is suggested that turbulence is described by motion on a torus of infinite dimensions. This scenario is called the *Landau-Hopf picture.*

[1] In the following we shall speak of bifurcations, though it would be more precise to speak of nonequilibrium phase transitions because of the role of fluctuations. We shall adopt a mathematician's attitude and neglect fluctuations.

8.11.2 The Ruelle and Takens Picture

Quite a different scenario has been derived by Ruelle and Takens. To explain their mathematical approach we must first discuss the meaning of the word "generic", which nowadays is frequently used in mathematics.

Let us consider a whole class of differential equations $\dot{q} = N(q)$, where N fulfills certain differentiability conditions. One then looks for those properties of the solutions $q(t)$ which are the rule and not the exception. Such properties are called "generic". Instead of trying to sharpen this definition let us illustrate it by a simple example taken from physics. Let us consider a central force which is continuous. If we denote the distance from the center as r, the class of functions $K(r)$, where K is a continuous function, is "generic". On the other hand, the force described by Coulomb's law, $K \propto 1/r^2$, is not generic; it is quite a special case[2]. Ruelle and Takens studied what happens in the generic case with respect to the bifurcations of tori into higher-dimensional tori.

Their analysis shows that after the two-dimensional torus has been reached, the next bifurcation should not lead to a three-dimensional torus, i.e., not to quasiperiodic motion at three basic frequencies, but rather to a new kind of attractor, a "strange attractor". A strange attractor can be characterized as follows: All the trajectories of the attractor are embedded in a certain region of q space. Trajectories outside that region but close enough are attracted into that region. Trajectories within that region will remain in it. The "strangeness" of the attractor consists in the fact that it is neither a fixed point nor a limit cycle nor a torus, and it does not form a manifold. Clearly, if a strange attractor is defined in this way, a large variety of them may exist, and it will be an important task for future research to attempt a classification. One such classification, which is espoused in particular by Ruelle, consists in the use of Lyapunov exponents, but of course there are other possibilities. In what follows it is important to note that Ruelle and Takens assume in their approach that the frequencies are close to rational numbers.

8.11.3 Bifurcations of Tori. Quasiperiodic Motions

Thus far we have very carefully studied the properties of quasiperiodic motion, and especially bifurcations from one torus to another, including bifurcations from two-dimensional to three-dimensional tori. The reason why we have put so much emphasis on this approach is the following: Experimentally not only are transitions from a two-dimensional torus to chaos found to occur, but also transitions to a three-dimensional torus. Therefore it is important to know why the *Ruelle and Takens picture* holds in some cases but fails in others. Our detailed discussion in the preceding section revealed that if a KAM condition is fulfilled, i.e., if the frequencies possess a certain kind of irrationality with respect to each other, bifurcation from, say, a two-dimensional to a three-dimen-

[2] This nearly trivial example illustrates another important aspect: sometimes one has to be cautious when applying the concept of "genericity" to physics (or other fields). Here, due to symmetry, conservation laws, or for other reasons, the phenomena may correspond to "nongeneric" solutions (such as Coulomb's law).

sional torus is possible. We drew the conclusion that probability arguments must be applied in order to decide whether a real system will show such a transition. Our approach solves the puzzle of why systems may show this kind of bifurcation despite the fact that the corresponding solutions are not "generic" in the sense of Ruelle and Takens. Indeed, for a given system there may be regions of the control parameter or of other parameters in which a scenario of subsequent bifurcations of tori is possible. However, when proceeding to higher-dimensional tori, such transitions become more and more improbable, so that the Landau-Hopf picture loses its validity and chaos sets in.

There are at least two further scenarios for the routes to turbulence; these are discussed in the next sections.

8.11.4 The Period-Doubling Route to Chaos. Feigenbaum Sequence

In some experiments two frequencies may lock together to a single frequency. The corresponding limit cycle can then undergo a sequence of period doubling bifurcations which eventually lead to chaos. According to my interpretation, in such a case the system is governed by few order parameters, and the period-doubling sequence takes place in the low-dimensional space of the corresponding order parameters, whose number is at least three. Such period doublings can be described by discrete maps, as explained in the introduction, but they can also be understood by means of differential equations, for which the Duffing equation (1.14.14) may stand as an example. Of course, in autonomous systems the driving force in the Duffing equation stems from a mode which oscillates at that driving frequency and drives two other nonlinearly coupled modes (or one nonlinear oscillator). Whether the Feigenbaum sequence is completely exhibited by real systems is an open question insofar as only the first bifurcation steps, say up to $n = 6$, have been observed experimentally. An important reason why no higher bifurcations can be observed is the presence of noise. In addition, as is known from specific examples, at higher bifurcations other frequencies, e.g., with period tripling, can also occur.

8.11.5 The Route via Intermittency

The last route to turbulence to be mentioned here is via *intermittency* according to Pomeau and Manneville. Here outbursts of turbulent motion are interrupted by quiet regimes. For a discussion of this phenomenon in terms of the logistic map, see [1], where further references are given.

In systems other than fluids, similar routes have been found. For instance in the laser, we find at the first laser threshold a Hopf bifurcation, and when laser pulses break into ultrashort pulses, bifurcation of a limit cycle into a torus. Motion of a limit cycle may also switch to chaotic motion under different circumstances where, more precisely speaking, we have periodic motion being modulated by chaotic motion. An important task of future research will be to study scenarios for general classes of systems and devise methods to obtain an overall picture.

9. Spatial Patterns

In the introduction we got acquainted with a number of systems in which spatial patterns evolve in a self-organized fashion. Such patterns may arise in continuous media such as fluids, or in cell assemblies in biological tissues. In this chapter we want to show how the methods introduced in the preceding chapters allow us to cope with the formation of such patterns. We note that such patterns need not be time independent but may be connected with oscillations or still more complicated time-dependent motions. Throughout this chapter we shall consider continuous media or problems in which a discrete medium, e. g., a cell assembly, can be well approximated by a continuum model.

9.1 The Basic Differential Equations

We denote the space vector (x, y, z) by $\boldsymbol{x}$. The state of the total system is then described by a state vector

$$\boldsymbol{q}(\boldsymbol{x}, t) \tag{9.1.1}$$

which depends on space and time.

Take for example a chemical reaction in a reactor. In this case the vector $\boldsymbol{q}$ is composed of components $q_1(\boldsymbol{x}, t)$, $q_2(\boldsymbol{x}, t), \ldots,$ where $q_j \equiv n_j(\boldsymbol{x}, t)$ is the concentration of a certain chemical j at space point $\boldsymbol{x}$ at time t. Because in continuously extended media we have to deal with diffusion or wave propagation or with streaming terms, spatial derivatives may occur. We shall denote such derivatives by means of the nabla-operator

$$\nabla = (\partial/\partial x, \partial/\partial y, \partial/\partial z) . \tag{9.1.2}$$

The temporal evolution of a system will be described by equations of the general form

$$\dot{\boldsymbol{q}}(\boldsymbol{x}, t) = \boldsymbol{N}(\boldsymbol{q}(\boldsymbol{x}, t), \nabla, \alpha, \boldsymbol{x}, t) . \tag{9.1.3}$$

These equations depend in a nonlinear fashion on $\boldsymbol{q}(\boldsymbol{x}, t)$. They contain, at least in general, spatial derivatives and they describe the impact of the surrounding by control parameters α. Furthermore, in spatially inhomogeneous media they may depend on the space vector $\boldsymbol{x}$ explicitly, and they may depend on time. Even in a stationary process such a time dependence must be included if the system is subject to internal or external fluctuations.

From the mathematical point of view, equations (9.1.3) are coupled nonlinear stochastic partial differential equations. Of course, they comprise an enormous class of processes and we shall again focus our attention on those situations where the system changes its macroscopic features dramatically. We mention a few explicit examples for (9.1.3). A large class used in chemistry consists of reaction diffusion equations of the form

$$\dot{\boldsymbol{q}}(\boldsymbol{x},t) = \boldsymbol{R}(\boldsymbol{q}(\boldsymbol{x},t),\boldsymbol{x},t) + D\Delta\boldsymbol{q}(\boldsymbol{x},t)\,. \tag{9.1.4}$$

Here $\boldsymbol{R}$ describes the reactions between the chemical substances. In general, $\boldsymbol{R}$ is a polynomial in $\boldsymbol{q}$ or it may consist of the sum of ratios of polynomials, for instance if Michaelis-Menton terms occur [1]. In homogeneous media the coefficients of the individual powers of $\boldsymbol{q}$ may depend on the space coordinate $\boldsymbol{x}$, and if controls are changed in time, $\boldsymbol{R}$ may also depend on time t. In most cases, however, we may assume that these coefficients are independent of space and time. In the second term of (9.1.4)

$$\nabla^2 \equiv \Delta \equiv \frac{\partial^2}{\partial x^2} + \frac{\partial^2}{\partial y^2} + \frac{\partial^2}{\partial z^2} \tag{9.1.5}$$

is the Laplace-operator describing a diffusion process. The diffusion matrix is

$$D = \begin{pmatrix} D_1 & & 0 \\ & \ddots & \\ 0 & & D_N \end{pmatrix}. \tag{9.1.6}$$

It takes into account that chemicals may diffuse with different diffusion constants.

Another broad class of nonlinear equations of the form (9.1.3) occurs in hydrodynamics. The most frequently occurring nonlinearity stems from the streaming term. This term arises when the streaming of a fluid is described in the usual way by local variables. This description is obtained by a transformation of the particle velocity from the coordinates of the streaming particles to the local coordinates of a liquid. Denoting the coordinate of a single particle by $\boldsymbol{x}(t)$ and putting

$$v = \dot{\boldsymbol{x}} \tag{9.1.7}$$

the just-mentioned transformation is provided by

$$\frac{dv(\boldsymbol{x}(t),t)}{dt} = \frac{\partial v}{\partial x}v_x + \frac{\partial v}{\partial y}v_y + \frac{\partial v}{\partial z}v_z + \frac{\partial v}{\partial t}\,. \tag{9.1.8}$$

At rhs of (9.1.8) we must then interpret the arguments $\boldsymbol{x}$ as that of a fixed space point which is no longer moving along with the fluid. The right-hand side of (9.1.8) represents the well-known streaming term, which is nonlinear in $\boldsymbol{v}$. Of course, other nonlinearities may also occur, for instance, when the density, which occurs in the equations of fluid dynamics, becomes temperature dependent and temperature itself is considered as part of the state vector $\boldsymbol{q}$. Note that in the equations of fluid dynamics $\boldsymbol{v}$ is part of the state vector $\boldsymbol{q}$

$$\boldsymbol{q}, \boldsymbol{v} \rightarrow \boldsymbol{q}(\boldsymbol{x}, t) . \tag{9.1.9}$$

As is well known from mathematics, the solutions of (9.1.3) are determined only if we fix appropriate initial and boundary conditions. In the following we shall assume that these equations are of such a type that the time evolution is determined by initial conditions, whereas the space dependence is governed by boundary conditions. We list a few typical boundary conditions, though it must be decided in each case which one of these conditions must be used and whether other types should also be taken into account. In general, this must be done on physical grounds, though some general theorems of mathematics may be helpful.

Some examples of boundary conditions: 1) The state vector $\boldsymbol{q}$ must vanish at the surface, i.e.,

$$\boldsymbol{q}(\boldsymbol{s}) = \boldsymbol{0} , \tag{9.1.10}$$

($\boldsymbol{s}$: surface), for a generalization see below. 2) The derivative normal to the surface must vanish. This condition is sometimes called the nonflux boundary condition

$$\frac{\partial q_j(\boldsymbol{s})}{\partial \boldsymbol{n}} = \boldsymbol{0} . \tag{9.1.11}$$

3) Another boundary condition, which is somewhat artificial but is quite useful if no other boundary conditions are given, is the periodic boundary condition

$$\boldsymbol{q}(\boldsymbol{x}) = \boldsymbol{q}(\boldsymbol{x} + \boldsymbol{L}) . \tag{9.1.12}$$

Here it is required that the solution is periodic along its coordinates x, y, z with periods L_x, L_y, L_z, respectively. 4) It there is no boundary condition within finite dimensions, in general one requires that

$$|\boldsymbol{q}| \ \text{bounded for} \ |\boldsymbol{x}| \rightarrow \infty . \tag{9.1.13}$$

Of course, within a state vector $\boldsymbol{q}$, different components may be subjected to different boundary conditions, e.g., mixtures of (9.1.10 and 11) may occur. Another boundary condition which generalizes (9.1.10) is given by

$$\boldsymbol{q}(\boldsymbol{s}) = \bar{\boldsymbol{q}}(\boldsymbol{s}) , \tag{9.1.14}$$

where $\bar{q}$ is a prescribed function at the surface. For instance one may require that the concentrations of certain chemicals are kept constant at the boundary.

There are still other boundary conditions which are less obvious, namely when $\boldsymbol{q}(x)$ is a function on a manifold, e.g., on a sphere or on a torus. Such examples arise in biology (evolution of morula, blastula, gastrula, and presumably many other cases of biological pattern formation), as well as in astrophysics. In such a case we require that $\boldsymbol{q}(x)$ is uniquely defined on the manifold, i.e., if we go along a meridian we must find the same values of $\boldsymbol{q}$ after one closed path.

9.2 The General Method of Solution

From a formal point of view our method of approach is a straightforward extension of the methods presented before. We assume that for a certain range of control parameter we have found a stable solution $\boldsymbol{q}_0(\boldsymbol{x}, t)$ so that

$$\dot{\boldsymbol{q}}_0(\boldsymbol{x}, t) = \boldsymbol{N}(\boldsymbol{q}_0(\boldsymbol{x}, t), \nabla, \alpha, \boldsymbol{x}, t) . \tag{9.2.1}$$

We assume that the solution $\boldsymbol{q}_0$ can be extended into a new region of control parameter values α but where linear stability is lost. To study the stability of the solution of (9.1.3) we put

$$\boldsymbol{q}(\boldsymbol{x}, t) = \boldsymbol{q}_0(\boldsymbol{x}, t) + \boldsymbol{w}(\boldsymbol{x}, t) \tag{9.2.2}$$

and insert it in (9.1.3). We assume that $\boldsymbol{w}$ obeys the linearized equation

$$\dot{\boldsymbol{w}}(\boldsymbol{x}, t) = L(\boldsymbol{q}_0(\boldsymbol{x}, t), \nabla, \alpha, \boldsymbol{x}, t) \boldsymbol{w}(\boldsymbol{x}, t) . \tag{9.2.3}$$

The right-hand side of (9.2.3) is sometimes called the linearization of (9.1.3) and L is sometimes called the Fréchêt derivative. If $\boldsymbol{N}$ is a functional of q_j (e.g., in the form of integrals over functions of q_j at different space points) the components of the matrix L are given by functional derivatives

$$L_{ij} = \delta N_i / \delta q_j(\boldsymbol{x}, t) . \tag{9.2.4}$$

However, in order no to overload our presentation we shall not enter the problem of how to define (9.2.3) in abstract mathematical terms but rather illustrate the derivation of (9.2.3) by explicit examples. In the case of a reaction diffusion equation (9.1.4), we obtain

$$L = L^{(1)} + L^{(2)} , \qquad \text{where} \tag{9.2.5}$$

$$L_{ik}^{(1)} = \left. \frac{\partial R_i}{\partial q_k} \right|_{q = q_0} \tag{9.2.6}$$

is just the usual derivative, whereas $L^{(2)}$ is given by

$$L^{(2)} = D\Delta \, . \tag{9.2.7}$$

We give here only one additional example of how to obtain the linearization (9.2.3). To this end we consider as a specific term the streaming term which may occur in N,

$$q_j(\boldsymbol{x}) \frac{\partial}{\partial x_k} q_l(\boldsymbol{x}) \, . \tag{9.2.8}$$

Making the replacement (9.2.2) we have to consider

$$[q_{0,j}(\boldsymbol{x}) + w_j(\boldsymbol{x})] \frac{\partial}{\partial x_k} [q_{0,l}(\boldsymbol{x}) + w_l(\boldsymbol{x})] \, . \tag{9.2.9}$$

Multiplying all terms with each other and keeping only the terms which are linear in w_j we readily obtain

$$q_{0,j}(\boldsymbol{x}) \frac{\partial}{\partial x_k} w_l(\boldsymbol{x}) + w_j(\boldsymbol{x}) \frac{\partial}{\partial x_k} q_{0,l}(\boldsymbol{x}) \, . \tag{9.2.10}$$

Let us consider the solutions of (9.2.3) in more detail. Because L depends on $\boldsymbol{q}_0$, we have to specify the different kinds of $\boldsymbol{q}_0$ from which we start. We begin with a $\boldsymbol{q}_0$ which is a constant vector, i.e., independent of space and time, and consider as example reaction diffusion equations. In this case, L is of the form (9.2.5), where $L^{(1)}$, (9.2.6), is a constant matrix and $L^{(2)}$ is given by (9.2.7).

In order to solve

$$L\boldsymbol{w} = [L^{(1)} + D\Delta]\boldsymbol{w} = \dot{\boldsymbol{w}} \, , \tag{9.2.11}$$

we make the hypothesis

$$\boldsymbol{w}(\boldsymbol{x}, t) = \boldsymbol{w}_1(t)\chi_k(\boldsymbol{x}) \, . \tag{9.2.12}$$

Under the assumption that $\chi_k(\boldsymbol{x})$ obeys the equation

$$\Delta\chi_k = -\lambda_k' \chi_k \tag{9.2.13}$$

(under given boundary conditions) we may transform the equation

$$\dot{\boldsymbol{w}} = L(\nabla, \alpha)\boldsymbol{w} \qquad \text{into} \tag{9.2.14}$$

$$\dot{\boldsymbol{w}}_1 = [L^{(1)} - D\lambda_k']\boldsymbol{w}_1 \, . \tag{9.2.15}$$

This equation represents a set of coupled linear ordinary differential equations with constant coefficients. From Sect. 2.6 we know the general form of its solution,

$$\boldsymbol{w}_1(t) = \mathrm{e}^{\lambda_k t} \boldsymbol{v}_k(\boldsymbol{x}, t) \, , \tag{9.2.16}$$

where v_k is independent of time if the characteristic exponents λ_k are non-degenerate. Otherwise v_k may contain a finite number of powers of t. For certain geometries the solutions of (9.2.13) are well known, e.g., in the case of rectangular geometry they are plane or standing waves. In the case of a sphere they are Bessel functions multiplied by spherical harmonics.

The next case concerns a $\boldsymbol{q}_0$ which is independent of space but a periodic or quasiperiodic function of time. In addition, we confine our considerations to reaction diffusion equations. Then $L^{(1)}$ acquires the same time dependence as $\boldsymbol{q}_0$. Hypothesis (9.2.12) is still valid and we have to seek the solutions of (9.2.15), a class of equations studied in extenso in Chaps. 2 and 3.

We now discuss a class of problems in which $\boldsymbol{q}_0$ depends on the space coordinate $\boldsymbol{x}$. In such a case, the solutions of (9.2.3) are, at least in general, not easy to construct analytically. In most cases we must resort to computer solutions. So far not very much has been done in this field, but I think that such a treatment will be indispensable, since this problem cannot be eventually circumvented, e.g., by specific forms of singularity theory. In spite of this open problem there are a few classes in which general statements can still be made. Let us assume that $\boldsymbol{q}_0(\boldsymbol{x})$ is periodic in $\boldsymbol{x}$ with periods

$$(a_1, a_2, a_3)\,. \tag{9.2.17}$$

If N is independent of $\boldsymbol{x}$, $L(\boldsymbol{x})$ possesses the same periodicity as $\boldsymbol{q}_0(\boldsymbol{x})$. By means of the hypothesis

$$\boldsymbol{w} = \mathrm{e}^{\lambda t}\boldsymbol{v}(\boldsymbol{x}) \tag{9.2.18}$$

we transform (9.2.14) into

$$L(\boldsymbol{q}_0(\boldsymbol{x}), \nabla, \alpha)\,\boldsymbol{v}(\boldsymbol{x}) = \lambda\,\boldsymbol{v}(\boldsymbol{x})\,. \tag{9.2.19}$$

Because $L(\boldsymbol{x})$ is periodic in each component of $\boldsymbol{x}$ we may apply the results of Sect. 2.7. If the boundary condition requires that $\boldsymbol{v}(\boldsymbol{x})$ is bounded for $|\boldsymbol{x}| \to \infty$, the general form of the solution $\boldsymbol{v}$ reads

$$\boldsymbol{v}(\boldsymbol{x}) = \mathrm{e}^{\mathrm{i}\boldsymbol{k}\cdot\boldsymbol{x}}\boldsymbol{z}(\boldsymbol{x})\,, \tag{9.2.20}$$

where $\boldsymbol{k}$ is a real vector and $\boldsymbol{z}$ is periodic with periods (9.2.17). In case of other boundary conditions in finite geometries, standing waves may be formed from (9.2.20) in order to fulfill the adequate boundary conditions. Here it is assumed that the boundary conditions are consistent with the requirement that $L(\boldsymbol{x})$ is periodic with (9.2.17).

9.3 Bifurcation Analysis for Finite Geometries

We now wish to show how to solve equations of the type (9.1.3). We study the bifurcation from a node or focus which are stable for a certain range of parameter values α and which become unstable beyond a critical value α_c. We assume that the corresponding old solution is given by

$$q_0(x) . \tag{9.3.1}$$

We assume that the solutions of the linearized equations (9.2.3) have the form

$$w_k(x,t) = e^{\lambda_k t} v_k(x) . \tag{9.3.2}$$

As is well known, when dealing with finite boundary conditions, at least in a standard problem, we may assume that the characteristic exponents λ_k are discrete and also that the index k is a discrete index. We decompose the required solution of (9.1.3) into the by now well-known superposition

$$q(x,t) = q_0(x) + \underbrace{\sum_k \xi_k(t) v_k(x)}_{W} . \tag{9.3.3}$$

The essential difference between our present procedure and that of the Chap. 8 (especially Sects. 8.1 – 5) consists in the fact that the index k covers an infinite set while in former chapters k was a finite set. The other difference consists in the space dependence of q, q_0 and v. We insert (9.3.3) into (9.1.3) where we expand the rhs of N into a power series of W. Under the assumption that (9.3.3) obeys (9.1.3) we obtain

$$\sum_k \dot{\xi}_k(t) v_k(x) = \sum_k \xi_k(t) L v_k(x) + H(W) \quad \text{with} \tag{9.3.4}$$

$$H(W) = \sum_{k'k''} \xi_{k'} \xi_{k''} \hat{N}^{(2)}(q_0(x)) : v_{k'}(x) : v_{k''}(x) + \dots . \tag{9.3.5}$$

We now multiply (9.3.4) by $\bar{v}_k(x)$ and integrate over the space within the given boundaries. According to the definition of $\bar{v}_k$ we obtain

$$\int d^3x\, \bar{v}_k(x) v_{k'}(x) = \delta_{kk'} . \tag{9.3.6}$$

Using in addition

$$L v_k(x) = \lambda_k v_k(x) , \tag{9.3.7}$$

we can cast (9.3.4) into the form

$$\dot{\xi}_k = \lambda_k \xi_k + \langle \bar{v}_k H(W) \rangle \quad \text{with} \tag{9.3.8}$$

$$\langle \bar{v}_k H(W) \rangle = \sum_{k'k''} \xi_{k'} \xi_{k''} \underbrace{\int d^3x (\bar{v}_k(x) \hat{N}^{(2)}(q_0(x)) : v_{k'}(x) : v_{k''}(x))}_{A^{(2)}_{kk'k''}} + \dots . \tag{9.3.9}$$

If N and q_0 are independent of space variables, the coefficients A, which occur in (9.3.9), can be cast into the form

$$A^{(2)}_{kk'k''} = \sum_{ll'l''} \hat{N}^{(2)}_{ll'l''} \underbrace{\int d^3x\, \bar{v}_{k,l}(x)\, v_{k,l'}(x)\, v_{kl''}(x)}_{\mathrm{I}} . \tag{9.3.10}$$

The remaining integral I depends on the solutions of the linearized equation (9.3.7) only. If L is invariant under symmetry operations, the v's can be chosen as irreducible representations of the corresponding transformation group. As is shown in group theory, on account of the transformation properties of the v's and $\bar{v}$'s, selection rules for the integral I in (9.3.10) arise. We mention that such group theoretical ideas are most useful to simplify the solution of (9.3.8). From the formal point of view, the set of (9.3.8) is identical with that of (8.1.17). Therefore, close to critical points we may apply the slaving principle, which allows us to reduce the set of (9.3.8) to a set of finite and in general even very few dimensions. The leading terms in (9.3.3) are those which contain the order parameters u_k only instead of all ξ_k. Close to the instability point the other remaining terms are comparatively small and give rise to small changes only. Therefore, close to instability points the evolving pattern is determined by superpositions of a finite number of terms of the form

$$u_k(t)\, v_k(x) . \tag{9.3.11}$$

What kind of combinations of (9.3.11) must be taken is then determined by the solutions of the order parameter equations alone. Here, particularly, we may have cases of competition in which only one u survives so that the evolving pattern is determined by a single v_{k_0}. In other cases, by cooperation, specific combinations of the u's stabilize each other. An example for the former is provided by the formation of rolls in the Bénard instability, while an example for the latter is given by hexagonal patterns in the same problem.

9.4 Generalized Ginzburg-Landau Equations

The spectrum of the operator L will be, at least in general, continuous if there are no boundary conditions in infinitely extended media. Let us consider the special case in which q_0 is space and time independent. We treat a linear operator L of the form (9.2.11), where L is space and time independent. According to (9.2.13) we may choose χ in the form of plane waves

$$\chi_k(x) = \mathrm{e}^{\mathrm{i}k \cdot x} . \tag{9.4.1}$$

Using hypothesis (9.2.12), we have to deal again with a coupled set of linear differential equations with constant coefficients (9.2.15). If w_1 is a finite dimensional vector, the eigensolutions w_1 can be characterized by a discrete set of indices which we call j. On the other hand, k is a continuous variable. Neglecting degeneracies of the eigenvalues of (9.2.15), the solutions of the linearized equation (9.2.14) are

$$\exp(\lambda_{k,j}t)\, v_{k,j}(x) = \exp(\lambda_{k,j}t)\, v_{k,j}(0) \exp(\mathrm{i}k \cdot x) . \tag{9.4.2}$$

We note that the eigenvalues can be written as

$$\lambda_{k,j} \equiv \lambda_j(\mathrm{i}k) . \tag{9.4.3}$$

The general hypothesis for q can be formulated analogously to (9.3.3) but we must take into account that k is continuous by a corresponding integral

$$q(x,t) = q_0(x) + \sum_j \int \xi_{k,j}(t)\, v_{k,j}(x)\, d^3k . \tag{9.4.4}$$

Such a continuous spectrum can cause some difficulties when applying the slaving principle. For this reason we resort to an approach well known to quantum mechanics and which amounts to the formation of wave packets. To this end we split k into a discrete set of vectors k' and remaining continuous parts $\hat{k}$

$$k = k' + \hat{k} . \tag{9.4.5}$$

We now consider an expression of the form

$$\int \xi_{k,j}(t)\, v_{k,j}(0)\, \mathrm{e}^{\mathrm{i}k \cdot x} d^3k , \tag{9.4.6}$$

where we split the integral into a sum over k'

$$\sum_{k'} \underbrace{\int_{k'_x-\delta/2}^{k'_x+\delta/2} \int_{k'_y-\delta/2}^{k'_y+\delta/2} \int_{k'_z-\delta/2}^{k'_z+\delta/2} d^3\hat{k}\, \xi_{k'+\hat{k},j}(t)\, \mathrm{e}^{\mathrm{i}\hat{k} \cdot x}\, v_{k'+\hat{k},j}(0)}_{\xi_{k',j}(x,t)}\, \mathrm{e}^{\mathrm{i}k' \cdot x} . \tag{9.4.7}$$

Introducing the abbreviation $\xi_{k',j}(x,t)$ as indicated in relation (9.4.7) and making for each wave packet the approximation

$$v_{k,j}(0) \to v_{k',j}(0) , \tag{9.4.8}$$

we may cast (9.4.7) into the form

$$\sum_{k'} \xi_{k',j}(x,t)\, v_{k',j}(x) . \tag{9.4.9}$$

Under these assumptions we may write (9.4.4) in the form

$$q(x,t) = q_0(x) + \sum_{k,j} \xi_{k,j}(x,t)\, v_{k,j}(x) . \tag{9.4.10}$$

The orthogonality relation (9.3.6) must now be taken in a somewhat different way, namely in the relation

$$\langle \bar{v}_{k',j'}(\boldsymbol{x}) v_{k,j}(\boldsymbol{x}) \rangle = \int d^3x \sum_l \bar{v}_{k',j',l}(\boldsymbol{x}) v_{k,j,l}(\boldsymbol{x}) \tag{9.4.11}$$

we shall integrate over a space region which contains many oscillations of $\boldsymbol{v}(\boldsymbol{x})$ but over which ξ_k does not change appreciably. Under these assumptions the relations

$$\langle \bar{v}_{k',j'} v_{k,j} \rangle_x = \delta_{k',k} \delta_{j'j} \quad \text{and} \tag{9.4.12}$$

$$\langle \bar{v}_{k',j'}(\boldsymbol{x}) \xi_{k,j}(\boldsymbol{x},t) v_{k,j}(\boldsymbol{x}) \rangle_x \approx \xi_{k,j}(\boldsymbol{x},t) \delta_{k',k} \delta_{j',j} \tag{9.4.13}$$

hold. In order to proceed further we must perform some intermediate steps. We consider the expression

$$\int \lambda_j(\mathrm{i}\boldsymbol{k}) \xi_{k,j}(t) v_{k,j}(0) \,\mathrm{e}^{\mathrm{i}\boldsymbol{k}\cdot\boldsymbol{x}} d^3k \,, \tag{9.4.14}$$

which can be written in the form

$$\lambda_j(\nabla) \int \xi_{kj}(t) v_{k,j}(0) \,\mathrm{e}^{\mathrm{i}\boldsymbol{k}\cdot\boldsymbol{x}} d^3k \,, \tag{9.4.15}$$

where $\lambda_j(\nabla)$ yields $\lambda_j(\mathrm{i}\boldsymbol{k})$ if $\lambda_j(\nabla)$ is followed by $\exp(\mathrm{i}\boldsymbol{k}\cdot\boldsymbol{x})$. The integral following λ_j in (9.4.15) can be split into the sum (9.4.9) as before, so that we obtain

$$\lambda_j(\nabla) \sum_{k'} \underbrace{\int \xi_{k'+\hat{k},j}(t) \,\mathrm{e}^{\mathrm{i}\hat{\boldsymbol{k}}\cdot\boldsymbol{x}} d^3\hat{k}}_{\xi_{k',j}(\boldsymbol{x},t)} \, v_{k,j}(0) \,\mathrm{e}^{\mathrm{i}\boldsymbol{k}'\cdot\boldsymbol{x}} \,. \tag{9.4.16}$$

We now consider the effect of $\lambda_j(\nabla)$ on ξ. To this end we treat

$$\lambda_j(\nabla) \xi_{k',j}(\boldsymbol{x},t) \,\mathrm{e}^{\mathrm{i}\boldsymbol{k}'\cdot\boldsymbol{x}} \,. \tag{9.4.17}$$

Performing the differentiation with respect to $\exp(\mathrm{i}\boldsymbol{k}'\cdot\boldsymbol{x})$ we readily obtain for (9.4.17)

$$\mathrm{e}^{\mathrm{i}\boldsymbol{k}'\cdot\boldsymbol{x}} \lambda_j(\mathrm{i}\boldsymbol{k}' + \nabla) \xi_{k',j}(\boldsymbol{x},t) \,, \tag{9.4.18}$$

where formulas well known in quantum physics have been used. It follows from (9.4.7) that $\xi_{k',j}$ contains only small $\hat{\boldsymbol{k}}$-vectors, enabling us to expand λ_j into a power series with respect to ∇. We readily obtain

$$\begin{aligned} &\lambda_j(\mathrm{i}\boldsymbol{k}' + \nabla) \xi_{k',j}(\boldsymbol{x},t) \\ &\quad = \lambda_j(\mathrm{i}\boldsymbol{k}') \xi_{k',j}(\boldsymbol{x},t) + \lambda_j^{(1)}(\mathrm{i}\boldsymbol{k}') : \nabla \, \xi_{k',j}(\boldsymbol{x},t) + \lambda_j^{(2)}(\mathrm{i}\boldsymbol{k}') : \nabla : \nabla \, \xi_{k',j}(\boldsymbol{x},t) + \dots \,. \end{aligned} \tag{9.4.19}$$

To complete the intermediate steps we have to evaluate

$$\langle \bar{v}_{k',j'}(\boldsymbol{x})\,\lambda_j(\mathrm{i}\boldsymbol{k})\,\xi_{k,j}(\boldsymbol{x},t)\,v_{k,j}(\boldsymbol{x})\rangle_x\,, \tag{9.4.20}$$

which leads to an expression of the form

$$\lambda_j(\mathrm{i}\boldsymbol{k}+\nabla)\,\xi_{k,j}(\boldsymbol{x},t)\,\delta_{k',k}\,\delta_{j',j}\,, \tag{9.4.21}$$

where we have made use of the result (9.4.13). After all these steps we are now in a position to treat the nonlinear equation (9.1.3). To this end we insert hypothesis (9.4.10) into (9.1.3), expand rhs of (9.1.3) into a power series of N with respect to $\boldsymbol{q}$, and make use of the linearized equation (9.2.3). We then readily obtain

$$\dot{\xi}_{k,j}(\boldsymbol{x},t)=\lambda_j(\mathrm{i}\boldsymbol{k}+\nabla)\,\xi_{k,j}(\boldsymbol{x},t)+H_{k,j}(\xi_{k,j}(\boldsymbol{x},t)) \quad \text{with} \tag{9.4.22}$$

$$H_{k,j}=\sum_{\substack{k'k''\\ j'j''}}\xi_{k',j'}(\boldsymbol{x},t)\,\xi_{k'',j''}(\boldsymbol{x},t)\,\langle \bar{v}_{k,j}(\boldsymbol{x})\hat{N}\colon v_{k',j'}(\boldsymbol{x})\colon v_{k'',j''}(\boldsymbol{x})\rangle_x+\dots\,. \tag{9.4.23}$$

Except for the fact that λ_j is an operator, (9.4.22) is of the same form as all the equations we have studied in a similar context before.

To make use of the slaving principle we must distinguish between the unstable and stable modes. To this end we distinguish between such λ's for which

$$\lambda_u=\mathrm{Re}\{\lambda_j(\mathrm{i}\boldsymbol{k})\}>-|C| \quad \text{and} \tag{9.4.24}$$

$$\lambda_s=\mathrm{Re}\{\lambda_j(\mathrm{i}\boldsymbol{k})\}\leqslant C<0 \tag{9.4.25}$$

hold. We must be aware of the fact that the indices j and $\boldsymbol{k}$ are not independent of each other when we define the unstable and stable modes according to (9.4.24, 25). We shall denote those modes for which (9.4.24) holds by u_{kj}, whereas those modes for which (9.4.25) is valid by s_{kj}. Note that in the following summations j and $\boldsymbol{k}$ run over such restricted sets of values implicitly defined by (9.4.24, 25). The occurrence of the continuous spectrum provokes a difficulty, because the modes go over continuously from the slaved modes to the undamped or unslaved modes. Because the slaving principle requires that the real part of the spectrum λ_s has an upper bound [cf. (9.4.25)], we make a cut between the two regions at such C. Consequently, we also have to treat modes with a certain range of negative real parts of λ_j as unstable modes. As one may convince oneself the slaving principle is still valid under the present conditions provided the amplitudes of u_{kj} are small enough.

Since the λ's are operators, the corresponding λ_u and λ_s are also operators with respect to spatial coordinates. Some analysis shows that the slaving principle also holds in this more general case provided the amplitudes ξ are only slowly varying functions of space so that $\lambda_s(\boldsymbol{k}+\nabla)$ does not deviate appreciably from $\lambda_s(\boldsymbol{k})$. After these comments it is easy to apply to formalisms of the slaving principle. Keeping in the original equation (9.1.3) only terms of N up to third order in

q, we obtain the following set of equations for the order parameters $u_{k,j} = u_{k,j}(x,t)$ in the form of

$$\left[\frac{d}{dt} - \lambda_{u,j}(ik+\nabla)\right] u_{k,j} = \sum_{\substack{k'k'' \\ j'j''}} A_{kk'k'',jj'j''} u_{k',j'} u_{k'',j''}$$

$$+ \sum_{\substack{k'k''k''' \\ j'j''j'''}} B_{k'k''k''',j'j''j'''} u_{k',j'} u_{k'',j''} u_{k''',j'''} + \ldots (+F_{k,j}(t)) . \tag{9.4.26}$$

Precisely speaking, the coefficients A and B may contain derivatives with respect to spatial coordinates. However, in most cases of practical importance, $\lambda_s(ik+\nabla)$ which occurs in the denominators may be well approximated by $\lambda_s(ik)$ because u depends only weakly on x and the denominators are bounded from below. If the original equation (9.1.3) contained fluctuations, the corresponding fluctuating forces reappear in (9.4.26) and are added here in the form of $F_{k,j}$. I have called these equations, which I derived some time ago, "generalized Ginzburg-Landau equations" because if we use λ_j in the approximation (9.4.19) and drop the indices and sums over k and j, the equations (9.4.26) reduce to equations originally established by Ginzburg and Landau in the case of equilibrium phase transitions, especially in the case of superconductivity. Besides the fact that the present equations are much more general, two points should be stressed. The equations are derived here from first principles and they apply in particular to systems far from thermal equilibrium.

9.5 A Simplification of Generalized Ginzburg-Landau Equations. Pattern Formation in Bénard Convection

In a number of cases of practical interest, the generalized Ginzburg-Landau equations which we derived above can be considerably simplified. In order to make our procedure as transparent as possible, we will use the Bénard instability of fluid dynamics as an example. (The relevant experimental results were presented in Sect. 1.2.1.) The procedure can be easily generalized to other cases. In the Bénard problem the order parameters depend on the two horizontal space coordinates x and y, which we lump together in the vector $\boldsymbol{x} = (x, y)$. Correspondingly, plane waves in the horizontal plane are described by wave vectors $k_\perp = (k_x, k_y)$. In this case the eigenvalues of the unstable modes can be written in the form [Ref. 1, Sect. 8.13]:

$$\lambda = a - (k_0^2 - k_\perp^2)^2 . \tag{9.5.1}$$

To make contact with the λ which occurs in the order parameter equations (9.4.26), we must transform (9.5.1) by means of

$$e^{ik_c \cdot x} \tag{9.5.2}$$

[where $\boldsymbol{k}_c$ corresponds to $\boldsymbol{k}$ in (9.4.26)], and in order to take into account finite bandwidth excitations we must introduce derivatives as in the above section. Therefore the λ to be used in (9.4.26) reads

$$\lambda = a - (k_0^2 + (\mathrm{i}\boldsymbol{k}_c + \nabla)^2)^2 . \tag{9.5.3}$$

In the following we shall use the order parameter equations (9.4.26) in the form

$$\begin{aligned} \dot{u}_{\boldsymbol{k}_c}(\boldsymbol{x}) = {} & [a - (k_0^2 + (\mathrm{i}\boldsymbol{k}_c + \nabla)^2)^2] u_{\boldsymbol{k}_c}(\boldsymbol{x}) \\ & + \sum_{\boldsymbol{k}_c', \boldsymbol{k}_c''} A_{\boldsymbol{k}_c, \boldsymbol{k}_c', \boldsymbol{k}_c''} \delta_{\boldsymbol{k}_c, \boldsymbol{k}_c' + \boldsymbol{k}_c''} u_{\boldsymbol{k}_c'}(\boldsymbol{x}) u_{\boldsymbol{k}_c''}(\boldsymbol{x}) \\ & - \sum_{\boldsymbol{k}_c', \boldsymbol{k}_c'', \boldsymbol{k}_c'''} B_{\boldsymbol{k}_c, \boldsymbol{k}_c', \boldsymbol{k}_c'', \boldsymbol{k}_c'''} \delta_{\boldsymbol{k}_c, \boldsymbol{k}_c' + \boldsymbol{k}_c'' + \boldsymbol{k}_c'''} u_{\boldsymbol{k}_c'}(\boldsymbol{x}) u_{\boldsymbol{k}_c''}(\boldsymbol{x}) u_{\boldsymbol{k}_c'''}(\boldsymbol{x}) , \end{aligned} \tag{9.5.4}$$

where the Kronecker symbols under the sums ensure the conservation of wave numbers. This is the case if we do not include boundary conditions with respect to the horizontal plane, and use plane-wave solutions for $v_{\boldsymbol{k},j}(\boldsymbol{x})$. It is evident that the specific form of λ (9.5.1) or λ (9.5.3) favors those $|\boldsymbol{k}_\perp|$ which are close to $|\boldsymbol{k}_c|$. We now introduce a new function ψ according to

$$\psi(\boldsymbol{x}) = \sum_{\boldsymbol{k}_c} \mathrm{e}^{\mathrm{i}\boldsymbol{k}_c \cdot \boldsymbol{x}} u_{\boldsymbol{k}_c}(\boldsymbol{x}) . \tag{9.5.5}$$

The sum runs over the critical $\boldsymbol{k}$ vectors, which all have the same absolute value k_0 but point in different horizontal directions. We now make our basic assumption, namely that we may put

$$A_{\boldsymbol{k}_c, \boldsymbol{k}_c', \boldsymbol{k}_c''} \approx A \quad \text{and} \tag{9.5.6}$$

$$B_{\boldsymbol{k}_c, \boldsymbol{k}_c', \boldsymbol{k}_c'', \boldsymbol{k}_c'''} \approx B \tag{9.5.7}$$

for the $\boldsymbol{k}$ vectors, which are selected through the δ-functions and the condition that $|\boldsymbol{k}_c| = k_0$.

We multiply (9.5.4) by $\exp(\mathrm{i}\boldsymbol{k}_c \cdot \boldsymbol{x})$ and sum over $\boldsymbol{k}_c$. To elucidate the further procedure we start with the thus resulting expression which stems from the last term in (9.5.4), and therefore consider

$$\sum_{\boldsymbol{k}_c} \mathrm{e}^{\mathrm{i}\boldsymbol{k}_c \cdot \boldsymbol{x}} \sum_{\boldsymbol{k}_c', \boldsymbol{k}_c'', \boldsymbol{k}_c'''} \delta_{\boldsymbol{k}_c, \boldsymbol{k}_c' + \boldsymbol{k}_c'' + \boldsymbol{k}_c'''} u_{\boldsymbol{k}_c'}(\boldsymbol{x}) u_{\boldsymbol{k}_c''}(\boldsymbol{x}) u_{\boldsymbol{k}_c'''}(\boldsymbol{x}) . \tag{9.5.8}$$

Since

$$-\boldsymbol{k}_c + \boldsymbol{k}_c' + \boldsymbol{k}_c'' + \boldsymbol{k}_c''' = 0 , \tag{9.5.9}$$

we can insert the factor

$$\exp[\mathrm{i}\{-\boldsymbol{k}_c + \boldsymbol{k}_c' + \boldsymbol{k}_c'' + \boldsymbol{k}_c'''\}\boldsymbol{x}] \equiv 1 \tag{9.5.10}$$

into (9.5.8). Exchanging the sequence of summations in (9.5.8) we obtain

$$\sum_{k_c',k_c'',k_c'''} \exp[\mathrm{i}\{\boldsymbol{k}_c'+\boldsymbol{k}_c''+\boldsymbol{k}_c'''\}\boldsymbol{x}]\, u_{k_c'}(\boldsymbol{x})\, u_{k_c''}(\boldsymbol{x})\, u_{k'''}(\boldsymbol{x}) \sum_{k_c} \delta_{k_c,k_c'+k_c''+k_c'''} \qquad (9.5.11)$$

and note that we may drop the last sum due to

$$\sum_{k_c} \delta_{k_c,k_c'+k_c''+k_c'''} = 1\ . \qquad (9.5.12)$$

Because of (9.5.5) the whole expression (9.5.11) then simplifies to

$$(9.5.11) = \psi^3(\boldsymbol{x})\ . \qquad (9.5.13)$$

The term with A can be treated in analogous fashion.

Finally, we use the transformation

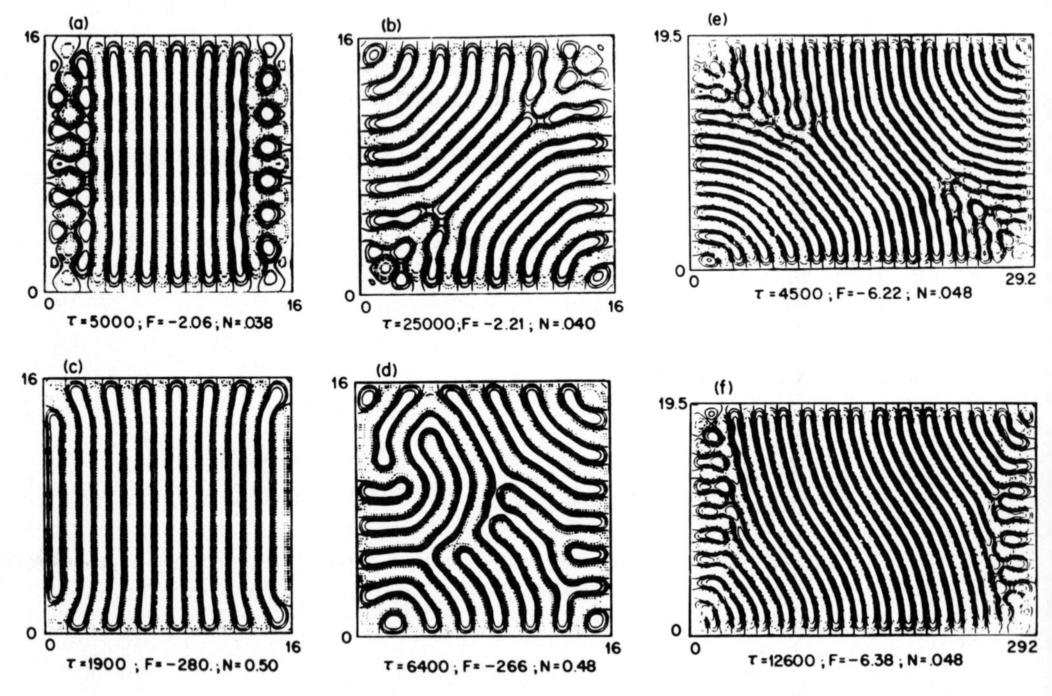

Fig. 9.1a – f. Contour plots of the amplitude field $\psi(x,y,t)$ at various times $t \equiv \tau$, for **(a, b, e, f)** $a = 0.10$ and **(c, d)** $a = 0.90$, where $A = 0$, $B = 1$. The cells in **(a – d)** have an aspect ratio of 16 while the cells in **(e, f)** have aspect ratios of 29.2 and 19.5. The initial conditions were parallel rolls for **(a, c)** and random domains of opposite sign for **(b, e)**. The solid and dotted contours represent positive and negative values at $\frac{1}{2}$, $\frac{1}{4}$, $\frac{1}{8}$, and $\frac{1}{16}$ the maximum amplitude. The contours correspond to vertical velocity contours in optical experiments. The values of the time at which equilibrium was reached (τ), the Lyapunov functional (F), and the Nusselt number (N) are given. The state in **(e)** has not reached equilibrium and is evolving into the equilibrium state, **(f)**. Here equilibrium is defined to occur when $d\ln(F)/d\tau$ is smaller than 10^{-8}. [After H. S. Greenside, W. M. Coughran, Jr., N. L. Schryer: Phys. Rev. Lett. **49**, 726 (1982)]

$$e^{i\boldsymbol{k}_c\cdot\boldsymbol{x}}(k_0^2 + (i\boldsymbol{k}_c + \nabla)^2)^2 = (k_0^2 + \nabla^2)^2 e^{i\boldsymbol{k}_c\cdot\boldsymbol{x}} . \tag{9.5.14}$$

Because the whole procedure yields the expression $\dot{\psi}(\boldsymbol{x})$ for the lhs of (9.5.4), we find the new equation

$$\dot{\psi}(\boldsymbol{x}) = \{a - (k_0^2 + \nabla^2)\}^2 \psi(\boldsymbol{x}) + A\,\psi^2(\boldsymbol{x}) - B\,\psi^3(\boldsymbol{x}) . \tag{9.5.15}$$

This equation has been solved numerically for $A = 0$ by H. S. Greenside, W. M. Coughran, Jr., and N. L. Schryer, and a typical result is shown in Fig. 9.1. Note the resemblance of this result to Figs. 1.2.6, 8, except for the fact that hexagons are now lacking. Our own results show that such hexagons can be obtained if the term with A is included.

10. The Inclusion of Noise

In the introduction we pointed out that noise plays a crucial role especially at instability points. Here we shall outline how to incorporate noise into the approach developed in the previous chapters. In synergetics we usually start from equations at a mezoscopic level, disregarding the microscopic motion, for instance of molecules or atoms. The equations of fluid dynamics may stand as an example for many others. Here we are dealing with certain macroscopic quantities such as densities, macroscopic velocities, etc. Similarly, in biological morphogenesis we disregard individual processes below the cell level, for instance metabolic processes. On the other hand, these microscopic processes cannot be completely neglected as they give rise to fluctuating driving forces in the equations for the state variables $\boldsymbol{q}$ of the system under consideration. We shall not derive these noise sources. This has to be done in the individual cases depending on the nature of noise, whether it is of quantum mechanical origin, or due to thermal fluctuations, or whether it is external noise, produced by the action of reservoirs to which a system is coupled. Here we wish rather to outline the general approach to deal with given noise sources. We shall elucidate our approach by explicit examples.

10.1 The General Approach

Adding the appropriate fluctuating forces to the original equations we find equations of the form

$$\dot{\boldsymbol{q}} = \boldsymbol{N}(\boldsymbol{q}, \alpha) + \boldsymbol{F}(t) , \tag{10.1.1}$$

which we shall call Langevin equations. If the fluctuating forces depend on the state variables we have to use stochastic differential equations of the form

$$d\boldsymbol{q} = \boldsymbol{N}(\boldsymbol{q}, \alpha)dt + d\boldsymbol{F}(t, \boldsymbol{q}) , \tag{10.1.2}$$

which may be treated according to the Îto or Stratonovich calculus (Chap. 4). In the present context we wish to study the impact of fluctuating forces on the behavior of systems close to instability points. We shall assume that $\boldsymbol{F}$ is comparatively small in a way that it does not change the character of the transition appreciably. This means that we shall concentrate on those problems in which the instability is not induced by fluctuations but by the deterministic part $\boldsymbol{N}$. Qualitative changes may occur in the case of multiplicative noise where $\boldsymbol{F}$ depends in certain ways on $\boldsymbol{q}$. Since these aspects are covered by *Horsthemke* and *Lefever* (cf. the references), we shall not enter the discussion on these problems here.

We now proceed as follows. We first disregard $\boldsymbol{F}$ or $d\boldsymbol{F}$, assume that

$$\dot{\boldsymbol{q}} = \boldsymbol{N}(\boldsymbol{q}, \alpha) \tag{10.1.3}$$

possesses a solution for a given range of the control parameter α, and study as before the stability of

$$\boldsymbol{q}_0(t, \alpha) . \tag{10.1.4}$$

The hypothesis

$$\boldsymbol{q} = \boldsymbol{q}_0 + \boldsymbol{w} \tag{10.1.5}$$

leads us to the by now well-known linearized equations

$$\dot{\boldsymbol{w}}_k(t) = L(t, \alpha)\, \boldsymbol{w}_k(t) , \tag{10.1.6}$$

where we shall assume the solutions in the form

$$\boldsymbol{w}_k(t) = \mathrm{e}^{\lambda_k t} \boldsymbol{v}_k(t) . \tag{10.1.7}$$

Again amplitudes ξ_k and phase angles ϕ_k may be introduced. Assuming that one or several eigenvalues λ's belonging to (10.1.7) acquire a positive real part, we shall apply the slaving principle. In Sect. 7.6 we have shown that the slaving principle can be applied to stochastic differential equations of the Langevin-Îto or -Stratonovich type. According to the slaving principle we may reduce the original set of equations (10.1.2) to a set of order parameter equations for the corresponding ξ_k and ϕ_k. The resulting equations are again precisely of the form (10.1.2), though, of course, the explicit form of $\boldsymbol{N}$ and $d\boldsymbol{F}$ has changed. To illustrate the impact of fluctuations and how to deal with it, we shall consider a specific example first.

10.2 A Simple Example

Let us study the order parameter equation

$$\dot{u} = \lambda u - b u^3 + F(t) , \tag{10.2.1}$$

where we assume the properties

$$\langle F(t) \rangle = 0 , \tag{10.2.2}$$

$$\langle F(t)F(t')\rangle = Q\delta(t-t') \tag{10.2.3}$$

for the fluctuating forces. If we are away from the instability point $\lambda = 0$, in general it suffices to solve (10.2.1) approximately by linearization. Because for $\lambda < 0$, u is a small quantity close to the stationary state, we may approximate (10.2.1) by

$$\dot{u} \approx \lambda u + F(t)\,. \tag{10.2.4}$$

For $\lambda > 0$ we make the replacement

$$u = u_0 + \eta\,, \quad \text{where} \quad \lambda - b u_0^2 = 0\,, \tag{10.2.5}$$

which transforms (10.2.1) into

$$\dot{\eta} \approx -2\lambda\eta + F(t)\,, \tag{10.2.6}$$

where we have neglected higher powers of η. Because (10.2.6) is of the same form as (10.2.4) and η is small, it suffices to study the solutions of (10.2.6) which can be written in the form

$$\eta = \int_0^t \exp[-2\lambda(t-\tau)]F(\tau)d\tau\,. \tag{10.2.7}$$

(We do not take care of the homogeneous solution because it drops out in the limit performed below.) The correlation function can be easily evaluated by means of (10.2.3) and yields in a straightforward fashion for $t \geqq t'$

$$\begin{aligned}
\langle \eta(t)\eta(t')\rangle &= \int_0^t\int_0^{t'} \exp[-2\lambda(t-\tau) - 2\lambda(t'-\tau')]Q\delta(\tau-\tau')d\tau d\tau' \\
&= \int_0^{t'} \exp[-2\lambda(t+t') + 4\lambda\tau']Qd\tau' \\
&= \exp[-2\lambda(t+t')]Q\frac{(e^{4\lambda t'}-1)}{4\lambda}\,.
\end{aligned} \tag{10.2.8}$$

Assuming that t and t' are big, but $t-t'$ remains finite, (10.2.8) reduces to the stationary correlation function

$$\langle \eta(t)\eta(t')\rangle = \frac{Q}{4\lambda}\exp(-2\lambda|t-t'|)\,, \tag{10.2.9}$$

which is valid for $t \lesseqgtr t'$.

Close to the instability point $\lambda = 0$ the linearization procedure breaks down. This can be easily seen from the result (10.2.9), because in such a case the rhs diverges for $\lambda \to 0$. This effect is well known in phase transition theory and is called "critical fluctuation". However, in physical systems far from thermal equilibrium and many other systems these fluctuations are limited, which, mathematically speaking, is due to the nonlinear term $-bu^3$ in (10.2.1). To take care of this term the approach via the Fokker-Planck equation is most elegant. To this end we assume that $F(t)$ has the properties (10.2.2) and (10.2.3) and that it is Gaussian [1]. According to Sect. 4.2 the Fokker-Planck equation for the probability density f belonging to the Langevin equation (10.2.1) reads

$$\dot{f} = -\frac{\partial}{\partial u}[(\lambda u - bu^3)f] + \frac{Q}{2}\frac{\partial^2}{\partial u^2}f, \quad \text{where} \tag{10.2.10}$$

$$\langle F(t)F(t')\rangle = Q\delta(t-t'). \tag{10.2.11}$$

As shown in [1], the stationary solution of (10.2.10) is given by

$$f_0(u) = \mathcal{N}\exp\left[\frac{2}{Q}(\lambda u^2/2 - bu^4/4)\right], \tag{10.2.12}$$

where $\mathcal{N}$ is a normalization factor. The branching of the solution of the deterministic equation with $F(t) \equiv 0$ from $u = 0$ for $\lambda \leqslant 0$ into $u_\pm = \pm\sqrt{\lambda/b}$ for $\lambda > 0$ is now replaced by the change of shape of the distribution function (10.2.12), in which the single peak for $\lambda < 0$ is replaced by two peaks for $\lambda > 0$. Some care must be exercised in interpreting this result, because f_0 is a probability distribution. Therefore, in reality the system may be at any point u but with given probability (10.2.12). Clearly, for $\lambda > 0$ the probability is a maximum for $u_\pm = \pm\sqrt{\lambda/b}$. But, of course, at a given moment the system can be in a single state only. From this point of view an important question arises. Let us assume that at time $t = 0$ we have prepared (or measured) the system in a certain initial state $u = u_i$. What is the probability of finding the system at a later time t in some other final given state $u = u_f$? This question can be answered by the time-dependent solution $f(u,t)$ of the Fokker-Planck equation with the initial condition $f(u,0) = \delta(u - u_i)$. Even for Fokker-Planck equations containing several variables these solutions can be found explicitly if the drift coefficients are linear in the variables and the diffusion coefficients constants. We shall present the results for the one-variable Fokker-Planck equation at the end of this section and the corresponding general theorem in Sect. 10.4.1. If, for example, the drift coefficients are nonlinear, even in the case of a single variable, computer solutions are necessary. We shall give an outline of the corresponding results in Sect. 10.3.

A further important problem is the following. Let us assume that the system was originally prepared in the state u_+. How long will it take for the system to reach the state u_- for the first time? This is a special case of the so-called first passage time problem which can be formally solved by a general formula. We shall deal with the first passage time problem in the context of the more general Chapman-Kolmogorov equation for discrete maps in Sect. 11.6. Now, let us take up the problem of finding the time-dependent solutions of (10.2.1) in an approximation, namely by *linearization*. It suffices to treat $\lambda > 0$, because the case of $\lambda < 0$ can be treated similarly. Using the decomposition

$$u = u_0 + \eta \tag{10.2.13}$$

and neglecting nonlinear terms, we transform (10.2.10) into

$$\dot{\tilde{f}} = -\frac{\partial}{\partial \eta}(-2\lambda \eta \tilde{f}) + \frac{Q}{2}\frac{\partial^2}{\partial \eta^2}\tilde{f} \tag{10.2.14}$$

with $\tilde{f} \equiv \tilde{f}(\eta) = f(u_0 + \eta)$.

Taking as initial condition

$$\tilde{f}(t=0) = \delta(\eta - \eta_0)\,, \tag{10.2.15}$$

the solution reads [1]

$$\tilde{f}(\eta, t) = [\pi a(t)]^{-1/2} \exp\{-[\eta - b(t)]^2/a(t)\} \tag{10.2.16}$$

where a and b are given by

$$a(t) = \frac{Q}{2\lambda}[1 - \exp(-4\lambda t)] \quad \text{and} \tag{10.2.17}$$

$$b(t) = b(0)\exp(-2\lambda t)\,. \tag{10.2.18}$$

10.3 Computer Solution of a Fokker-Planck Equation for a Complex Order Parameter

In this section we study the branching from a time-independent solution when a complex eigenvalue acquires a positive real part [cf. Sect. 8.4, in particular (8.4.9)]. Making the substitution

$$u(t) \rightarrow u(t)\,e^{i\omega t}, \quad \omega \equiv \lambda'', \tag{10.3.1}$$

we assume the order parameter equation in the form

$$\dot{u} = \beta(\bar{n} - |u|^2)u + F(t)\,, \tag{10.3.2}$$

where $F(t)$ fulfills (10.2.2, 3). Decomposing u according to

$$u = r e^{-i\varphi}, \tag{10.3.3}$$

we may derive the following Fokker-Planck equation

$$\frac{\partial W}{\partial t}+\beta\frac{1}{r}\frac{\partial}{\partial r}[(\bar{n}-r^2)rW]=Q\left[\frac{1}{r}\frac{\partial}{\partial r}\left(r\frac{\partial W}{\partial r}\right)+\frac{1}{r^2}\frac{\partial^2 W}{\partial\varphi^2}\right]. \tag{10.3.4}$$

[This can best be done by writing down the Fokker-Planck equation for the real and imaginary parts of u and transforming that equation to polar coordinates according to (10.3.3).]

In order to get rid of superfluous constants we introduce new variables or constants by means of

$$\hat{r}=\sqrt[4]{\beta/Q}\,r\,,\quad \hat{t}=\sqrt{\beta Q}\,t\,,\quad a=\sqrt{\beta/Q}\,\bar{n}\,. \tag{10.3.5}$$

The Fokker-Planck equation (10.3.4) then takes the form

$$\frac{\partial W}{\partial\hat{t}}+\frac{1}{\hat{r}}\frac{\partial}{\partial\hat{r}}[(a-\hat{r}^2)\hat{r}W]=\frac{1}{\hat{r}}\frac{\partial}{\partial\hat{r}}\left(\hat{r}\frac{\partial W}{\partial\hat{r}}\right)+\frac{1}{\hat{r}^2}\frac{\partial^2 W}{\partial\varphi^2}. \tag{10.3.6}$$

The stationary solution is given by ($\mathcal{N}$: normalization constant)

$$W(\hat{r})=\frac{\mathcal{N}}{2\pi}\exp\left(-\frac{\hat{r}^4}{4}+a\frac{\hat{r}^2}{2}\right),\quad \frac{1}{\mathcal{N}}=\int_0^\infty\hat{r}\exp\left(-\frac{\hat{r}^4}{4}+a\frac{\hat{r}^2}{2}\right)d\hat{r}\,. \tag{10.3.7}$$

In order to obtain correlation functions, e.g., of the type (10.2.8), the nonstationary solutions of the Fokker-Planck equation must be used. Since general analytical expressions are not known, either approximation methods (like a variational method) or painstaking computer calculations were performed, the latter yielding results of great accuracy, of which those relevant are represented.

First (10.3.6) is reduced to a one-dimensional Schrödinger equation. The hypothesis

$$W(\hat{r},\varphi,\hat{t})=\sum_{m=0}^{\infty}\sum_{n=-\infty}^{\infty}A_{nm}\left[\frac{1}{\sqrt{\hat{r}}}\exp\left(\frac{\hat{r}^4}{8}+a\frac{\hat{r}^2}{4}\right)\Psi_{nm}(\hat{r})\right]\exp(\mathrm{i}n\varphi-\lambda_{nm}\hat{t}) \tag{10.3.8}$$

leads to

$$\Psi''_{nm}+[\lambda_{nm}-V_n(\hat{r})]\,\Psi_{nm}=0\qquad\text{with}\quad \Psi''\equiv\frac{d^2}{d\hat{r}^2}\Psi \tag{10.3.9}$$

for the eigenfunctions $\Psi_{nm}=\Psi_{-nm}$ and the eigenvalues $\lambda_{nm}=\lambda_{-nm}$. For explicit values of the λ's see Table 10.3.1.

Table 10.3.1. Eigenvalues and moments for different pump parameters, a

a	λ_{01}	M_1	λ_{02}	M_2	λ_{03}	M_3	λ_{04}	M_4
10	19.1142	0.4614	19.1237	0.4885	34.5184	0.0212	35.3947	0.0226
8	14.6507	0.4423	14.9666	0.4622	23.6664	0.0492	28.3894	0.0344
7	12.0787	0.4085	13.0891	0.4569	20.0129	0.0895	26.4382	0.0354
6	9.4499	0.4061	11.5823	0.4459	18.0587	0.1132	25.6136	0.0287
5	7.2368	0.4717	10.6059	0.4095	17.3876	0.0980	25.8079	0.0179
4	5.6976	0.5925	10.2361	0.3344	17.6572	0.0634	26.9004	0.0086
3	4.8564	0.7246	10.4763	0.2387	18.6918	0.0330	28.7914	0.0033
2	4.6358	0.8284	11.2857	0.1553	20.3871	0.0151	31.3963	0.0011
0	5.6266	0.9370	14.3628	0.0601	25.4522	0.0028	38.4621	0.0001

The potential $V_n(\hat{r})$ is given by (Fig. 10.3.1)

$$V_n(\hat{r}) = \frac{n^2}{\hat{r}^2} + \frac{\Psi_{00}''}{\Psi_{00}}$$

$$= \left(n^2 - \frac{1}{4}\right)\frac{1}{\hat{r}^2} + a + \left(\frac{a^2}{4} - 2\right)\hat{r}^2 - \frac{a}{2}\hat{r}^4 + \frac{\hat{r}^6}{4}. \qquad (10.3.10)$$

According to (10.3.7, 8) the eigenfunction Ψ_{00} belonging to the stationary eigenvalue $\lambda_{00} = 0$ is

$$\Psi_{00} = \sqrt{N\hat{r}}\exp(-\hat{r}^4/8 + a\hat{r}^2/4). \qquad (10.3.11)$$

The first five eigenfunctions, which are determined numerically, are plotted in Fig. 10.3.2. For a plot of eigenvalues consult Fig. 10.3.3.

If follows from (10.3.9) that the Ψ_{nm}'s are orthogonal for different m. If they are normalized to unity, we have

$$\int_0^\infty \Psi_{nm}(\hat{r})\,\Psi_{nm'}(\hat{r})\,d\hat{r} = \delta_{mm'}. \qquad (10.3.12)$$

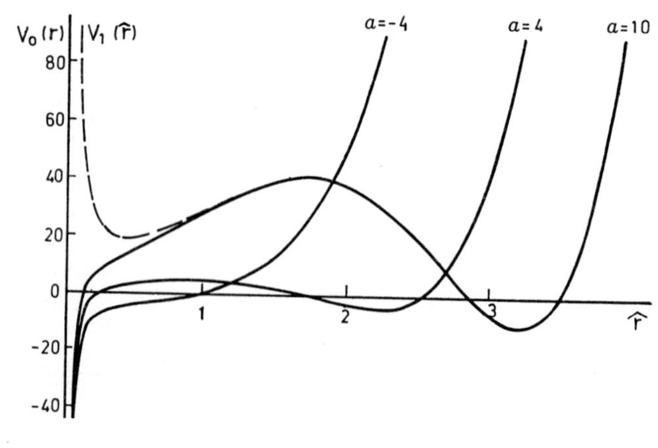

Fig. 10.3.1. The potential $V_0(\hat{r})$ of the Schrödinger equation (10.3.9) for three pump parameters (solid line) and $V_1(\hat{r})$ for $a = 10$ (broken line). [After H. Risken, H. D. Vollmer: Z. Physik **201**, 323 (1967)]

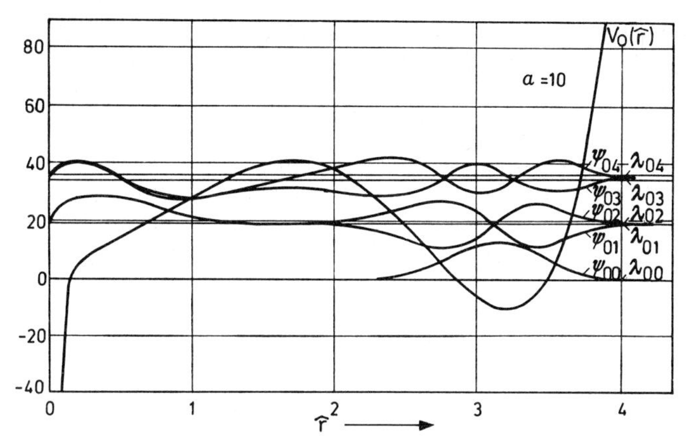

Fig. 10.3.2. The potential V_0 of the Schrödinger equation (10.3.9) and the first five eigenvalues and eigenfunctions for the pump parameter $a = 10$. [After H. Risken, H. D. Vollmer: Z. Phys. **201**, 323 (1967)]

The completeness relation

$$\delta(\hat{r}-\hat{r}') = \sum_{m=0}^{\infty} \Psi_{nm}(\hat{r})\,\Psi_{nm}(\hat{r}') \tag{10.3.13}$$

leads immediately to Green's function of the Fokker-Planck equation. It is obtained from the general solution (10.3.8) by putting

$$A_{nm} = \frac{1}{2\pi\sqrt{\hat{r}'}} \exp\left(\frac{\hat{r}'^4}{8} - a\frac{\hat{r}'^2}{4}\right) \Psi_{nm}(\hat{r}') \exp(-\mathrm{i}n\varphi')\,. \tag{10.3.14}$$

Thus the Green's function reads

$$G(\hat{r},\varphi;\hat{r}',\varphi',\hat{\tau}) = \frac{1}{2\pi\sqrt{\hat{r}\hat{r}'}} \exp\left(-\frac{\hat{r}^4}{8} + a\frac{\hat{r}^2}{4} + \frac{\hat{r}'^4}{8} - a\frac{\hat{r}'^2}{4}\right)$$
$$\times \sum_{m=0}^{\infty} \sum_{n=-\infty}^{\infty} \Psi_{nm}(\hat{r})\,\Psi_{nm}(\hat{r}') \exp[\mathrm{i}n(\varphi-\varphi') - \lambda_{nm}\hat{\tau}]\,. \tag{10.3.15}$$

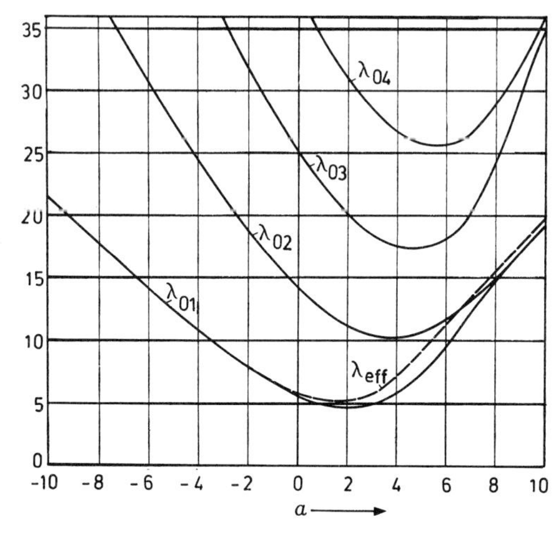

Fig. 10.3.3. The first four nonzero eigenvalues λ_{om} and the effective eigenvalue λ_{eff} (10.3.24) as functions of the pump parameter a. [After H. Risken, H. D. Vollmer: Z. Phys. **201**, 323 (1967)]

For the calculation of stationary two-time correlation functions the joint distribution function $F(\hat{r},\varphi;\hat{r}',\varphi';\hat{\tau})$ is needed. $F(\hat{r},\varphi;\hat{r}',\varphi';\hat{\tau})\hat{r}\,d\hat{r}\,d\varphi\,\hat{r}'\,d\hat{r}'\,d\varphi'$ is the probability that $\hat{r}(t+\hat{\tau})$, $\hat{\varphi}(t+\hat{\tau})$ lie in the interval $\hat{r},\dots,\hat{r}+d\hat{r};\ \varphi,\dots,\varphi+d\varphi$ and that $\hat{r}'(t)$, $\varphi'(t)$ lie in the interval $\hat{r}',\dots,\hat{r}'+d\hat{r}',\ \varphi',\dots,\varphi'+d\varphi'$. Moreover, F can be expressed by means of the Green's function $G(\hat{r},\varphi;\hat{r}',\varphi';\hat{\tau})$ and of $W(\hat{r}',\varphi')$, the latter describing the distribution at the initial time, t,

$$F(\hat{r},\varphi;\hat{r}',\varphi';\hat{\tau}) = G(\hat{r},\varphi;\hat{r}',\varphi';\hat{\tau})\,W(\hat{r}',\varphi'),\ \hat{\tau}>0\,. \tag{10.3.16}$$

The correlation function of the intensity fluctuations is obtained by

$$\begin{aligned} K(a,\ \hat{\tau}) &\equiv \langle(\hat{r}^2(\hat{t}+\hat{\tau})-\langle\hat{r}^2\rangle)(\hat{r}^2(\hat{t})-\langle\hat{r}^2\rangle)\rangle \\ &= \int\!\!\int\!\!\int\!\!\int \hat{r}\,d\hat{r}\,\hat{r}'\,d\hat{r}'\,d\varphi\,d\varphi'\,(\hat{r}^2-\langle\hat{r}^2\rangle)(\hat{r}'^2-\langle\hat{r}'^2\rangle)\,F(\hat{r},\varphi;\hat{r}',\varphi';\hat{\tau}) \\ &= K(a,0)\sum_{m=1}^{\infty} M_m \exp(-\lambda_{om}\hat{\tau})\,, \end{aligned} \tag{10.3.17}$$

where

$$M_m = \frac{\mathcal{N}}{K(a,0)}\left[\int_0^\infty \sqrt{\hat{r}}\,\hat{r}^2 \exp\left(-\frac{\hat{r}^4}{8}+a\frac{\hat{r}^2}{4}\right)\Psi_{om}(\hat{r})\,d\hat{r}\right]^2. \tag{10.3.18}$$

For a plot of the first four matrix elements M_m see Fig. 10.3.4, and for explicit values, Table 10.3.1. Here $K(a,0)$ is given by

$$\langle\hat{r}^4\rangle-\langle\hat{r}^2\rangle^2 = 2\pi\int_0^\infty \hat{r}^5\,W(\hat{r})\,d\hat{r} - \left[2\pi\int_0^\infty \hat{r}^3\,W(\hat{r})\,d\hat{r}\right]^2, \tag{10.3.19}$$

where $W(\hat{r})$ is defined in (10.3.7).

Both $\mathcal{N}$ and $K(a,0)$ may be reduced to the error integral

$$\Phi(y) = \frac{2}{\sqrt{\pi}}\int_0^y \exp(-x^2)\,dx\,. \tag{10.3.20}$$

Introducing the new variable $v=\hat{r}^2$ we define

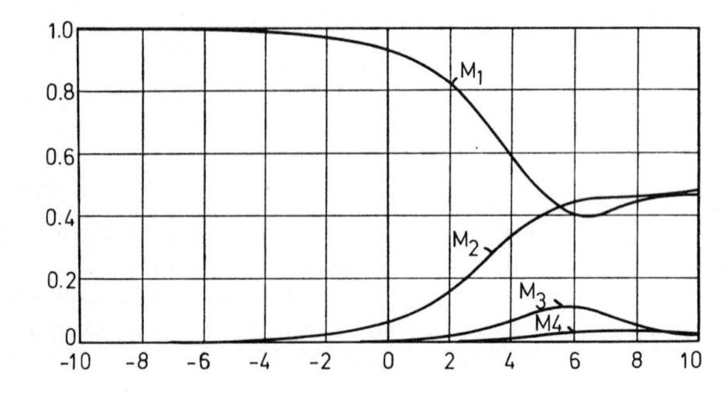

Fig. 10.3.4. The first four matrix elements M_m as functions of the pump parameter a. [After H. Risken, H. D. Vollmer: Z. Phys. **201**, 323 (1967)]

$$I_n(a) = \int_0^\infty v^n \exp\left(-\frac{v^2}{4} + a\frac{v}{2}\right) dv \,. \tag{10.3.21}$$

Then the following recurrence relations hold:

$$\begin{aligned} I_0(a) &= \sqrt{\pi}\exp(a^2/4)\,[1+\Phi(a/2)]\,,\\ I_1(a) &= 2 + a I_0(a)\,,\\ I_n(a) &= 2(n-1)I_{n-2}(a) + a I_{n-1}(a) \quad \text{for} \quad n \geqq 2\,. \end{aligned} \tag{10.3.22}$$

The spectrum $S(a, \hat{\omega})$ of the intensity fluctuations is given by the Fourier transform of the correlation function (10.3.17)

$$S(a, \hat{\omega}) = K(a,0) \sum_{n=1}^{\infty} M_m \,. \tag{10.3.23}$$

Although it is a sum of Lorentzian lines of widths λ_{0m}, $S(a, \hat{\omega})$ may be well approximated by an "effective" Lorentzian curve

$$S_{\text{eff}}(a, \hat{\omega}) = K(a,0)\frac{\lambda_{\text{eff}}}{\hat{\omega}^2 + \lambda_{\text{eff}}^2} \quad \text{with} \quad \frac{1}{\lambda_{\text{eff}}} = \sum_{m=1}^{\infty} \frac{M_m}{\lambda_{0m}} \tag{10.3.24}$$

which has the same area and the same maximum value (Fig. 10.3.5).

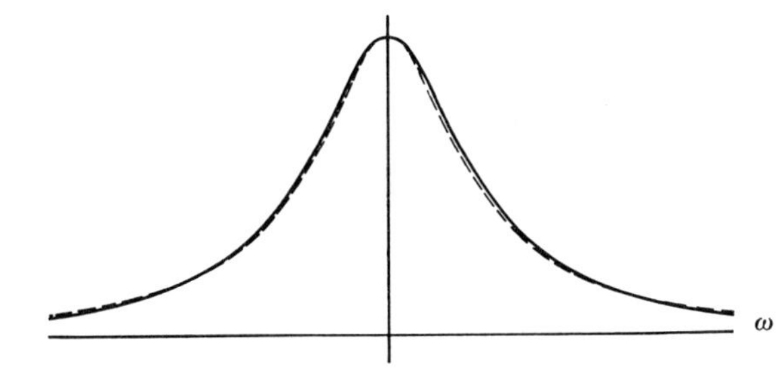

Fig. 10.3.5. A comparison between the exact noise spectrum and the effective Lorentzian line for $a = 5$. [After H. Risken, H. D. Vollmer: Z. Phys. **201**, 323 (1967)]

This effective width λ_{eff} is, however, about 25% larger than λ_{01} for $a \approx 5$. The eigenvalues and matrix elements were calculated numerically in particular for the threshold region $-10 \leqq a \leqq 10$. Similar calculations for the correlation function of the amplitude yield

$$\begin{aligned} g(a, \hat{\tau}) &= \langle \hat{r}(\hat{t}+\hat{\tau})\exp[\mathrm{i}\varphi(\hat{t}+\hat{\tau})]\,\hat{r}(\hat{t})\exp[-\mathrm{i}\varphi(\hat{t})]\rangle\\ &= \int\!\!\int\!\!\int\!\!\int \hat{r}\,d\hat{r}\,\hat{r}'\,d\hat{r}'\,d\varphi\,d\varphi'\,\hat{r}\hat{r}'\exp(\mathrm{i}\varphi - \mathrm{i}\varphi')F(\hat{r},\varphi;\hat{r}',\varphi';\,\hat{\tau})\,,\\ &= g(a,0)\sum_{m=0}^{\infty} V_m \exp(-\lambda_{1m}\,\hat{\tau})\,, \text{ where} \end{aligned} \tag{10.3.25}$$

$$V_m = \frac{N}{g(a,0)} \left[\int_0^\infty \sqrt{\hat{r}}\,\hat{r} \exp\left(-\frac{\hat{r}^4}{8} + a\frac{\hat{r}^2}{4} \right) \Psi_{1m}(\hat{r})\,d\hat{r} \right]^2 . \tag{10.3.26}$$

Further, $g(a,0)$ is given by

$$g(a,0) = \langle \hat{r}^2 \rangle = 2\pi \int_0^\infty \hat{r}^3 W(\hat{r})\,d\hat{r} \tag{10.3.27}$$

and can be reduced to the error integral by the same substitution leading to (10.3.21). A calculation of V_0 shows that

$$1 - V_0 = \sum_{m=1}^\infty V_m$$

is of the order of 2% near threshold and smaller farther away from threshold. Therefore the spectral profile is nearly Lorentzian with a linewidth (in unnormalized units)

$$\Delta\omega = \sqrt{\beta Q}\,\lambda_{10} = \alpha(a)\,Q/\langle n \rangle, \alpha = \lambda_{10}(a)\langle \hat{r}^2(a) \rangle . \tag{10.3.28}$$

The factor α is plotted in Fig. 10.3.6.

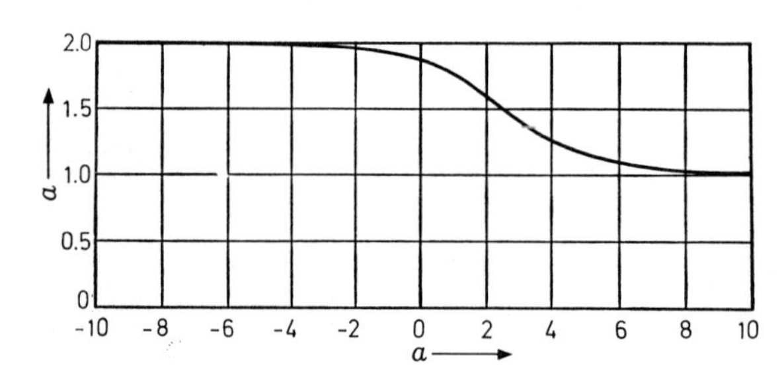

Fig. 10.3.6. The linewidth factor $\alpha(a) = \lambda_{10}$ as a function of the pump parameter a. [After H. Risken: Z. Phys. **191**, 302 (1966)]

Transient Solution. The general transient solution of (10.3.6) is given by

$$W(\hat{r}, \varphi, \hat{t}) = \int_0^\infty \int_0^{2\pi} G(\hat{r}, \varphi; \hat{r}', \varphi'; \hat{t})\, W(\hat{r}', \varphi', 0)\, \hat{r}'\, d\hat{r}'\, d\varphi' , \tag{10.3.29}$$

where G is the Green's function (10.3.15) and $W(\hat{r}', \varphi', 0)$ is the initial distribution.

After this specific example we now turn to the presentation of several useful general theorems.

10.4 Some Useful General Theorems on the Solutions of Fokker-Planck Equations

10.4.1 Time-Dependent and Time-Independent Solutions of the Fokker-Planck Equation, if the Drift Coefficients are Linear in the Coordinates and the Diffusion Coefficients Constant

In certain classes of applications we may assume that drift coefficients can be linearized around certain stable values of the coordinates and that the diffusion coefficients are independent of the coordinates. If we denote the elongation from the equilibrium positions by q_j, the corresponding Fokker-Planck equation reads

$$\frac{\partial f}{\partial t} + \sum_{ij} C_{ij} \frac{\partial}{\partial q_i}(q_j f) = \frac{1}{2} \sum_{ij} Q_{ij} \frac{\partial^2 f}{\partial q_i \partial q_j}. \tag{10.4.1}$$

We abbreviate $q_1, \ldots, q_N$ by $\boldsymbol{q}$. The Green's function of (10.4.1) must fulfill the initial condition

$$G(\boldsymbol{q}, \boldsymbol{q}', 0) = \prod_j \delta(q_j - q_j'). \tag{10.4.2}$$

The solution of (10.4.1) with (10.4.2) reads explicitly

$$G(\boldsymbol{q}, \boldsymbol{q}', t) = [\pi^n \det\{\sigma(t)\}]^{-1/2} \times \exp\left\{ - \sum_{ij} (\sigma^{-1})_{ij} [q_i - \sum_k b_{ik}(t) q_k'] [q_j - \sum_l b_{jl}(t) q_l'] \right\}, \tag{10.4.3}$$

where

$$\sigma = (\sigma_{ij}), \quad \sigma_{ij}(t) = \sum_{sr} [\delta_{is}\delta_{jr} - b_{is}(t) b_{jr}(t)] \sigma_{sr}(\infty). \tag{10.4.4}$$

The functions b_{is} which occur in (10.4.3, 4) obey the equations

$$\dot{b}_{is} = \sum_j C_{ij} b_{js}, \tag{10.4.5}$$

with the initial conditions

$$b_{js}(0) = \delta_{js}. \tag{10.4.6}$$

Here $\sigma(\infty)$ is determined by

$$C\sigma(\infty) + \sigma(\infty) C^T = -2Q, \tag{10.4.7}$$

where we have used the abbreviations

$$\begin{aligned} C &= (C_{ij}), \\ Q &= (Q_{ij}), \end{aligned} \tag{10.4.8}$$

and the superscript T denotes the transposed matrix. In particular the stationary solution reads

$$f(q) = G(q, q', \infty) = [\pi^n \det\{\sigma(\infty)\}]^{-1/2} \times \exp\left[-\sum_{ij} (\sigma^{-1})_{ij}(\infty)\} q_i q_j\right]. \tag{10.4.9}$$

10.4.2 Exact Stationary Solution of the Fokker-Planck Equation for Systems in Detailed Balance

In this section we mainly demonstrate two things.

1) We derive sufficient and necessary conditions for the drift and diffusion coefficients of the Fokker-Planck equation so that the "principle of detailed balance" is fulfilled.
2) We show that under the condition of detailed balance the stationary solution of the Fokker-Planck equation may be found explicitly by quadratures.

While the principle of detailed balance is expected to hold for practically all systems in thermal equilibrium, this need not be so in systems far from thermal equilibrium. Thus each individual case requires a detailed discussion (e.g., by symmetry considerations) as to whether this principle is applicable. Also the inspection of the structure of the Fokker-Planck equation will enable us to decide whether detailed balance is present.

a) Detailed Balance

We denote the set of variables $q_1, \ldots, q_N$ by $\boldsymbol{q}$ and the set of the variables under time reversal by

$$\tilde{q} = \{\varepsilon_1 q_1, \ldots, \varepsilon_N q_N\}, \tag{10.4.10}$$

where $\varepsilon_i = -1\ (+1)$ depending on whether the coordinate q_i changes sign (does not change sign) under time reversal. Furthermore, $\boldsymbol{\lambda}$ stands for a set of externally determined parameters. The time reversed quantity is denoted by

$$\tilde{\lambda} = \{v_1 \lambda_1, \ldots, v_M \lambda_M\}, \tag{10.4.11}$$

where $v_i = -1\ (+1)$ depends on the inversion symmetry of the external parameters under time reversal. We denote the joint probability of finding the system at t_1 with coordinates $\boldsymbol{q}$ and at t_2 with coordinates $\boldsymbol{q}'$ by

$$f_2(\boldsymbol{q}', \boldsymbol{q}; t_2, t_1). \tag{10.4.12}$$

In the following we consider a stationary system so that the joint probability depends only on the time difference $t_2 - t_1 = \tau$. Thus (10.4.12) may be written as

$$f_2(\boldsymbol{q}', \boldsymbol{q}; t_2, t_1) = W(\boldsymbol{q}', \boldsymbol{q}; \tau). \tag{10.4.13}$$

We now formulate the principle of detailed balance. The following two definitions are available.

1) The principle of detailed balance (first version)

$$W(q',q;\tau,\lambda) = W(\tilde{q},\tilde{q}';\tau,\tilde{\lambda}) . \tag{10.4.14}$$

The joint probability may be expressed by the stationary distribution $f(q)$ multiplied by the conditional probability P, where stationarity is exhibited by writing

$$P = P(q'|q;\tau,\lambda) . \tag{10.4.15}$$

Therefore, we may reformulate (10.4.14) as follows:

$$P(q'|q;\tau,\lambda)f(q,\lambda) = P(\tilde{q}|\tilde{q}';\tau,\tilde{\lambda})f(\tilde{q}',\tilde{\lambda}) . \tag{10.4.16}$$

Here and in the following we assume that the Fokker-Planck equation possesses a unique stationary solution. One may then show directly that

$$f(q,\lambda) = f(\tilde{q},\tilde{\lambda}) \tag{10.4.17}$$

holds. We define the transition probability per second by

$$w(q',q;\lambda) = [(d/d\tau)P(q'|q;\tau,\lambda)]_{\tau=0} . \tag{10.4.18}$$

Taking the derivative with respect to τ on both sides of (10.4.16) and putting $\tau = 0$ (but $q \neq q'$), we obtain

2) the principle of detailed balance (second version)

$$w(q',q;\lambda)f(q,\lambda) = w(\tilde{q},\tilde{q}';\tilde{\lambda})f(\tilde{q}',\tilde{\lambda}) . \tag{10.4.19}$$

It has obviously a very simple meaning. The left-hand side describes the total transition rate out of the state q into a new state q'. The principle of detailed balance then requires that this transition rate is equal to the rate in the reverse direction for q' and q with reverse motion, e.g., with reverse momenta.

b) The Required Structure of the Fokker-Planck Equation and Its Stationary Solution

We now derive necessary and sufficient conditions on the form of a Fokker-Planck equation so that the principle of detailed balance in its second (and first) versions is satisfied. Using the conditional probability P (which is nothing but the Green's function) we write the Fokker-Planck equation (or generalized Fokker-Planck equation having infinitely many derivatives) in the form of the equation

$$\frac{d}{d\tau}P(q'|q;\tau,\lambda) = L(q',\lambda)P(q'|q;\tau,\lambda) . \tag{10.4.20}$$

Note that, if not otherwise stated, L may also be an integral operator. The solution of (10.4.20) is subject to the initial condition

$$P(q'|q;0,\lambda) = \delta(q'-q) . \tag{10.4.21}$$

The formal solution of (10.4.20) with (10.4.21) reads

$$P(q'|q;\tau,\lambda) = \exp[L(q',\lambda)\tau]\,\delta(q'-q) . \tag{10.4.22}$$

Putting (10.4.22) into (10.4.20) and taking $\tau = 0$ on both sides, (10.4.20) acquires the form

$$w(q',q;\lambda) = L(q',\lambda)\,\delta(q'-q) . \tag{10.4.23}$$

The backward equation (backward Kolmogorov equation) is defined by

$$(d/d\tau)P(q'|q;\tau,\lambda) = L^+(q,\lambda)P(q'|q;\tau,\lambda) , \tag{10.4.24}$$

where L^+ is the operator adjoint to L. Again specializing (10.4.24) for $\tau = 0$ we obtain

$$w(q',q;\lambda) = L^+(q,\lambda)\,\delta(q'-q) . \tag{10.4.25}$$

Proceeding in (10.4.25) to time-inverted coordinates and then inserting this and (10.4.23) into (10.4.19), we obtain

$$L(q',\lambda)\,\delta(q'-q)f(q,\lambda) = \{L^+(\tilde{q}',\tilde{\lambda})\,\delta(\tilde{q}'-\tilde{q})\}f(q',\lambda) , \tag{10.4.26}$$

where we have used (10.4.17).

We now demonstrate how one may derive an operator identity to be fulfilled by L, L^+ which is a consequence of (10.4.19). On the lhs of (10.4.26) we replace q by q' in f. On the rhs we make the replacement

$$\delta(\tilde{q}'-\tilde{q}) = \delta(q'-q) . \tag{10.4.27}$$

With these substitutions and bringing rhs to lhs, (10.4.26) acquires the form

$$L(q',\lambda)f(q',\lambda)\,\delta(q'-q) - f(q',\lambda)L^+(\tilde{q}',\tilde{\lambda})\,\delta(q'-q) = 0 . \tag{10.4.28}$$

Because the δ-function is an arbitrary function if we let q accept all values, (10.4.28) is equivalent to the following operator equation

$$L(q',\lambda)f(q',\lambda) - f(q',\lambda)L^+(\tilde{q}',\tilde{\lambda}) = 0 . \tag{10.4.29}$$

In (10.4.29), L acts in the usual sense of an operator well known from operators in quantum mechanics, so that Lf is to be interpreted as $L(f\ldots)$, the points indicating an arbitrary function. So far we have seen that the condition of detailed balance has the consequence (10.4.29).

We now demonstrate that if (10.4.29) is fulfilled the system even has the property of the first-version principle of detailed balance (which appears to be stronger). First we note that (10.4.29) may be iterated yielding

$$[L(q',\lambda)]^n f(q',\lambda) = f(q',\lambda)[L^+(\tilde{q}',\tilde{\lambda})]^n \,. \tag{10.4.30}$$

We multiply (10.4.30) by $\tau^n(1/n!)$ and sum up over n from $n = 0$ to $n = \infty$. Now making all steps which have led from (10.4.26 to 29) in the reverse direction, and using (10.4.22) and its analogous form for L^+, we obtain (10.4.16) and thus (10.4.14). We now exploit (10.4.29) to determine the explicit form of the Fokker-Planck equation if the system fulfills the condition of detailed balance. Because (10.4.29) is an operator identity each coefficient of all derivatives with respect to q_i must vanish. Though in principle the comparison of coefficients is possible for arbitrarily high derivatives, we confine ourselves to the usual Fokker-Planck equation with an operator L of the form

$$L(q) = -\sum_i \frac{\partial}{\partial q_i} K_i(q,\lambda) + \frac{1}{2}\sum_{ik} \frac{\partial^2}{\partial q_i \partial q_k} K_{ik}(q,\lambda) \tag{10.4.31}$$

and its adjoint

$$L^+(q) = \sum_i K_i(q,\lambda) \frac{\partial}{\partial q_i} + \frac{1}{2}\sum_{ik} K_{ik}(q,\lambda) \frac{\partial^2}{\partial q_i \partial q_k} \,. \tag{10.4.32}$$

We may always assume that the diffusion coefficients are symmetric

$$K_{ik} = K_{ki} \,. \tag{10.4.33}$$

It is convenient to define the following new coefficients:

a) the irreversible drift coefficients

$$D_i(q,\lambda) = \tfrac{1}{2}[K_i(q,\lambda) + \varepsilon_i K_i(\tilde{q},\tilde{\lambda})] \equiv D_i^{ir} \,; \tag{10.4.34}$$

b) the reversible drift coefficients

$$J_i(q,\lambda) = \tfrac{1}{2}[K_i(q,\lambda) - \varepsilon_i K_i(\tilde{q},\tilde{\lambda})] \equiv D_i^{r} \,. \tag{10.4.35}$$

For applications it is important to note that J_i transforms as q_i under time reversal. We then explicitly obtain the necessary and sufficient conditions for K_{ik}, D_i and J_i so that the principle of detailed balance holds. We write the stationary solution of the Fokker-Planck equation in the form

$$f(q,\lambda) = \mathcal{N} \mathrm{e}^{-\Phi(q,\lambda)} \,, \tag{10.4.36}$$

where $\mathcal{N}$ is the normalization constant and Φ may be interpreted as a generalized thermodynamic potential. The conditions read

$$K_{ik}(q,\lambda) = \varepsilon_i \varepsilon_k K_{ik}(\tilde{q},\tilde{\lambda}) , \tag{10.4.37}$$

$$D_i - \frac{1}{2}\sum_k \frac{\partial K_{ik}}{\partial q_k} = -\frac{1}{2}\sum_k K_{ik}\frac{\partial \Phi}{\partial q_k} , \tag{10.4.38}$$

$$\sum_i \left(\frac{\partial J_i}{\partial q_i} - J_i \frac{\partial \Phi}{\partial q_i} \right) = 0 . \tag{10.4.39}$$

If the diffusion matrix K_{ik} possesses an inverse, (10.4.38) may be solved with respect to the gradient of Φ

$$\frac{\partial \Phi}{\partial q_i} = \sum_k (K^{-1})_{ik} \left(\sum_l \frac{\partial K_{kl}}{\partial q_l} - 2D_k \right) \equiv A_i . \tag{10.4.40}$$

This shows that (10.4.40) implies the integrability condition

$$(\partial/\partial q_j)A_i = (\partial/\partial q_i)A_j , \tag{10.4.41}$$

which is a condition on the drift and diffusion coefficients as defined by rhs of (10.4.40). Substituting A_i, A_j by (10.4.40), the condition (10.4.39) acquires the form

$$\sum_i \left[\frac{\partial J_i}{\partial q_i} - J_i \sum_k (K^{-1})_{ik} \left(\sum_l \frac{\partial K_{kl}}{\partial q_l} - 2D_k \right) \right] = 0 . \tag{10.4.42}$$

Thus the conditions for detailed balance to hold are given finally by (10.4.37, 41, 42). Equation (10.4.38) or equivalently (10.4.40) then allows us to determine Φ by pure quadratures, i.e., by a line integral. Thus the stationary solution of the Fokker-Planck equation may be determined explicitly.

10.4.3 An Example

Let us consider the following Langevin equations:

$$\dot{q}_1 = -\alpha q_1 + \omega q_2 + F_1(t) , \tag{10.4.43}$$

$$\dot{q}_2 = -\alpha q_2 - \omega q_1 + F_2(t) , \tag{10.4.44}$$

where the fluctuating forces F_j give rise to the diffusion coefficients

$$Q_{ik} = \delta_{ik}\bar{Q} = \text{const} . \tag{10.4.45}$$

In order to find suitable transformation properties of q_1, q_2 with respect to time reversal, we establish a connection between (10.4.43, 44) and the equations of the harmonic oscillator, i.e.,

$$\dot{x} = \omega p\,, \tag{10.4.46}$$

$$\dot{p} = -\omega x\,, \tag{10.4.47}$$

where we have chosen an appropriate scaling of the momentum p and the coordinate x. As is well known from mechanics, these variables transform under time reversal as

$$\tilde{x} = x\,, \tag{10.4.48}$$

$$\tilde{p} = -p\,. \tag{10.4.49}$$

A comparison between (10.4.43, 44) and (10.4.48, 49) suggests an identification (at least in the special case $\alpha = 0$)

$$x = q_1\,, \tag{10.4.50}$$

$$p = q_2\,. \tag{10.4.51}$$

Retaining this identification also for $\alpha \neq 0$, we are led to postulate the following properties of q_1, q_2,

$$\tilde{q}_1 = +q_1\,, \tag{10.4.52}$$

$$\tilde{q}_2 = -q_2\,, \tag{10.4.53}$$

so that $\varepsilon_1 = +1$, $\varepsilon_2 = -1$.

Using definition (10.4.34) of the irreversible drift coefficient D_1 we obtain

$$D_1 = \tfrac{1}{2}[-\alpha q_1 + \omega q_2 + (-\alpha q_1 - \omega q_2)] \tag{10.4.54}$$

so that

$$D_1 = -\alpha q_1\,. \tag{10.4.55}$$

In an analogous fashion we obtain

$$D_2 = \tfrac{1}{2}[-\alpha q_2 - \omega q_1 - (+\alpha q_2 - \omega q_1)] \tag{10.4.56}$$

so that

$$D_2 = -\alpha q_2\,. \tag{10.4.57}$$

Let us now calculate the reversible drift coefficients by means of (10.4.35). We obtain

$$J_1 = \tfrac{1}{2}[-\alpha q_1 + \omega q_2 - (-\alpha q_1 - \omega q_2)]$$
$$= +\omega q_2 \tag{10.4.58}$$

and similarly

$$J_2 = \tfrac{1}{2}[-\alpha q_2 - \omega q_1 + (\alpha q_2 - \omega q_1)]$$
$$= -\omega q_1 . \tag{10.4.59}$$

Inserting (10.4.55, 57), respectively, into (10.4.40) we obtain

$$A_i = -\frac{2\alpha}{\bar{Q}} q_i , \quad i = 1, 2 . \tag{10.4.60}$$

Using (10.4.55, 57 – 59), we readily verify (10.4.42). Clearly, (10.4.60) can be derived from the potential function

$$\Phi = \frac{\alpha}{\bar{Q}}(q_1^2 + q_2^2) \tag{10.4.61}$$

so that we have constructed the stationary solution of the Fokker-Planck equation belonging to (10.4.43, 44). (For an exercise see the end of Sect. 10.4.4.)

10.4.4 Useful Special Cases

We mention two special cases which have turned out to be extremely useful for applications.

1) $J_i = 0$ yields the so-called potential conditions in which case (10.4.37, 39) are fulfilled identically so that only (10.4.40, 41) remain to be satisfied.

2) In many practical applications one deals with complex variables (instead of the real ones) and the Fokker-Planck equation has the following explicit form

$$\frac{\partial f}{\partial t} = \left[\sum_j \left(\frac{\partial}{\partial u_j} C_j + \frac{\partial}{\partial u_j^*} \tilde{C}_j \right) + \sum_{kj} Q_{kj} \frac{\partial^2}{\partial u_k \partial u_j^*} \right] f \tag{10.4.62}$$

with

$$Q_{kj} = \delta_{kj} Q_j , \quad Q_j = Q . \tag{10.4.63}$$

Then the above conditions reduce to the following ones: C_j, $\tilde{C}_j$ must have the form

$$C_j = \partial B / \partial u_j^* + I_j^{(1)} , \tag{10.4.64}$$

$$\tilde{C}_j = \partial B/\partial u_j + I_j^{(2)}, \tag{10.4.65}$$

and the following conditions must be satisfied

$$\sum_j \left(\frac{\partial B}{\partial u_j} I_j^{(1)} + \frac{\partial B}{\partial u_j^*} I_j^{(2)} \right) = 0, \tag{10.4.66}$$

$$\sum_j \left(\frac{\partial I_j^{(1)}}{\partial u_j} + \frac{\partial I_j^{(2)}}{\partial u_j^*} \right) = 0. \tag{10.4.67}$$

As a result the stationary solution of (10.4.62) reads

$$f = \mathcal{N} e^{-\Phi}, \quad \text{where} \tag{10.4.68}$$

$$\Phi = 2B/Q. \tag{10.4.69}$$

Exercise. Determine the stationary solution of the Fokker-Planck equation belonging to the following coupled Langevin equations:

$$\dot{q}_1 = -\alpha q_1 + \omega q_2 - \beta q_1(q_1^2 + q_2^2) + F_1(t) \tag{10.4.70}$$

$$\dot{q}_2 = -\alpha q_2 - \omega q_1 - \beta q_2(q_1^2 + q_2^2) + F_2(t), \tag{10.4.71}$$

where $Q_{ik} = \delta_{ik}\bar{Q}$ independent of q_1, q_2.

10.5 Nonlinear Stochastic Systems Close to Critical Points: A Summary

After all the detailed representations of sometimes rather heavy mathematical methods it may be worthwhile to relax a little and to summarize the individual steps. One starting point would be nonlinear equations containing fluctuating forces. We assumed that in a first step we may neglect these forces. We then studied these systems close to instability points. Close to such points the behavior of a system is generally governed by few order parameters, and the slaving principle allows us to eliminate all "slaved" variables. The elimination procedure works also when fluctuating forces are present. This leads us to order parameter equations including fluctuating forces. Such order parameter equations may be of the Langevin-Îto or -Stratonovich type. In general these equations are nonlinear, and close to instability points we must not neglect nonlinearity. On the other hand, it is often possible to keep only the leading terms of the nonlinearities. The most elegant way to cope with the corresponding problem consists in transforming the corresponding order parameter equations of the Langevin-Îto or -Stratonovich type into a Fokker-Planck equation. Over the past decades we have applied this "program" to various systems. In quite a number of cases in which spatial patterns are formed it turned out that on the level of the order parameters the principle of detailed balance holds due to symmetry relations. In such a case one may get an overview of the probability with which individual configurations of the order parameters u_j can be realized. This allows one to determine the probability with which certain spatial patterns can be formed, and it allows us to find their stable configurations by looking for the minima of $V(\boldsymbol{u})$ in

$$f(\boldsymbol{u}) = \mathcal{N} \mathrm{e}^{-V(\boldsymbol{u})},$$

where $V(\equiv \Phi)$ is the potential defined in (10.4.40).

11. Discrete Noisy Maps

In this chapter we shall deal with discrete noisy maps, which we got to know in the introduction. In the first sections of the present chapter we shall study in how far we can extend previous results on differential equations to such maps. In Sects. 7.7 – 7.9 we showed how the slaving principle can be extended. We are now going to show how an analog to the Fokker-Planck equation (Sect. 10.2 – 4) can be found. We shall then present a path-integral solution and show how the analog of solutions of time-dependent Fokker-Planck equations can be found.

11.1 Chapman-Kolmogorov Equation

We consider an n-dimensional mapping where the state vectors $\boldsymbol{q}_k$ have n components. From what follows it appears that the whole procedure can be extended to continuously distributed variables. We assume that $\boldsymbol{q}_k$ obeys the equation

$$\boldsymbol{q}_{k+1} = \boldsymbol{f}(\boldsymbol{q}_k) + \boldsymbol{G}(\boldsymbol{q}_k)\boldsymbol{\eta}_k\,, \tag{11.1.1}$$

where $\boldsymbol{f}$ is a nonlinear function and the matrix G can be decomposed into

$$G = A + M(\boldsymbol{q}_k)\,, \tag{11.1.2}$$

where A is independent of the variables $\boldsymbol{q}_k$ and thus jointly with the random vector $\boldsymbol{\eta}_k$ represents additive noise. Here M is a function of the state vector $\boldsymbol{q}_k$ and thus jointly with $\boldsymbol{\eta}_k$ gives rise to multiplicative noise. We wish to find an equation for the probability density at "time" $k+1$ which is defined by

$$P(\boldsymbol{q}, k+1) = \langle \delta(\boldsymbol{q} - \boldsymbol{q}_{k+1}) \rangle\,. \tag{11.1.3}$$

The average goes over the random paths of the coordinates $\boldsymbol{q}_{k'}$ up to the index k, due to the systematic motion and the fluctuations $\boldsymbol{\eta}_{k'}$. Therefore we transform (11.1.3) into

$$P(\boldsymbol{q}, k+1) = \int d^n\xi \int d^n\eta\, \delta(\boldsymbol{q} - \boldsymbol{f}(\boldsymbol{\xi}) - G(\boldsymbol{\xi})\boldsymbol{\eta})\, W(\boldsymbol{\eta}) P(\boldsymbol{\xi}, k)\,, \tag{11.1.4}$$

where $W(\eta)$ is an arbitrary distribution function of the noise amplitude η and may even depend on each step k. For simplicity, we neglect the latter dependence, however.

To evaluate the integrals in (11.1.4) we introduce the variable η'

$$G(\xi)\eta = \eta' \,, \tag{11.1.5}$$

by which we can replace (11.1.4) by

$$P(q,k+1) = \int d^n\xi \int d^n\eta' D(\xi)^{-1}\delta(q-f(\xi)-\eta')\, W(G^{-1}(\xi)\eta')P(\xi,k)\,, \tag{11.1.6}$$

where $D = \det G$. The integrals can now be evaluated by use of the properties of the δ-function, yielding the final result

$$P(q,k+1) = \int d^n\xi D(\xi)^{-1}\, W(G^{-1}(\xi)[q-f(\xi)])P(\xi,k)\,, \tag{11.1.7}$$

which has the form of the Chapman-Kolmogorov equation.

11.2 The Effect of Boundaries. One-Dimensional Example

In our above treatment we have implicitly assumed that either the range of the components of q goes from $-\infty$ to $+\infty$ or that the spread of the distribution function W is small compared to the intervals on which the mapping is executed. If these assumptions do not hold, boundary effects must be taken into account. Since the presentation is not difficult in principle but somewhat clumsy, we show how this problem can be treated by means of the one-dimensional case.

We start from the equation

$$P(q,k+1) = \int_a^b d\xi \int d\eta\, \delta(q-f(\xi)-\eta)P(\xi,k)\, W(\eta)\,, \tag{11.2.1}$$

where

$$a \leqslant q \leqslant b \quad \text{and} \tag{11.2.2}$$

$$q = f(\xi) + \eta \tag{11.2.3}$$

hold. Equations (11.2.2, 3) imply

$$a - f(\xi) \leqslant \eta \leqslant b - f(\xi)\,. \tag{11.2.4}$$

When integrating over η in (11.2.1), we have to observe the boundary values (11.2.4). In order to guarantee the conservation of probability, i.e.,

$$\int_a^b P(q,k)\, dq = 1 \tag{11.2.5}$$

for all k, we must normalize $W(\eta)$ on the interval (11.2.4). This requires the introduction of a normalization factor $\mathcal{N}$ which is explicitly dependent on $f(\xi)$

$$\int_{a-f(\xi)}^{b-f(\xi)} W(\eta)\, d\eta = \mathcal{N}^{-1}(f(\xi)) . \tag{11.2.6}$$

Therefore, in (11.2.1) we are forced to introduce instead of $W(\eta)$ the function

$$W(\eta; f(\xi)) = \mathcal{N}(f(\xi))\, W(\eta) . \tag{11.2.7}$$

With this in mind we can eventually transform (11.2.1) into

$$P(q,k+1) = \int_a^b d\xi\, W(q-f(\xi))\, \mathcal{N}(f(\xi)) P(\xi,k) , \tag{11.2.8}$$

which is again a Chapman-Kolmogorov equation.

11.3 Joint Probability and Transition Probability. Forward and Backward Equation

In the following we use the Chapman-Kolmogorov equation for the probability density $P(\boldsymbol{q},k)$ in the form

$$P(\boldsymbol{q},k+1) = \int_{\mathcal{D}} d^n\xi\, K(\boldsymbol{q},\xi) P(\xi,k) . \tag{11.3.1}$$

The kernel K has the general form

$$K(\boldsymbol{q},\xi) = D^{-1}(\xi)\, \mathcal{N}(f(\xi))\, W(G^{-1}[q-f(\xi)]) \tag{11.3.2}$$

and $D = \det G$. Here W is the probability distribution of the random vector $\boldsymbol{\eta}$ and $\mathcal{N}$ is a properly chosen normalization factor which secures that the fluctuations do not kick the system away from its domain $\mathcal{D}$ and that probability is conserved, i.e.,

$$\int_{\mathcal{D}} d^n q\, K(\boldsymbol{q},\xi) = 1 . \tag{11.3.3}$$

In order to obtain a complete description of the underlying Markov process corresponding to (11.1.1), we have to consider the joint probability

$$P_2(\boldsymbol{q},k;\xi,k'), \quad (k>k') . \tag{11.3.4}$$

Here the "time" index k is related to $\boldsymbol{q}$, whereas k' belongs to ξ. Further, P_2 may be expressed in terms of the transition probability $p(\boldsymbol{q},k\,|\,\xi,k')$ and $P(\boldsymbol{q},k)$ via

$$P_2(q,k;\xi,k') = p(q,k\,|\,\xi,k')P(\xi,k') \,. \tag{11.3.5}$$

In the case of a stationary Markov process where the kernel (11.3.2) does not depend on the time index k explicitly, we have furthermore

$$p(q,k\,|\,\xi,k') = p(q\,|\,\xi;k-k') \,. \tag{11.3.6}$$

In order to derive the so-called forward equation for the transition probability p we perform an integration over ξ in (11.3.5) to obtain

$$P(q,k) = \int_{\mathscr{D}} d^n\xi p(q,k\,|\,\xi,k')P(\xi,k') \,. \tag{11.3.7}$$

Equations (11.3.7, 1) jointly with the fact that the initial distribution may be chosen arbitrarily yield the desired equation

$$p(q,k+1\,|\,\xi,k') = \int_{\mathscr{D}} K(q,z)p(z,k\,|\,\xi,k')d^nz \,. \tag{11.3.8}$$

Furthermore, we note that jointly with (11.3.7)

$$P(q,k) = \int_{\mathscr{D}} d^n\xi p(q,k\,|\,\xi,k'+1)P(\xi,k'+1) \tag{11.3.9}$$

holds. Subtracting (11.3.9) from (11.3.7) and using the fact that $P(\xi,k'+1)$ obeys (11.3.1) we arrive at

$$0 = \int_{\mathscr{D}} d^n\xi[p(q,k\,|\,\xi,k')P(\xi,k') - \int_{\mathscr{D}} d^nzp(q,k\,|\,\xi,k'+1)K(\xi,z)P(z,k')] \,. \tag{11.3.10}$$

Renaming now the variables in the second term, i.e., $z \leftrightarrow \xi$, and changing the order of integration we obtain

$$p(q,k\,|\,\xi,k') = \int_{\mathscr{D}} d^nzp(q,k\,|\,z,k'+1)K(z,\xi) \,. \tag{11.3.11}$$

Again we made use of the arbitrariness of the initial distribution $P(\xi,k')$. Equation (11.3.11) is the backward equation for the transition probability p. Equations (11.3.8 and 11) complete the description of process (11.1.1) in terms of probability distributions. Moments, correlation functions, etc., may now be defined in the standard way.

11.4 Connection with Fredholm Integral Equation

For the steady-state distribution we put

$$P(q,k) = P_s(q) \tag{11.4.1}$$

which transforms (11.1.7) into

$$P_s(q) = \int d^n\xi K(q,\xi) P_s(\xi) , \tag{11.4.2}$$

where the kernel is defined by [compare (11.3.2)]

$$K(q,\xi) = D(\xi)^{-1} W(G^{-1}(\xi)[q-f(\xi)]) . \tag{11.4.3}$$

Equation (11.4.2) is a homogeneous Fredholm integral equation. In order to treat transients we seek the corresponding eigenfunctions making the hypothesis

$$P_\lambda(q,k) = \lambda^{-k} P_\lambda(q) . \tag{11.4.4}$$

It transforms (11.1.7) again into a Fredholm integral equation

$$P_\lambda(q) = \lambda \int d^n\xi K(q,\xi) P_\lambda(\xi) . \tag{11.4.5}$$

The k-dependent solution can be expressed in the form

$$P(q,k) = \sum c_\lambda \lambda^{-k} P_\lambda(q) . \tag{11.4.6}$$

11.5 Path Integral Solution

We start from (11.1.7) in the form

$$P(q,k+1) = \int d^n\xi K(q,\xi) P(\xi,k) . \tag{11.5.1}$$

Iterating it for $k = 1, 2, \ldots$ leads to

$$P(q,k+1) = \int d^n q_k \ldots d^n q_1 K(q,q_k) K(q_k,q_{k-1}) \ldots K(q_2,q_1) P(q_1,1) , \tag{11.5.2}$$

which can be considered as a path integral. To make contact with other formulations we specialize W to a Gaussian distribution which we may take, without restriction of generality, in the form

$$W(\eta) = \mathcal{N} \exp(-\tilde{\eta} A \eta) , \tag{11.5.3}$$

where A is a diagonal matrix with diagonal elements α_j^2. Introducing the Fourier transform of (11.5.3) we may cast (11.5.2) into the form

$$P(q,k+1) = \int Dq \int Ds \prod_{l=1}^{k} D(q_l)^{-1} \exp[\cdots] \, P(q_1,1) , \tag{11.5.4}$$

where

$$
\begin{aligned}
Dq &= \prod_{l=1}^{k} d^n q_l\,, \\
Ds &= \prod_{l=1}^{k} [(1/(2\pi)^n d^n s_l]\,, \\
[\cdots] &= \sum_{l=1}^{k} \mathrm{i} s_l \{G^{-1}(q_l)[q_{l+1} - f(q_l)]\} - \tfrac{1}{4}\sum_{j=1}^{k} \tilde{s}_j^2\,, \\
q_{k+1} &\equiv q\,, \quad \tilde{s}_j = (s_{j1}/|\alpha_1|, s_{j2}/|\alpha_2|, \dots)\,.
\end{aligned}
\tag{11.5.5}
$$

11.6 The Mean First Passage Time

An important application of the backward equation (11.3.11) is given by the mean first passage time problem. In the following we shall derive an inhomogeneous integral equation for the conditional first passage time which will be introduced below. Here we assume that the process is stationary [compare (11.3.6)]. In order to formulate the problem precisely we consider some closed subdomain $\mathscr{V}$ of $\mathscr{D}$ and assume that the system is initially concentrated within that region $\mathscr{V}$ with probability 1, i.e.,

$$\int_{\mathscr{V}} d^n q P(q,0) = 1\,. \tag{11.6.1}$$

At this stage it becomes advantageous to define the probability $\tilde{P}(q,k)$, where $\tilde{P}(q,k)\,d^n q$ measures the probability to meet the system under consideration in the volume element $d^n q$ at q, without having reached the boundary of $\mathscr{V}$ before. We introduce

$$\tilde{P}(k) = \int_{\mathscr{V}} d^n q \tilde{P}(q,k)\,, \tag{11.6.2}$$

where $\tilde{P}(k)$ is the probability of finding the system within $\mathscr{V}$ without having reached the boundary up to time k. Combining (11.6.1 and 2) we obtain the probability that the system has reached the boundary of $\mathscr{V}$ during k, namely $1 - \tilde{P}(k)$. Finally, the probability that the system reaches the boundary of $\mathscr{V}$ between k and $k+1$ is given by $\tilde{P}(k) - \tilde{P}(k+1)$. It now becomes a simple matter to obtain the mean first passage time by

$$\langle \tau \rangle = \sum_{k=0}^{\infty} (k+1)[\tilde{P}(k) - \tilde{P}(k+1)]\,. \tag{11.6.3}$$

At this stage it should be noted that the mean first passage time not only depends on $\mathscr{V}$ but also on the initial distribution $P(q,0)$ [compare (11.6.1)]. It is this fact which suggests the introduction of the conditional first passage time $\langle \tau(q) \rangle$. This is the mean first passage time for a system which has been at q at time $k=0$ with certainty. Obviously we have

$$\langle\tau(\xi)\rangle = \sum_{k=0}^{\infty}(k+1)[\tilde{p}(q\,|\,\xi,k)-\tilde{p}(q\,|\,\xi,k+1)]\,, \tag{11.6.4}$$

where $\tilde{p}(q\,|\,\xi,k)$ is the corresponding transition probability. The relation between (11.6.3 and 4) is given by

$$\langle\tau\rangle = \int_{\mathscr{V}} d^n\xi\langle\tau(\xi)\rangle P(\xi,0)\,. \tag{11.6.5}$$

In the following we shall use the fact that within $\mathscr{V}$ the transition probability $\tilde{p}(q\mid\xi,k)$ obeys the backward equation (11.3.11), which allows us to rewrite (11.6.4)

$$\langle\tau(\xi)\rangle = \sum_{k=0}^{\infty}(k+1)\int_{\mathscr{V}} d^nq\int_{\mathscr{V}} d^nz[\delta(\xi-z)-K(z,\xi)]\tilde{p}(q\,|z,k)\,. \tag{11.6.6}$$

Equation (11.6.6) may now be considerably simplified by adding and subtracting the expression

$$\tilde{p}(q\,|z,k+1)K(z,\xi) \tag{11.6.7}$$

under the integral. Using the definition of $\langle\tau(q)\rangle$ according to (11.6.4) and applying the backward equation again we arrive at

$$\langle\tau(\xi)\rangle = -\int_{\mathscr{V}} d^nzK(z,\xi)\langle\tau(z)\rangle + R\,, \tag{11.6.8}$$

where R is given by

$$R = \sum_{k=0}^{\infty}(k+1)\int_{\mathscr{V}} d^nq[\tilde{p}(q\,|\,\xi,k)-\tilde{p}(q\,|\,\xi,k+2)]\,. \tag{11.6.9}$$

It is now simple to evaluate the expression for R. Adding and subtracting

$$\tilde{p}(q\,|\,\xi,k+1) \tag{11.6.10}$$

in the sum, using the obvious relation

$$\int_{\mathscr{V}} d^nq\sum_{k=0}^{\infty}[\tilde{p}(q\,|\,\xi,k)-\tilde{p}(q\,|\,\xi,k+1)] = 1\,, \tag{11.6.11}$$

performing the summation over k, and replacing ξ by q, our final result reads

$$\langle\tau(q)\rangle = \int_{\mathscr{V}} d^nzK(z,q)\langle\tau(z)\rangle + 1\,. \tag{11.6.12}$$

Equation (11.6.12) contains the result announced in the beginning of this section: We find an inhomogeneous integral equation for the conditional first passage time $\langle\tau(q)\rangle$ for the discrete time process (11.1.1).

11.7 Linear Dynamics and Gaussian Noise. Exact Time-Dependent Solution of the Chapman-Kolmogorov Equation

We consider the linear version of (11.1.1), i.e.,

$$f(q_k) = A q_k \,, \tag{11.7.1}$$

where A is a matrix depending on the external parameters only. Furthermore, we shall assume $G = 1$ and that the probability density of the random vector $\boldsymbol{\eta}$ is of the Gaussian type

$$W(\eta_k) = \left(\frac{\det \beta}{(2\pi)^n}\right)^{1/2} \exp\left(-\frac{1}{2}\bar{\eta}_k \beta \eta_k\right), \tag{11.7.2}$$

where β denotes a symmetric positive matrix. We note that (11.7.1) together with (11.7.2) may be visualized as a linearized version of (11.1.1) around a fixed point. If in addition the fluctuations are small, we are allowed to neglect the effect of the boundaries in the case of a finite domain $\mathscr{D}$ when the fixed point is far enough from the boundaries. Using (11.7.1 and 2), the kernel (11.3.2) obviously has the form

$$K(q,\xi) = \left(\frac{\det \beta}{(2\pi)^n}\right)^{1/2} \exp\left[-\frac{1}{2}(\bar{q} - \bar{\xi}A^T)\beta(q - A\xi)\right]. \tag{11.7.3}$$

Equation (11.1.7) may now be solved by the hypothesis

$$P(\xi,k) = \left(\frac{\det B}{(2\pi)^n}\right)^{1/2} \exp\left[-\frac{1}{2}(\bar{\xi} - \bar{\xi}_0)B(\xi - \xi_0)\right], \tag{11.7.4}$$

where ξ_0 denotes the center of the probability distribution a time k and B again is a positive symmetric matrix. Indeed, inserting both (11.7.4, 3) into (11.1.7) we obtain

$$P(q,k+1) = \frac{(\det B \cdot \det \beta)^{1/2}}{(2\pi)^n} \int_{-\infty}^{+\infty} d^n\xi \exp\{\cdots\}, \tag{11.7.5}$$

where

$$\{\cdots\} = \tfrac{1}{2}[(\bar{q} - \bar{\xi}A^T)\beta(q - A\xi) + (\bar{\xi} - \bar{\xi}_0)B(\xi - \xi_0)] \,. \tag{11.7.6}$$

Shifting ξ by a constant vector $\boldsymbol{a}$

$$\xi = \xi' + \boldsymbol{a} \tag{11.7.7}$$

and choosing

$$\boldsymbol{a} = (A^T\beta A + B)^{-1}(A^T\beta \boldsymbol{q} + B\xi_0)\,, \tag{11.7.8}$$

we are able to perform the ξ'integration and find

$$P(\boldsymbol{q},k+1) = \tilde{\mathcal{N}}\exp\left[-\tfrac{1}{2}(\bar{q}-\bar{q}_0)\tilde{B}(\boldsymbol{q}-\boldsymbol{q}_0)\right]. \tag{11.7.9}$$

Identifying

$$\begin{aligned}
\tilde{B} &= B_{k+1};\; B = B_k \\
\tilde{\mathcal{N}} &= \mathcal{N}_{k+1};\; \mathcal{N}_k = (2\pi)^{-n/2}(\det\beta)^{1/2} \\
\boldsymbol{q}_0 &= \boldsymbol{q}_{k+1};\; \xi_0 = \boldsymbol{q}_k
\end{aligned} \tag{11.7.10}$$

and comparing (11.7.4) with (11.7.9), we immediately find the recursion relations

$$\boldsymbol{q}_{k+1} = A\boldsymbol{q}_k\,, \tag{11.7.11}$$

$$B_{k+1} = \beta - \beta A(A^T\beta A + B_k)^{-1}A^T\beta\,, \tag{11.7.12}$$

$$\mathcal{N}_{k+1} = \mathcal{N}_k\cdot\left(\frac{\det\beta}{\det(A^T\beta A + B_k)}\right)^{1/2}. \tag{11.7.13}$$

The stationary solution can be obtained from the condition

$$B_{k+1} = B_k\,, \tag{11.7.14}$$

etc. In the case of a diagonal matrix β, which fulfills the condition

$$\beta_{ik} = \tilde{\beta}\delta_{ik}\,, \tag{11.7.15}$$

we may solve (11.7.14) if additionally the following situation is met

$$A^TA = AA^{T-1}\,. \tag{11.7.16}$$

We then simply have

$$B = \tilde{\beta}(1 - A^TA)\,. \tag{11.7.17}$$

Finally we mention that the Gaussian case also reveals instabilities of the system. Indeed, as soon as an eigenvalue a of the matrix A crosses $|a| = 1$ due to the variation of an external parameter, a corresponding divergence in the variance of the probability distribution indicates the instability.

12. Example of an Unsolvable Problem in Dynamics

When looking back at the various chapters of this book we may note that even by a seemingly straightforward extension of a problem (e.g., from periodic to quasi-periodic motion) the solution of the new problem introduces qualitatively new difficulties. Here we want to demonstrate that seemingly simple questions exist in dynamic systems which cannot be answered even in principle. Let us consider a dynamic system whose states are described by state vectors $\boldsymbol{q}$. Then the system may proceed from any given state to another one $\boldsymbol{q}'$ via transformations, A, B, C within a time interval τ,

$$\boldsymbol{q}' = A\boldsymbol{q}, \quad \text{or} \quad \boldsymbol{q}' = B\boldsymbol{q}, \quad \dots . \tag{12.1}$$

We assume that the inverse operators A^{-1}, B^{-1}, ... exist. Obviously, A, B, C, ... form a group. We may now study all expressions (words) formed of A, A^{-1}, B, B^{-1}, etc., e.g., $BA^{-1}C$. We may further define that for a number of specific words, $W(A,B,C\dots) = 1$, e.g., $BC = 1$. This means that after application of C and B any initial state $\boldsymbol{q}$ of the dynamic system is reached again. Then we may ask the following question: given two words $W_1(A,B,\dots)$ and $W_2(A,B,\dots)$ can we derive a general procedure by which we can decide in finitely many steps whether the dynamic system has reached the same end points $\boldsymbol{q}_1 = \boldsymbol{q}_2$, if it started from the same initial arbitrary point $\boldsymbol{q}_0$. That is, we ask whether

$$\boldsymbol{q}_1 \equiv W_1(A,B,\dots)\boldsymbol{q}_0 = \boldsymbol{q}_2 \equiv W_2(A,B,\dots)\boldsymbol{q}_0 , \tag{12.2}$$

or, equivalently, whether

$$W_1(A,B,\dots) = W_2(A,B,\dots) , \qquad \text{or} \tag{12.3}$$

$$W_1 W_2^{-1} = 1 . \tag{12.4}$$

This is a clearcut and seemingly simple task. Yet it is unsolvable in principle. There is no *general* procedure available. The problem we have stated is the famous *word problem* in group theory. By means of Gödel's theorem (which we are not going to present here) the unsolvability of this problem can be shown. On the other hand, if *specific* classes of words $W_j = 1$ as defining relations are given, the word problem can be solved.

This example shows clearly that some care must be exercised when problems are formulated. This is especially so if *general* solutions are wanted. Rather, one should be aware of the fact that some questions can be answered only with respect to *restricted classes* of equations (or problems). It is quite possible that such a cautious approach is necessary when dealing with self-organizing systems.

13. Some Comments on the Relation Between Synergetics and Other Sciences

In the introduction we presented *phenomena* which are usually treated within different disciplines, so that close links between synergetics and other disciplines could be established at that level. But the present book is primarily devoted to the *basic concepts* and *theoretical methods* of synergetics, and therefore in the following the relations between synergetics and other disciplines are discussed at this latter level. Since synergetics has various facets, a scientist approaching it from his own discipline will probably notice those aspects of synergetics first which come closest to the basic ideas of his own field. Based on my discussions with numerous scientists, I shall describe how links can be established in this way. Then I shall try to elucidate the basic differences between synergetics and the other disciplines.

When physicists are dealing with synergetics, quite often *thermodynamics* is brought to mind. Indeed, one of the most striking features of thermodynamics is its *universality*. Its laws are valid irrespective of the different components which constitute matter in its various forms (gases, liquids, and solids). Thermodynamics achieves this universal validity by dealing with *macroscopic quantities* (or "observables") such as volume, pressure, temperature, energy or entropy. Clearly these concepts apply to large ensembles of molecules, but not to individual molecules. A closely related approach is adopted by *information theory*, which attempts to make *unbiased estimates* about systems on which only limited information is available. Other physicists recognize common features between synergetics and *irreversible thermodynamics*. At least in the realm of physics, chemistry, and biology, synergetics and irreversible thermodynamics deal with systems driven away from thermal equilibrium.

Chemists and physicists are struck by the close analogy between the various macroscopic transitions of synergetic systems and *phase transitions* of systems in thermal equilibrium, such as the liquid – gas transition, the onset of ferromagnetism, and the occurrence of superconductivity. Synergetic systems may undergo continuous or discontinuous transitions, and they may exhibit features such as symmetry breaking, critical slowing down, and critical fluctuations, which are well known in phase transition theory.

The appropriate way to cope with fluctuations, which are a necessary part of any adequate treatment of phase transitions, is provided by *statistical mechanics*. Scientists working in that field are delighted to see how typical equations of their field, such as Langevin equations, Fokker-Planck equations and master equations, are of fundamental importance in synergetics. *Electrical engineers* are immediately familiar with other aspects of synergetics, such as networks, positive and negative feedback, and nonlinear oscillations, while *civil* and *mechanical engineers* probably consider synergetics to be a theory of static or dynamic instabilities, postbuckling phenomena of solid structures, and nonlinear oscillations. Synergetics studies the behavior of systems when controls are changed; clearly, scientists working in *cybernetics* may consider synergetics from the point of view of *control theory.*

From a more general point of view, both *dynamic systems theory* and synergetics deal with the temporal evolution of systems. In particular, mathematicians dealing with *bifurcation theory* observe that synergetics – at least in its present stage – focusses attention on qualitative changes in the dynamics (or statics) of a system, and in particular on bifurcations. Finally, synergetics may be considered part of *general systems theory*, because in both fields scientists are searching for the *general principles* under which systems act.

Quite obviously, each of the above-mentioned disciplines (and probably many others) have good reason to consider synergetics part of them. But at the same time in each case, synergetics introduces features, concepts, or methods which are alien to each specific field. *Thermodynamics* acts at its full power only if it deals with systems in thermal equilibrium, and *irreversible thermodynamics* is confined to systems close to thermal equilibrium. Synergetic systems in physics, chemistry, and biology are driven *far from thermal equilibrium* and can exhibit new features such as oscillations. While the concept of *macroscopic variables* retains its importance in synergetics, these variables, which we have called *order parameters,* are quite different in nature from those of thermodynamics. This becomes especially clear when thermodynamics is treated with the aid of information theory, where numbers of realizations are computed under given constraints. In other words, information theory and thermodynamics are *static* approaches, whereas synergetics deals with *dynamics.*

The *nonequilibrium phase transitions* of synergetic systems are much more varied than phase transitions of systems in thermal equilibrium, and include oscillations, spatial structures, and chaos. While phase transitions of systems in thermal equilibrium are generally studied in their thermodynamic limit, where the volume of the sample is taken to be infinite, in most nonequilibrium phase transitions the geometry of the sample plays a crucial role, leading to quite different structures. Electrical engineers are quite familiar with the concepts of nonlinearity and noise, which also play a fundamental role in synergetics. But synergetics also offers other insights. Not only can synergetic processes be realized on quite different substrates (molecules, neurons, etc.), but synergetics also deals with spatially extended media, and the concept of phase transitions is alien to electrical engineering. Similar points may be made with respect to mechanical engineering, where in general, fluctuations are of lesser concern. Though in cybernetics and synergetics the concept of *control* is crucial, the two disciplines have quite different goals. In cybernetics, procedures are devised for controlling a system so that it performs in a prescribed way, whereas in synergetics we change controls in a more or less unspecified manner and study the *self-organization* of the system, i.e. the various states it acquires under the newly imposed control.

The theory of dynamic systems and its special (and probably most interesting) branch, bifurcation theory, ignore fluctuations. But, as is shown in synergetics, fluctuations are crucial at precisely those points where bifurcations occur (and bifurcation theory should work best in the absence of fluctuations). Or, in other words, the transition region can be adequately dealt with only if fluctuations are taken into account. In contrast to traditional bifurcation theory (e.g. of the Lyapunov-Schmidt type), which derives the branching solutions alone, in synergetics we study the entire stochastic dynamics in the subspace spanned by the time-dependent order parameters. This is necessary in order to take fluctuations into account. At the same time our approach allows us to study the stability of the newly evolving branches and the temporal growth of patterns. Thus there is close contact with phase transition theory, and it is possible to introduce concepts new to bifurcation theory, such as critical slowing down, critical fluctuations, symmetry breaking, and the restoration of broken symmetry via fluctuations. In addition, our methods cover bifurcation sequences within such a subspace, e.g. period-doubling sequences and frequency locking. In most cases a number of (noisy) components are necessary to establish a coherent state, and consequently synergetics deals with systems composed of *many components*; this in turn requires a stochastic approach.

While bifurcation theory as yet excludes fluctuations, in some of its recent developments it does consider the neighborhood of branching solutions. As experts of dynamic systems theory and bifurcation theory will notice, this book advances to the frontiers of modern research and offers these fields new results. One such result concerns the form of the solutions (analogous to Floquet's theorem) of linear differential equations with quasiperiodic coefficients, where we treat a large class of such equations by means of embedding. Another result concerns the bifurcation of an n-dimensional torus into other tori. Finally, the slaving principle contains a number of important theorems as special cases, such as the center manifold theorem, the slow manifold theorem, and adiabatic elimination procedures.

With respect to general systems theory, synergetics seems to have entered virgin land. By focussing its attention on situations in which the macroscopic behavior of systems undergoes dramatic changes, it has enabled us to make general statements and to cover large classes of systems.

In conclusion, a general remark on the relation between synergetics and mathematics is in order. This relation is precisely the same as that between the natural sciences and mathematics. For instance, quantum mechanics is not just an application of the theory of matrices or of the spectral theory of linear operators. Though quantum mechanics uses these mathematical tools, is has developed its own characteristic system of concepts. This holds a fortiori for synergetics. Its concepts of order parameters and slaving can be applied to sciences which have not yet been mathematized and to ones which will probably never be mathematized, e.g. the theory of the development of science.

Appendix A: Moser's Proof of His Theorem

A.1 Convergence of the Fourier Series

Lemma A.1.1

We assume that the vector $\boldsymbol{F}$ (6.3.27) (with its subvectors $\boldsymbol{f}$, $\boldsymbol{g}$, $\boldsymbol{G}$) is a real analytic function of period 2π in $\psi_1, \ldots, \psi_n$. For a given $r > 0$ we introduce the norm

$$\|\boldsymbol{F}\|_r = \sup_{|\mathrm{Im}\{\psi_\nu\}| < r} \|\boldsymbol{F}\|, \quad \|\boldsymbol{F}\| = \|\boldsymbol{f}\| + \|\boldsymbol{g}\| + \|\boldsymbol{G}\|, \tag{A.1.1}$$

which is finite for some positive r. Here $\|\cdots\|$ denotes sum over the modulus of the individual vector components.

For any real analytic periodic function the Fourier coefficients decay exponentially. More precisely, if

$$\boldsymbol{F} = \sum_j \boldsymbol{F}_j \exp[\mathrm{i}(\boldsymbol{j}, \boldsymbol{\psi})], \tag{A.1.2}$$

then

$$\|\boldsymbol{F}_j\| \leqq \|\boldsymbol{F}\|_r \exp(-\|\boldsymbol{j}\| r). \tag{A.1.3}$$

To prove this inequality we represent $\boldsymbol{F}_j$ in the form

$$\boldsymbol{F}_j = \frac{1}{(2\pi)^n} \int \ldots \int \boldsymbol{F} \exp[-\mathrm{i}(\boldsymbol{j}, \boldsymbol{\psi})] \, d^n \psi, \tag{A.1.4}$$

where the integration is taken over $0 \leq \psi_\nu \leq 2\pi$. Now shifting the integration domain into the complex to $\mathrm{Im}\{\psi_\nu\} = -\rho \operatorname{sign}\{j_\nu\}$ gives

$$\|\boldsymbol{F}_j\| \leqq \|\boldsymbol{F}\|_\rho \exp(-\|\boldsymbol{j}\| \rho), \tag{A.1.5}$$

and since this holds for every $\rho < r$ we obtain (A.1.3).

In Sect. 6.3 we introduced the null-space of the operator L. In the following we shall denote this null-space by $\mathcal{N}$. We may decompose the space of functions $\boldsymbol{F}$ into that null-space $\mathcal{N}$ and a residual space, which we shall assume real and denote by $\mathcal{R}$. We then formulate the following.

Lemma A.1.2

Let $\boldsymbol{F} \in \mathscr{R}$ and $\boldsymbol{F}$ be analytic in $|\mathrm{Im}\{\psi_\nu\}| < r$. Then the unique solution of

$$L\boldsymbol{U} = \boldsymbol{F}, \qquad \boldsymbol{U} \in \mathscr{R}, \tag{A.1.6}$$

is analytic in the same domain. If $0 < \rho < r < 1$, one has for $\boldsymbol{U}$ the estimate

$$\|\boldsymbol{U}\|_\rho \leqq \frac{c}{K(r-\rho)^\sigma}\|\boldsymbol{F}\|_r, \tag{A.1.7}$$

where c depends on n, m, τ only, and $\sigma = \tau + 1$.

Proof: It is sufficient for the present purposes to prove this lemma only with $\sigma = \tau + n$, $n \geqslant 2$.

Since the operator L acts component-wise on the various terms in the Fourier expansion, it suffices to verify the convergence of the Fourier series so obtained. For this purpose we write

$$\boldsymbol{F} = \sum_j \boldsymbol{F}_j \exp[\mathrm{i}(\boldsymbol{j}, \psi)], \qquad \boldsymbol{U} = \sum_j \boldsymbol{U}_j \exp[\mathrm{i}(\boldsymbol{j}, \psi)], \tag{A.1.8}$$

and have from

$$L\boldsymbol{U} = \sum [\mathrm{i}(\omega, \boldsymbol{j})\boldsymbol{U}_j + L\boldsymbol{U}_j] \exp[\mathrm{i}(\boldsymbol{j}, \psi)] \tag{A.1.9}$$

the condition

$$[\mathrm{i}(\omega, \boldsymbol{j}) + L]\boldsymbol{U}_j = \boldsymbol{F}_j. \tag{A.1.10}$$

Since L can be diagonalized one finds $\boldsymbol{U}_j$ by dividing the various eigenfunctions $\in \mathscr{R}$ of L by the eigenvalues which are nonzero. On account of (6.2.6) we can estimate $\boldsymbol{U}_j$ by

$$\|\boldsymbol{U}_j\| \leqq \frac{c'}{K}(\|\boldsymbol{j}\|^r + 1)\|\boldsymbol{F}_j\| \leqq \frac{c'}{K}(\|\boldsymbol{j}\|^\tau + 1)\exp(-\|\boldsymbol{j}\|r)\|\boldsymbol{F}\|_r \tag{A.1.11}$$

according to (A.1.3). Hence

$$\|\boldsymbol{U}\|_\rho \leqq \|\boldsymbol{F}\|_r \frac{c'}{K}\sum_j (\|\boldsymbol{j}\|^r + 1)\exp[\|\boldsymbol{j}\|(\rho - r)]. \tag{A.1.12}$$

The latter sum always converges and can be estimated as follows. With $\delta = r - \rho \leqq 1$ we have

$$\delta^{r+n}\sum_j (\|\boldsymbol{j}\|^r + 1)\exp(-\|\boldsymbol{j}\|\delta) \leqq \sum_j (\delta^r\|\boldsymbol{j}\|^r + 1)\exp(-\|\boldsymbol{j}\|\delta)\delta^n, \tag{A.1.13}$$

which is bounded by a constant independent of δ since it can be estimated by the integral

$$\int \ldots \int (\|x\|^r + 1) \exp(-\|x\|) d^n x \,, \tag{A.1.14}$$

which it approximates. Hence we have from (A.1.12)

$$\|U\|_\rho \leqq \|F\|_r \frac{c}{K} \frac{1}{(r-\rho)^{\tau+n}} \,, \tag{A.1.15}$$

which proves lemma A.1.2 with $\sigma = \tau + n$.

Summarizing the results of this appendix and Sect. 6.3, we have established the following theorem.

Theorem A.1.1. There exist unique formal expansions in powers of ε for $\Delta(\varepsilon)$, $d(\varepsilon)$, $D(\varepsilon)$ and for $u(\psi, \varepsilon)$, $v(\psi, \varepsilon)$, $V(\psi, \varepsilon)$ which formally satisfy the conditions of theorem 6.2.1 and the normalization that all coefficients in this expansion of u, v, V belong to $\mathscr{R}$. The proof is evident from the preceding discussion. Comparison of coefficients leads at each step to equations of the type (6.3.33), which by lemma A.1.2 admit a unique solution with the normalization $U \in \mathscr{R}$. The more difficult proof of the convergence of the series expansion in ε is settled in Sect. A.3. However, in Sect. A.3 we shall not be able to ensure this normalization and therefore we investigate now the totality of *all* formal solutions.

A.2 The Most General Solution to the Problem of Theorem 6.2.1

We discuss what the meaning of the arbitrary constants (of the null-space) is. To find the most general solution, $U \in \mathscr{R}$, and $\mathscr{N}$ to the problem of Moser's theorem we let

$$\mathscr{U}: \begin{cases} \varphi = \psi + u(\psi, \varepsilon) \\ \xi = \chi + v(\psi, \chi, \varepsilon) \end{cases} \tag{A.2.1}$$

be the unique particular solution which satisfies the normalization that all vectors lying in the null-space vanish, i.e., $U \in \mathscr{R}$. It transforms the differential equations (6.3.1 and 2) into a system whose linearization is

$$\left.\begin{aligned} \dot{\psi} &= \omega \\ \dot{\chi} &= \Lambda \chi \end{aligned}\right\} . \tag{A.2.2}$$

Clearly, any transformation

$$\hat{\mathscr{U}}: \begin{cases} \psi' = \psi + \varepsilon \hat{u}(\psi, \varepsilon) \\ \chi' = \chi + \varepsilon[\hat{v}(\psi, \varepsilon) + \hat{V}(\psi, \varepsilon)\chi] \,, \end{cases} \tag{A.2.3}$$

which transforms the system (A.2.2) into itself will give rise to another solution

$$\mathscr{U} \circ \hat{\mathscr{U}} \tag{A.2.4}$$

to theorem 6.2.1. Here (A.2.4) denotes the composition of $\mathscr{U}$ and $\hat{\mathscr{U}}$. For this reason we determine the self-transformations $\hat{\mathscr{U}}$ of (A.2.2). Inserting (A.2.3) into (A.2.2) and requiring that the differential equation in the new variables ψ', χ' has the same form as before, we find that [L has the shape (6.3.29) with $L_1 = L_2 = L_3$]

$$L\hat{\boldsymbol{U}} = \boldsymbol{0}, \qquad \text{where} \tag{A.2.5}$$

$$\hat{\boldsymbol{U}} = \begin{pmatrix} \hat{\boldsymbol{u}} \\ \hat{\boldsymbol{v}} \\ \hat{\boldsymbol{V}} \end{pmatrix}. \tag{A.2.6}$$

Equation (A.2.5) means that $\hat{\boldsymbol{U}}$ is a vector of the null-space, and these self-transformations have the form

$$\mathscr{C} \in \mathfrak{C}: \begin{cases} \psi' = \psi + \varepsilon \boldsymbol{a} \\ \chi' = \chi + \varepsilon(\boldsymbol{b} + B\chi), \end{cases} \tag{A.2.7}$$

where $\boldsymbol{a}$, $\boldsymbol{b}$, B are independent of ψ and satisfy

$$\Lambda \boldsymbol{b} = \boldsymbol{0}, \tag{A.2.8}$$

$$\Lambda B = B\Lambda. \tag{A.2.9}$$

We shall denote the group of self-transformation $\mathfrak{C}$. Thus, both $\mathscr{U}$ and $\mathscr{U} \circ \mathscr{C}$ are solutions to theorem 6.2.1. In fact, they form the most general solution provided $\boldsymbol{\Delta}$, $\boldsymbol{d}$, D are considered given. Indeed, one may show that these latter quantities are uniquely determined. Since this proof is more of a technical nature we do not present it here but rather formulate the corresponding lemma.

Lemma A.2.1

Let $\mathscr{U}$ in (A.2.1) be the unique (normalized) formal transformation and let

$$N = \begin{pmatrix} \boldsymbol{\Delta} \\ \boldsymbol{d} \\ D \end{pmatrix} \tag{A.2.10}$$

be the corresponding modifying term found in theorem A.1.1. Thus $\mathscr{U}$ transforms the system (6.3.1 and 2) into a system (6.1.26 and 27) whose linearization is given by

$$\left.\begin{aligned}\dot{\psi} &= \omega \\ \dot{\chi} &= \Lambda\chi\end{aligned}\right\}. \tag{A.2.11}$$

Let $\tilde{\mathscr{U}}$, $\tilde{N}$ (where $\tilde{\mathscr{U}}$ = identity, $\tilde{N} = \mathbf{0}$ for $\varepsilon = 0$) be any other formal expansion with this property. Then there exists a transformation $\mathscr{C} \in \mathfrak{C}$, cf. (A.2.7), such that $\tilde{\mathscr{U}} = \mathscr{U} \circ \mathscr{C}$ and $\tilde{N} = N$.

A.3 Convergent Construction

a) To prove the convergence of the series expansion obtained in the previous section one is tempted to use Cauchy's majorant method. But this method fails because of the presence of the small divisors. A crude estimate leads to majorants of the form $\Sigma(n!)^{2\tau}\varepsilon^n$ which diverge for all $|\varepsilon| > 0$. Namely, by lemma A.1.2 the solution of the linearized equation leads to multiplication of the coefficients by the factor $(r-r')^{-\tau}$ where $|\mathrm{Im}\{\varphi\}| < r$, $|\mathrm{Im}\{\varphi'\}| < r'$ are the complex domains in which the iterative estimates hold. Choosing a sequence of such domains $|\mathrm{Im}\{\varphi\}| < r_n$ with

$$r_n = \frac{r_0}{2}\left(1 + \frac{1}{n}\right) \rightarrow \frac{r_0}{2} \qquad \text{as } n \rightarrow \infty\,, \tag{A.3.1}$$

we obtain a factor $(r_{n-1} - r_n)^{-\tau} = O(n^{2\tau})$ going from the determination of the coefficients of ε^{n-1} to those of ε^n. This leads then to the series $\Sigma(n!)^{2\tau}\varepsilon^n$, indicating the divergence of the series. Still, the convergence of the series can be established by using an alternate construction which converges more rapidly (Sects. 5.2 and 5.3). This idea is from Kolmogorov, Moser's proof here being a generalization and sharpened version of that approach.

We shall describe an iteration method in which at each step the linear equations of lemma A.1.2 have to be solved but the precision increases with an exponent 3/2 so that the previous series can be replaced by

$$\Sigma(n!)^{2\tau}\varepsilon^{(3/2)^n}$$

which is obviously convergent. Below, we shall describe this iteration procedure with detailed estimates and show in the following section how this result can be used to prove the convergence of the series found in Sect. 6.3.

b) We consider again a family of differential equations

$$\left.\begin{aligned}\dot{\varphi} &= \boldsymbol{a} + \boldsymbol{f} \\ \dot{\xi} &= \boldsymbol{b} + B\xi + \boldsymbol{g}\end{aligned}\right\}, \tag{A.3.2}$$

where $\boldsymbol{a} = (a_1, \ldots, a_n)$ vary freely, while $\boldsymbol{b} = (b_1, \ldots, b_m)$ and the m by m matrix B are restricted by

$$\Lambda \boldsymbol{b} = \boldsymbol{0}, \qquad B\Lambda = \Lambda B. \tag{A.3.3}$$

Here $\omega = (\omega_1, \ldots, \omega_n)$ and the eigenvalues $\Lambda_1, \ldots, \Lambda_m$ of the matrix Λ satisfy the condition

$$|(\boldsymbol{j}, \omega) + \Sigma k_\mu \Lambda_\mu| \geqq (\|\boldsymbol{j}\|^\tau + 1)^{-1} K, \quad 0 < K \leqq 1, \tag{A.3.4}$$

for all integers, all vectors $\boldsymbol{j}$, and all k_μ, with

$$|\Sigma k_\mu| \leqq 1, \quad \Sigma |k_\mu| \leqq 2, \tag{A.3.5}$$

except the finitely many $(\boldsymbol{j}, \boldsymbol{k}) = (\boldsymbol{0}, \boldsymbol{k})$ for which the lhs of (A.3.4) vanishes. The number τ is chosen $> n-1$ to ensure the existence of such ω, Λ.

In the following it will be decisive to consider $\boldsymbol{a}$, $\boldsymbol{b}$, B as variables [under the linear restrictions (A.3.3)] and we shall specify the complex domain of these variables as well as that of φ, ξ, by

$$\mathscr{D}: \begin{cases} |\mathrm{Im}\{\varphi\}| < r \leqq 1, \quad |\xi| < s \\ \dfrac{|\boldsymbol{a} - \omega|}{r} + \dfrac{|\boldsymbol{b}|}{s} + |B - \Lambda| < q, \end{cases} \tag{A.3.6}$$

and we require with a fixed constant $c_0 \geqq 1$, that $q \geqq c_0 K$. In the following, all constants which depend on n, m, c_0, τ only will be denoted by $c_1, c_2, c_3, \ldots$.

Theorem A.3.1. Constants δ_0, c depending on c_0, n, m, τ only exist such that if $\delta < \delta_0$ and

$$\frac{|f|}{r} + \frac{|g|}{s} < r^\sigma K \delta \quad \text{in} \quad \mathscr{D}, \quad \sigma = \tau + 1 \tag{A.3.7}$$

then a transformation exists

$$\mathscr{U}: \begin{cases} \varphi = \psi + \boldsymbol{u}(\psi) \\ \xi = \chi + \boldsymbol{v}(\psi, \chi) \quad \text{linear in } \chi \end{cases} \tag{A.3.8}$$

and $\boldsymbol{a} = \hat{\boldsymbol{a}}$, $\boldsymbol{b} = \hat{\boldsymbol{b}}$, $B = \hat{B}$ in $\mathscr{D}$ such that (A.3.2) for this choice of $\boldsymbol{a} = \hat{\boldsymbol{a}}$, $\boldsymbol{b} = \hat{\boldsymbol{b}}$, $B = \hat{B}$ is transformed by (A.3.8) into a system

$$\left.\begin{aligned} \dot{\varphi} &= \omega + O(\chi) \\ \dot{\chi} &= \Lambda \chi + O(\chi^2) \end{aligned}\right\} \text{for } |\mathrm{Im}\{\psi\}| < r/2; |\chi| < s. \tag{A.3.9}$$

In particular

$$\varphi = \omega t + \psi + \boldsymbol{u}(\omega t + \varphi), \qquad \xi = \boldsymbol{v}(\omega t, \boldsymbol{0}) \tag{A.3.10}$$

is a quasiperiodic solution of (A.3.2) with characteristic numbers $\omega_1, \ldots, \omega_n$, $\Lambda_1, \ldots, \Lambda_m$.

Moreover, the $\hat{a}$, $\hat{b}$, $\hat{B}$ lie in

$$\frac{|\hat{a}-\omega|}{r} + \frac{|\hat{b}|}{s} + |\hat{B}-\Lambda| < cr^{\sigma}K\delta < q \tag{A.3.11}$$

and u, v are real analytic satisfying

$$\frac{|u|}{r} + \frac{|v|}{s} < c\delta, \quad \text{for } |\mathrm{Im}\{\psi\}| < r/2; \quad |\chi| < s. \tag{A.3.12}$$

c) The remainder of this section is devoted to the proof of this theorem. First we observe that replacing ξ by $s\xi$ we can assume $s=1$. Similarly, replacing t by λt, multiplying $a, b, B, \omega, \Lambda, f, g$ by λ^{-1} we can normalize K to $K=1$. However, r cannot be replaced by 1 by stretching the φ variable, since the angular variables $\varphi_1, \ldots, \varphi_n$ where chosen of period 2π. Therefore, we take $s=K=1$, $r \leqq 1$, $q \geqq c_0 \geqq 1$.

In the following construction the transformation $\mathscr{U}$ will be built up as an infinite product of transformations

$$\mathscr{U} = \lim_{\nu\to\infty} (\mathscr{U}_0 \circ \mathscr{U}_1 \circ \ldots \circ \mathscr{U}_\nu), \tag{A.3.13}$$

where each $\mathscr{U}_\nu$ refines the previous approximation further. We denote the given family of differential equations symbolically by $\mathscr{F} = \mathscr{F}_0$, and $\mathscr{F}_1$ denotes the system obtained by transforming $\mathscr{F}_0$ by the coordinate transformation $\mathscr{U}_0$, etc. Hence $\mathscr{F}_\nu$ is transformed by $\mathscr{U}_\nu$ into $\mathscr{F}_{\nu+1}$, and $\mathscr{F}_0$ by $\mathscr{U}_0 \circ \mathscr{U}_1 \circ \ldots \circ \mathscr{U}_\nu$ into $\mathscr{F}_{\nu+1}$.

It will be decisive in the following proof to describe precisely the domain $\mathscr{D}_\nu$ of validity of the transformation and the differential equations. In particular, we mention that the transformation $\mathscr{U}_\nu$ will involve a change of the parameters a, b, B, as well as a transformation of the variables φ, ξ. To make this change of variables $\mathscr{U}_\nu$ clear we drop the subscript ν and write $\mathscr{U}_\nu$ in the form

$$\mathscr{U}_\nu: \begin{cases} \varphi - \psi + u(\psi, \chi, \alpha, \beta, \tilde{B}) \\ \xi = \chi + v(\psi, \chi, \alpha, \beta, \tilde{B}) \\ \alpha = \alpha + w_1(\alpha, \beta, \tilde{B}) \\ b = \beta + w_2(\alpha, \beta, \tilde{B}) \\ B = \tilde{B} + w_3(\alpha, \beta, \tilde{B}). \end{cases} \tag{A.3.14}$$

The variables φ, χ, α, β, $\tilde{B}$ will be restricted to

$$\mathscr{D}_{\nu+1}: |\mathrm{Im}\{\psi\}| < r_{\nu+1}, \quad |\chi| < s_{\nu+1}, \quad \frac{|\alpha-\omega|}{r_{\nu+1}} + \frac{|\beta|}{s_{\nu+1}} + |B-\Lambda| < q_{\nu+1}, \tag{A.3.15}$$

where the sequence r_ν, s_ν, q_ν will be chosen in such a manner that $\mathscr{U}_\nu$ maps $\mathscr{D}_{\nu+1}$ into

$$\mathscr{D}_\nu: \quad |\mathrm{Im}\{\varphi\}| < r_\nu, \quad |\xi| < s_\nu, \quad \frac{|\alpha-\omega|}{r_\nu} + \frac{|b|}{s_\nu} + |\tilde{B} - \Lambda| < q_\nu \tag{A.3.16}$$

and such that the family of differential equations $\mathscr{F}_\nu$ is mapped into a system $\mathscr{F}_{\nu+1}$ which approximates (A.3.9) to a higher degree than $\mathscr{F}_\nu$. We shall drop the index ν and write $\mathscr{F}_\nu$ in the form (A.3.2) and $\mathscr{F}_{\nu+1}$ in Greek letters

$$\left.\begin{aligned} \dot{\psi} &= \alpha + \Phi \\ \chi &= \beta + \tilde{B}\chi + \Xi , \end{aligned}\right\} \text{ in } \mathscr{D}_{\nu+1} \tag{A.3.17}$$

where the aim will be to make $\boldsymbol{\Phi}$, Ξ very small.

To set up the estimates which will be proven inductively we introduce the sequences r_ν, s_ν, δ_ν, q_ν in the following manner:

$$\left\{\begin{aligned} r_\nu &= \frac{r_0}{2}(1+2^{-\nu}), \qquad 0 < r_0 \leqq 1, \\ \delta_\nu &= c^\nu \delta_{\nu-1}^{3/2}, \\ s_\nu &= c^{-\nu^2/2}\delta_\nu, \\ q_{\nu+1} &= r_\nu^\sigma \delta_\nu, \qquad \text{for} \quad \nu = 0, 1, 2, \ldots; \, q_0 = q \geqq c_0 \geqq 1, \end{aligned}\right. \tag{A.3.18}$$

where $c \geqq 1$ will be determined later. Notice that δ_ν tends to zero rapidly if $\delta_0 < c^{-6} \leqq 1$ and, similarly, s_ν, q_ν approach zero while $r_\nu \to r_0/2$.

We shall assume that $\mathscr{F}_\nu$ satisfies

$$\frac{|f|}{r_\nu} + \frac{|g|}{s_\nu} < r_\nu^\sigma \delta_\nu = q_{\nu+1} \text{ in } \mathscr{D}_\nu , \tag{A.3.19}$$

and construct $\mathscr{U}_\nu$ in such a manner that it maps $\mathscr{D}_{\nu+1}$ into $\mathscr{D}_\nu$ and that for the transformed system $\mathscr{F}_{\nu+1}$ we have the corresponding error estimate

$$\frac{|\boldsymbol{\Phi}|}{r_{\nu+1}} + \frac{|\Xi|}{s_{\nu+1}} < r_{\nu+1}^\sigma \delta_{\nu+1} = q_{\nu+2} \text{ in } \mathscr{D}_{\nu+1} . \tag{A.3.20}$$

For the mapping $\mathscr{U}_\nu$ [see (A.3.14)] we establish

$$\left.\begin{aligned} &\frac{|u|}{r_\nu} + \frac{|v|}{s_\nu} < c_4^{\nu+1}\delta_\nu \\ &\frac{|w_1|}{r_\nu} + \frac{|w_2|}{s_\nu} + |w_3| < c_4 q_{\nu+1} \end{aligned}\right\} \text{ in } \mathscr{D}_{\nu+1} \tag{A.3.21}$$

with an appropriate constant $c_4 > 1$.

d) If the above statement and the estimates (A.3.20, 21) are established, the above theorem follows readily as we will now show. Since $\mathscr{U}_\nu$ maps $\mathscr{D}_{\nu+1}$ into $\mathscr{D}_\nu$, the composite transformation $\tilde{\mathscr{U}}_\nu = \mathscr{U}_0 \circ \mathscr{U}_1 \circ \ldots \circ \mathscr{U}_\nu$ maps $\mathscr{D}_{\nu+1}$ into $\mathscr{D}_0$ and can be estimated by

$$\frac{|\tilde{\boldsymbol{u}}|}{r_0} + \frac{|\tilde{\boldsymbol{v}}|}{s_0} < c_4(\delta_0 + c_4\delta_1 + \ldots + c_4^\nu \delta_\nu) < 2c_4\delta_0 \tag{A.3.22}$$

there. This implies that $\tilde{\mathscr{U}}_\nu$ converges in

$$\mathscr{D}_\infty\colon\ |\mathrm{Im}\{\psi\}| < \frac{r_0}{2}, \quad \chi = 0, \quad \boldsymbol{\alpha} = \boldsymbol{\beta} = \mathbf{0}, \quad \tilde{B} = 0 \tag{A.3.23}$$

uniformly and the transformation $\tilde{\mathscr{U}}_\infty$ is analytic in $\mathscr{D}_\infty$. Moreover, the above inequality implies (A.3.12) for $\boldsymbol{\chi} = \mathbf{0}$ if c is chosen $> 2c_4$. Since $\tilde{\boldsymbol{u}}$ is independent of $\boldsymbol{\chi}$ and $\tilde{\boldsymbol{v}}$ depends linearly on $\boldsymbol{\chi}$, it remains to estimate $\partial\tilde{\boldsymbol{v}}/\partial\boldsymbol{\chi}$. The term $1 + \partial\tilde{\boldsymbol{v}}/\partial\boldsymbol{\chi}$ is the product of the corresponding terms $1 + \partial\boldsymbol{v}_\nu/\partial\boldsymbol{\chi}$ and, since $|\partial\boldsymbol{v}_\nu/\partial\boldsymbol{\chi}| < c_4\delta_\nu$, leads to the estimate

$$\left|\frac{\partial\tilde{\boldsymbol{v}}}{\partial\boldsymbol{\chi}}\right| \leqq \prod_{\mu=0}^{\nu} (1 + c_4\delta_\mu) - 1 \leqq 2c_4\delta_0 \,. \tag{A.3.24}$$

This proves (A.3.12) for an appropriate c.

The transformed system $\mathscr{F}_\infty$ has the property that $\boldsymbol{\Phi}_\infty$, Ξ_ν, $\partial\Xi_\infty/\partial\boldsymbol{\chi}$ vanish (for $\boldsymbol{\chi} = \mathbf{0}$) in $\mathscr{D}_\infty$. This follows from the estimate (A.3.20) and $q_{\nu+2} \to 0$ together with the fact that $\partial\Xi/\partial\boldsymbol{\chi}$ at $\boldsymbol{\chi} = \mathbf{0}$ can be estimated by $\sup|\Xi| s_{\nu+1}^{-1}$, which also tends to zero as $\nu \to \infty$. Hence, for $\mathscr{F}_\infty$ we have

$$\boldsymbol{\Phi}_\infty = O(\chi)\,, \qquad \Xi_\infty = O(\chi^2)\,, \tag{A.3.25}$$

as was to be proven.

Finally, the determination of $\hat{\boldsymbol{a}}$, $\hat{\boldsymbol{b}}$, $\hat{\boldsymbol{B}}$ is obtained as the image of $\tilde{\mathscr{U}}_\infty$ from the last three components in (A.3.14). Since $\mathscr{U}_\nu$ maps $\mathscr{D}_{\nu+1}$ into $\mathscr{D}_\nu$, the images

$$\mathscr{U}_0 \circ \mathscr{U}_1 \circ \ldots \circ \mathscr{U}_\nu \mathscr{D}_{\nu+1} = \mathscr{D}_0^{(\nu)} \tag{A.3.26}$$

form a sequence of nested domains in $\mathscr{D}_0$:

$$\mathscr{D}_0^{(\nu+1)} \subset \mathscr{D}_0^{(\nu)} \subset \mathscr{D}_0 \,. \tag{A.3.27}$$

In particular, the range of the last three components $\boldsymbol{\alpha}$, $\boldsymbol{\beta}$, $\tilde{B}$ in (A.3.14) tends to zero, as follows from (A.3.15) and $q_{\nu+1} \to 0$, which implies that the corresponding $\boldsymbol{a}$, $\boldsymbol{b}$, $B \in \mathscr{D}_0$ shrink to a point $\hat{\boldsymbol{a}}$, $\hat{\boldsymbol{b}}$, $\hat{B} \in \mathscr{D}_0$ as $\nu \to \infty$. This follows immediately from (A.3.21) and the convergence of $\sum_{\nu=0}^{\infty} q_{\nu+1}$:

$$\frac{|\boldsymbol{a} - \hat{\boldsymbol{a}}|}{r_0} + |\hat{\boldsymbol{b}}| + |\hat{B} - \Lambda| < c_4 \sum_{\nu=0}^{\infty} q_{\nu+1} < 2c_4 q_1 < q_0 \,, \tag{A.3.28}$$

if δ_0 is chosen small enough, proving (A.3.11) for $c > 2c_4$. One readily verifies that the conditions of the theorem imply those of (A.3.20) for $\nu = 0$ so that the induction can be started.

e) This reduces the proof of theorem A.3.1 to the construction of $\mathscr{U} = \mathscr{U}_\nu$ and the proof of the estimates (A.3.20, 21).

For this purpose, we truncate f, g to its linear part

$$\left.\begin{aligned} f_0 &= f(\varphi, 0, a, b, B) \\ g_0 &= g(\varphi, 0, a, b, B) + g_\xi(\varphi, 0, a, b, B)\xi \end{aligned}\right\} \tag{A.3.29}$$

and break up (f_0, g_0) into its components in $\mathscr{N}$, $\mathscr{R}$ (as in Sect. A.2), which we denote by $(f_{\mathscr{N}}, g_{\mathscr{N}})$ and $(f_{\mathscr{R}}, g_{\mathscr{R}})$. Then the transformation $\mathscr{U}_\nu$ will be obtained by solving the linearized equations

$$\left.\begin{aligned} u_\psi \omega &= f_{\mathscr{R}}(\psi, a, b, B) \\ v_\psi \omega + v_\chi \Lambda\chi - \Lambda v &= g_{\mathscr{R}}(\psi, a, b, B) \end{aligned}\right\}. \tag{A.3.30}$$

As we saw in the previous section, these equations can be uniquely solved if we require that

$$U \in \mathscr{R}\ . \tag{A.3.31}$$

This defines u, v. The transformation of a, b, B will be given implicitly by the equations

$$\left.\begin{aligned} \alpha &= a + f_{\mathscr{N}}(a, b, B) \\ \beta + \tilde{B}\chi &= b + B\chi + g_{\mathscr{N}}(a, b, B; \chi) \end{aligned}\right\}. \tag{A.3.32}$$

f) The relations (A.3.30 – 32) define the transformation $\mathscr{U}$ and we proceed to verify that it maps $\mathscr{D}_{\nu+1} = \mathscr{D}_+$ into $\mathscr{D}_\nu = \mathscr{D}$. (To simplify the notation, we denote quantities referring to $\nu+1$, such as $s_{\nu+1}$ by s_+, and those referring to ν without subscript.) For this purpose we have to check that (A.3.32) can be inverted for

$$\frac{|\alpha - \omega|}{r_+} + \frac{|\beta|}{s_+} + |\tilde{B} - \Lambda| < q_+ \tag{A.3.33}$$

and that the solution a, b, B falls into $\mathscr{D}$, see (A.3.6). We explain the argument for the first equation ignoring the dependence on b, B.

We use the implicit function theorem: In $|a - \omega| < rq$ we have through (A.3.19, 18)

$$|f_{\mathscr{N}}| \leqq |f_0| \leqq rq_+\ , \tag{A.3.34}$$

and using Cauchy's estimate in the sphere $|a - \omega| < R = c_4 q_+ < rq/2$, we find

$$\left|\frac{\partial}{\partial a} f_{\mathscr{N}}\right| \leqq c_2 \frac{rq_+}{R} = \frac{c_2}{c_4} < \frac{1}{2} \quad \text{with} \quad c_2 \geqq 1\ . \tag{A.3.35}$$

The last relation can be achieved by taking $c_4 > 2c_2$. In the sphere $|\boldsymbol{a}-\omega| < R$ we also have from (A.3.33)

$$|f_{\mathscr{R}}| + |\alpha-\omega| < rq_+ + r_+ q_+ < \frac{2}{c_4} R < \frac{1}{2} R \quad \text{for} \quad c_4 > 4 \tag{A.3.36}$$

and the standard implicit function theorem guarantees the unique existence of an analytic solution $\boldsymbol{a} = \boldsymbol{a}(\alpha)$ of the equation

$$\boldsymbol{a} - \omega = \alpha - \omega - f_{\mathscr{N}} \,. \tag{A.3.37}$$

By the same argument we verify the unique existence of a solution $\boldsymbol{a}$, $\boldsymbol{b}$, $\boldsymbol{B}$ of (A.3.32) in

$$\frac{|a-\omega|}{r} + \frac{|b|}{s} + |B - \Lambda| < c_4 q_+ < \frac{1}{2} q \,, \tag{A.3.38}$$

which verifies the second half of (A.3.21), if δ_0 is chosen sufficiently small.

To estimate the solution $\boldsymbol{u}$, $\boldsymbol{v}$ of (A.3.30) we make use of lemma A.1.2. The unique solution of (A.3.30) can be estimated by

$$\frac{|u|}{r} + \frac{|v|}{s} < c_3 (r - r_+)^{-\sigma} r^{\sigma} \delta \leqq c_4^{\nu+1} \delta \text{ in } \mathscr{D}_{\nu+1} \tag{A.3.39}$$

since by (A.3.18) we have

$$r - r_+ \geqq 2^{-\nu-1} r \,. \tag{A.3.40}$$

We may choose the same constant c_4 as before by enlarging the previous one if necessary. This proves the first half of (A.3.21), the second part having been verified already above.

g) Having found the transformation $\mathscr{U}$ we transform $\mathscr{F}_\nu$ [or (A.3.2)] into new variables (A.3.17) and estimate the remainder terms $\boldsymbol{\Phi}$, $\boldsymbol{\Xi}$ to complete the induction proof.

For this purpose we introduce the following symbolic notation:

$$F = \begin{pmatrix} f \\ g \end{pmatrix}, \quad \hat{\boldsymbol{\Phi}} = \begin{pmatrix} \boldsymbol{\Phi} \\ \boldsymbol{\Xi} \end{pmatrix} \tag{A.3.41}$$

where the arguments in F are ψ, χ, $\boldsymbol{a}$, $\boldsymbol{b}$, B (not φ, ξ!) and in $\hat{\boldsymbol{\Phi}}$ are ψ, χ, α, β, $\tilde{B}$. Corresponding to the transformation $\mathscr{U}$, we introduce the vector

$$W = \begin{pmatrix} \psi + u \\ \chi + v \end{pmatrix} \tag{A.3.42}$$

and its Jacobian matrix

$$W' = \begin{pmatrix} 1+u_\psi & 0 \\ v_\psi & 1+v_\psi \end{pmatrix}. \tag{A.3.43}$$

Finally let

$$A = \begin{pmatrix} a \\ b+B\chi \end{pmatrix}, \quad A_+ = \begin{pmatrix} \alpha \\ \beta+\tilde{B}\chi \end{pmatrix}. \tag{A.3.44}$$

Then the transformation equations which express that $\mathscr{U}$ takes (A.3.2) into (A.3.17) take the form

$$W'(A_+ + \hat{\Phi}) = (A+F) \circ \mathscr{U}, \tag{A.3.45}$$

where on the left side we have matrix multiplication and on the rhs the circle $\circ$ indicates composition, i.e., substitution of φ by $\psi + u$, etc.

We compare these equations with those satisfied by solving (A.3.30, 32). In the present notation these equations take the form

$$\left.\begin{aligned} W'A_\infty - A_\infty \circ \mathscr{U} &= F_{\mathscr{R}} \\ A_+ - A &= F_{\mathscr{N}}, \end{aligned}\right\} \tag{A.3.46}$$

where

$$A_\infty = \begin{pmatrix} \omega \\ \Lambda\chi \end{pmatrix} \tag{A.3.47}$$

and the meaning of $F_{\mathscr{R}}$, $F_{\mathscr{N}}$ is obvious. Adding these relations we have

$$W'A_\infty - A_\infty \circ \mathscr{U} + (A_+ - A) = F_0 \tag{A.3.48}$$

with

$$F_0 = \begin{pmatrix} f \\ g + g_\xi \chi \end{pmatrix} \tag{A.3.49}$$

and the arguments are ψ, χ, a, b, B. Subtracting (A.3.48) from (A.3.45) we find after a short calculation for Φ

$$W'\hat{\Phi} = -\mathrm{I} + (F \circ \mathscr{U} - F_0) = -\mathrm{I} + \mathrm{II} + \mathrm{III}, \tag{A.3.50}$$

where

$$\mathbf{I} = (W'-1)(A_+ - A_\infty) - (A - A_\infty) \circ U + (A - A_\infty)$$

$$= \begin{pmatrix} u_\psi(\alpha-\omega) \\ v_\psi(\alpha-\omega) + v_\chi[\tilde{\beta} + (\tilde{B}-\Lambda)\chi] - (B-\Lambda)v \end{pmatrix}, \tag{A.3.51}$$

$$\mathbf{II} = F \circ \mathcal{U} - F_0 \circ \mathcal{U}, \tag{A.3.52}$$

$$\mathbf{III} = F_0 \circ \mathcal{U} - F_0. \tag{A.3.53}$$

We proceed to estimate the three error terms. The second one is due to linearization of F, the third is due to evaluation of F – or F_0 – at a displaced argument and the first is due to the fact that α was replaced by ω, etc., when we solved (A.3.30).

h) To estimate the quantity

$$|\hat{\Phi}|_+ = \sup_{\mathcal{D}_+} \left\{ \frac{|\Phi|}{r_+} + \frac{|\Xi|}{s_+} \right\} \tag{A.3.54}$$

(the + in $|\Phi|_+$ refers to the new domain $\mathcal{D}_+$ as well as to r_+, s_+; for the success of this approach it is essential that the norm be changed during the iteration) it suffices to estimate the corresponding terms $|\mathbf{I}|_+, |\mathbf{II}|_+, |\mathbf{III}|_+$. This follows from the fact that the Jacobian W' is close to one. But since the two components Φ, Ξ are scaled differently one has to show that even

$$\begin{pmatrix} 1+u_\psi & 0 \\ v_\psi \dfrac{r_+}{s_+} & 1+v_\chi \end{pmatrix} \tag{A.3.55}$$

is close to the identity. For the diagonal elements this is obvious from (A.3.21), and for the remaining term we have by (A.3.21) and by Cauchy's estimate

$$\left| v_\psi \frac{r_+}{s_+} \right| \leqq c_4^{\nu+1} \delta 2 \frac{s}{r} \frac{r_+}{s_+} \leqq 2 c_4^{\nu+1} \delta \frac{s}{s_+}. \tag{A.3.56}$$

By our choice of s in (A.3.18) we have

$$\frac{s_+}{s} = \sqrt{c\delta} \tag{A.3.57}$$

and the term

$$\left| v_\psi \frac{r_+}{s_+} \right| \leqq 2 c_4^{\nu+1} \sqrt{c^{-1}\delta_\nu} \tag{A.3.58}$$

can be made arbitrarily small by the choice of δ_0. Thus we have

$$|\hat{\Phi}|_+ \leqq 2(|\mathbf{I}|_+ + |\mathbf{II}|_+ + |\mathbf{III}|_+)\,. \tag{A.3.59}$$

The estimation of these terms is now straightforward, but we shall show how the various scale factors enter.

i) First we shall show

$$|\mathbf{I}|_+ < c_5 c_4^{\nu+1} \frac{s}{s_+} \delta q_+\,. \tag{A.3.60}$$

To estimate a typical term in **I**, see (A.3.15), we consider

$$\frac{|v_\psi(\alpha-\omega)|}{s_+} < \frac{r_+ q_+}{s_+} \cdot 2c_4^{\nu+1} \frac{s}{r} \delta \leqq 2c_4^{\nu+1} \frac{s}{s_+} \delta q_+\,, \tag{A.3.61}$$

where we used (A.3.15 and 21). The other terms can be handled similarly, but the term $(B-\Lambda)\boldsymbol{v}$ requires special attention. It does not suffice to use the estimate $|B-\Lambda| < q$ from (A.3.52), but rather

$$|B-\Lambda| = |B-\tilde{B}| + |\tilde{B}-\Lambda| = |w_3| + q_+ < (c_4+1)q_+\,, \tag{A.3.62}$$

where we used (A.3.21). The statement (A.3.60) is then clear.

To turn to the expression **II** we have to estimate the remainder in the Taylor expansion, e.g.,

$$|f(\varphi,\xi,\ldots) - f(\varphi,0,\ldots)| \leq \max\{|f_\xi||\xi|\} \leq 2\frac{r}{s} q_+ |\xi|\,, \tag{A.3.63}$$

where by (A.3.21)

$$|\xi| \leqq |\chi| + |v| \leqq s_+ \left(1 + \frac{s}{s_+} c_4^{\nu+1} \delta\right) \leqq 2s_+\,, \tag{A.3.64}$$

since by (A.3.57) the second term in the brackets can be made small by choice of δ_0. Hence

$$|f(\varphi,\xi) - f(\varphi,0)| \leqq 4r\frac{s_+}{s} q_+\,. \tag{A.3.65}$$

Similarly

$$|g(\varphi,\xi) - g(\varphi,0) - g_\xi(\varphi,0)\xi| \leqq 2\frac{q_+}{s}|\xi|^2 \leqq 8\frac{s_+^2}{s} q_+\,, \tag{A.3.66}$$

hence

$$|\mathbf{II}|_+ \leqq c_6 \frac{s_+}{s} q_+\,. \tag{A.3.67}$$

Finally, in **III** we use the mean value theorem, e.g.,

$$|f(\psi+u,\chi+v)-f(\psi,\chi)|\leqq \max\{|f_\psi|\cdot|u|\}+\max\{|f_\chi|\cdot|v|\}$$
$$\leqq 2q_+ 2rc_4^{\nu+1}\delta\,. \qquad (A.3.68)$$

With the corresponding estimate for g we find

$$|\mathbf{III}|_+\leqq c_7c_4^{\nu+1}\frac{s}{s_+}q_+\delta\,. \qquad (A.3.69)$$

Combining the estimates (A.3.60, 67, 68) for **I, II, III** we have

$$|\hat{\Phi}|_+\leqq c_8c_4^{\nu+1}q_+\left(\frac{s}{s_+}\delta+\frac{s_+}{s}\right). \qquad (A.3.70)$$

Clearly, the optimal choice for s is found by making both terms in the bracket equal. This agrees approximatively with our choice, as is seen by (A.3.57). Thus, with (A.3.18) we get

$$|\hat{\Phi}|_+\leqq c_8c_4^{\nu+1}q_+2\sqrt{c\delta}=c_8c_4^{\nu+1}r_\nu^\sigma 2\sqrt{c}\,\delta^{3/2}\leqq\sqrt{c}\,c_9^{\nu+1}r_{\nu+1}^\sigma\delta^{3/2}\,. \qquad (A.3.71)$$

Taking c large enough ($c>c_9^2$) we have through (A.3.18)

$$|\hat{\Phi}|_+<c^{\nu+1}r_{\nu+1}^\sigma\delta^{3/2}=r_{\nu+1}^\sigma\delta_{\nu+1}\,, \qquad (A.3.72)$$

which was claimed in (A.3.20). This completes the proof of the theorem.

A.4 Proof of Theorem 6.2.1

a) In this section we show that the series expansion constructed in Sect. 6.3 actually converges. For this purpose we make use of the existence theorem A.3.1 of the previous section and establish the convergence of some power series solution. This solution, however, may not agree with that found by formal expansion (in theorem 6.2.1), since in Sect. 6.3 the normalization was imposed on the factors $\mathcal{U}_\nu$ and not on the products $\mathcal{U}_1\circ\mathcal{U}_2\circ\ldots\circ\mathcal{U}_\nu$. But with the help of lemma A.2.1 we shall be able to establish the convergence of the unique solution described in theorem 6.2.1.

For the first step we consider the equations (6.3.1 and 2) and assume that f, g are real analytic in a fixed domain

$$|\mathrm{Im}\{\varphi\}|<r\,,\qquad|\xi|<s\,,\qquad|\varepsilon|<\varepsilon_0\,, \qquad (A.4.1)$$

where

$$|\varepsilon|\left(\frac{|f|}{r}+\frac{|g|}{s}\right)<M\varepsilon_0 . \tag{A.4.2}$$

We apply theorem A.3.1 to these equations, where we replace f, g of Sect. A.3 by $\varepsilon f(\varphi,\xi,\varepsilon)$, $\varepsilon g(\varphi,\xi,\varepsilon)$. Although the proof of that theorem was carried out without such a complex parameter ε, one sees immediately that the solutions $\boldsymbol{u}$, $\boldsymbol{v}$, $\hat{\boldsymbol{a}}$, $\hat{\boldsymbol{b}}$, $\hat{\boldsymbol{B}}$ of theorem A.3.1 are analytic functions of ε. In fact, the approximation constructed in the proof of theorem A.3.1 turns out to be analytic, and since the final solution is obtained as the uniform limit (in a complex domain) of the approximations, the final solution is analytic there. It remains to be verified that the inequalities – required by theorem A.3.1 –

$$\frac{|\varepsilon f|}{r}+\frac{|\varepsilon g|}{s}<r^{\sigma}K\delta_0 \tag{A.4.3}$$

are satisfied. Clearly, (A.4.2) implies (A.4.3) if

$$\varepsilon_0 \leqq \frac{r^{\sigma}K\delta_0}{M} . \tag{A.4.4}$$

This gives a lower bound for the radius of convergence for the solutions $\boldsymbol{u}(\psi,\varepsilon)$, $\boldsymbol{v}(\psi,\chi,\varepsilon)$, $\hat{\boldsymbol{a}}(\varepsilon)$, $\hat{\boldsymbol{b}}(\varepsilon)$, $\hat{\boldsymbol{B}}(\varepsilon)$ which are analytic in $|\mathrm{Im}\{\psi\}|<r/2$, $|\varepsilon|<\varepsilon_0$ ($\boldsymbol{v}$ is linear in χ).

Moreover, for $\varepsilon=0$ the solutions constructed in Sect. A.3 reduce to $\boldsymbol{u}=\boldsymbol{v}=\mathbf{0}$, $\hat{\boldsymbol{a}}=\omega$, $\hat{\boldsymbol{b}}=\mathbf{0}$, $\hat{\boldsymbol{B}}=\Lambda$ as one sees by setting $\delta=0$ there. Hence $\boldsymbol{u}$, $\boldsymbol{v}$, $\hat{\boldsymbol{a}}$; ω, $\hat{\boldsymbol{b}}$, $\hat{\boldsymbol{B}}$; Λ can be expanded into power series, without constant terms, which converge in $|\varepsilon|<\varepsilon_0$.

This proves the existence of one analytic solution

$$\left.\begin{aligned} \boldsymbol{u} &= \boldsymbol{u}(\psi,\varepsilon)\\ \boldsymbol{v} &= \boldsymbol{v}(\psi,\xi,\varepsilon)\\ \boldsymbol{\Delta} &= \hat{\boldsymbol{a}}(\varepsilon)-\omega\\ \boldsymbol{d} &= \hat{\boldsymbol{b}}(\varepsilon)\\ \boldsymbol{D} &= \hat{\boldsymbol{B}}(\varepsilon)-\Lambda \end{aligned}\right\} \tag{A.4.5}$$

to our problem and theorem 6.2.1 is established. In fact, as was stated at the end of Sect. A.2, the power series for $\boldsymbol{\Delta}$, $\boldsymbol{d}$, $\boldsymbol{D}$ are uniquely determined and independent of normalizations.

b) We now turn to the proof of the convergence of the series expansion constructed in Sect. 6.3. It was normalized by the condition

$$U\in\mathscr{R} \tag{A.4.6}$$

which is not necessarily satisfied for the solution (A.4.5). We denote the transformation given by (A.4.5) by

$$\left.\begin{aligned}\varphi &= \psi + \boldsymbol{u}(\psi, \varepsilon)\\ \xi &= \chi + \boldsymbol{v}(\psi, \xi, \varepsilon)\end{aligned}\right\}. \tag{A.4.7}$$

By lemma A.2.1 the most general solution is given in the form $\mathscr{U} \circ \mathscr{C}$, where $\mathscr{C} \in \mathfrak{C}$, while Δ, $\boldsymbol{d}$, D is independent of the normalization. Therefore it suffices to show that a convergent series expansion for $\mathscr{C}$ can be found such that the expansion of $\mathscr{U} \circ \mathscr{C}$ agrees with that found in theorem A.1.1. Since $\mathfrak{C}$ is finite dimensional, this assertion follows from the implicit function theorem, as we now show. With

$$\mathscr{C}: \begin{cases}\psi' = \psi + \boldsymbol{a}\\ \chi' = \chi + \boldsymbol{b} + B\chi\end{cases} \quad \text{with } \Lambda \boldsymbol{b} = \boldsymbol{0},\ \Lambda B = B\Lambda\,, \tag{A.4.8}$$

we get for $\mathscr{U} \circ \mathscr{C}$

$$\left.\begin{aligned}\varphi &= \psi + \boldsymbol{a} + \boldsymbol{u}(\psi+\boldsymbol{a}, \varepsilon)\\ \xi &= \chi + \boldsymbol{b} + B\chi + \boldsymbol{v}(\psi+\boldsymbol{a}, \chi+\boldsymbol{b}+B\chi, \varepsilon)\end{aligned}\right\} \tag{A.4.9}$$

and it remains to find $\boldsymbol{a}$, $\boldsymbol{b}$, B in such a way that

$$\begin{pmatrix}\boldsymbol{a} + \boldsymbol{u}\\ \boldsymbol{b} + \boldsymbol{v} + B\chi\end{pmatrix} \in \mathscr{R} \tag{A.4.10}$$

is fulfilled.

We decompose

$$\begin{pmatrix}\boldsymbol{u}\\ \boldsymbol{v}\end{pmatrix}$$

into its components in $\mathscr{N}$ and $\mathscr{R}$ and denote by $P_{\mathscr{N}}$ the projection of our function space into $\mathscr{N}$.

Then we try to determine $\boldsymbol{a}$, $\boldsymbol{b}$, B in such a manner that

$$P_{\mathscr{N}}\begin{pmatrix}\boldsymbol{a} + \boldsymbol{u}\\ \boldsymbol{b} + B\chi + \boldsymbol{v}\end{pmatrix} = \begin{pmatrix}\boldsymbol{a}\\ \boldsymbol{b} + B\chi\end{pmatrix} + P_{\mathscr{N}}\begin{pmatrix}\boldsymbol{u}\\ \boldsymbol{v}\end{pmatrix} \tag{A.4.11}$$

vanishes. Here the arguments in $\boldsymbol{u}$, $\boldsymbol{v}$ are $\psi' = \psi + \boldsymbol{a}$, $\chi' = \chi + \boldsymbol{b} + B\chi, \varepsilon$. This gives rise to finitely many equations of equally many unknowns $(\boldsymbol{a}, \boldsymbol{b}, B)$ varying with $\mathscr{N}$. For $\varepsilon = 0$ the functions $\boldsymbol{u}$ und $\boldsymbol{v}$ vanish and the solution is $\boldsymbol{a} = \boldsymbol{b} = B = 0$. By the implicit function theorem there exist analytic functions $\boldsymbol{a}(\varepsilon)$, $\boldsymbol{b}(\varepsilon)$, $B(\varepsilon)$ without constant term annihilating (A.4.11). This defines the mapping $\mathscr{C}$ in (A.4.8) for which $\mathscr{U} \circ \mathscr{C}$ satisfies our normalization (A.4.6). Since $\mathscr{U}$, $\mathscr{C}$ are given by a convergent series, so is $\mathscr{U} \circ \mathscr{C}$ given by a convergent series, provided $|\varepsilon|$ is sufficiently small, as we wanted to prove.

c) This result frees one from the intricate construction in Sect. A.3 and ensures the convergence of the series obtained by formal expansion, at least if the above normalization is observed. Actually, the same statement holds if one requires instead that the free terms

$$P_{\mathscr{N}} \begin{pmatrix} u \\ v \end{pmatrix} \tag{A.4.12}$$

which remain in the expansion form a convergent series. To summarize:

Theorem A.4.1. The formal series expansions for $\boldsymbol{u}(\psi, \varepsilon)$, $\boldsymbol{v}(\psi, \chi, \varepsilon)$, $\Delta(\varepsilon)$, $\boldsymbol{d}(\varepsilon)$, $D(\varepsilon)$ of theorem A.1.1 are convergent for sufficiently small ε provided that

$$P_{\mathscr{N}} \begin{pmatrix} u \\ v \end{pmatrix}$$

is prescribed as a convergent series in ε.

Bibliography and Comments

Since the field of synergetics has ties to many disciplines, an attempt to provide a more or less complete list of references seems hopeless. Indeed, such a list would fill a whole volume. We therefore confine the references to those works which we used in the preparation of this book. In addition, we quote a number of papers, articles, or books which the reader might find useful for further study. We list the references and further reading material according to the individual chapters.

1. Introduction

1.1 What is Synergetics About?

H. Haken: *Synergetics, An Introduction*, 3rd ed. (Springer, Berlin, Heidelberg, New York 1983) This reference is referred to in the present book as [1]
H. Haken, R. Graham: Synergetik – Die Lehre vom Zusammenwirken. Umschau **6**, 191 (1971)
H. Haken (ed.): *Synergetics* (Proceedings of a Symposium on Synergetics, Elmau 1972) (Teubner, Stuttgart 1973)
H. Haken (ed.): *Cooperative Effects, Progress in Synergetics* (North-Holland, Amsterdam 1974)
H. Haken: Cooperative effects in systems far from thermal equilibrium and in nonphysical system. Rev. Mod. Phys. **47**, 67 (1975)
A further source of references is the Springer Series in Synergetics, whose individual volumes are listed in the front matter of this book.
For a popularisation see
H. Haken: *Erfolgsgeheimnisse der Natur* (Deutsche Verlagsanstalt, Stuttgart 1981)

1.2 Physics

The modern treatment of phase transitions of systems in thermal equilibrium rests on the renormalization group approach:
K. G. Wilson: Phys. Rev. **B4**, 3174; 3184 (1971)
K. G. Wilson, M. E. Fisher: Phys. Rev. Lett. **28**, 248 (1972)
F. J. Wegener: Phys. Rev. **B5**, 4529 (1972); **B6**, 1891 (1972)
T. W. Burkhardt, J. M. J. van Leeuwen (eds.): *Real-Space Renormalization*, Topics Curr. Phys., Vol. 30 (Springer, Berlin, Heidelberg, New York 1982)
Books and reviews on the subject are, for example,
K. G. Wilson, J. Kogut: Phys. Rep. **12C**, 75 (1974)
C. Domb, M. S. Green (eds.): *Phase Transitions and Critical Phenomena*. Internat. Series of Monographs in Physics, Vols. 1–6 (Academic, London 1972–76)
S. K. Ma: *Modern Theory of Critical Phenomena* (Benjamin, Reading, MA 1976)

1.2.1 Fluids: Formation of Dynamic Patterns

Taylor Instability

G. I. Taylor: Philos. Trans. R. Soc. London **A223**, 289 (1923)
For recent and more detailed studies see, for example,
R. P. Fenstermacher, H. L. Swinney, J. P. Gollub: J. Fluid Mech. **94**, 103 (1979)
R. C. DiPrima: In *Transition and Turbulence,* ed. by R. E. Meyer (Academic, New York 1981)

Bénard Instability

H. Bénard: Rev. Gén. Sci. Pures Appl. **11**, 1261, 1309 (1900)
Lord Rayleigh: Philos. Mag. **32**, 529 (1916)
For more recent theoretical studies on linear stability see, e.g.
S. Chandrasekhar: *Hydrodynamic and Hydromagnetic Stability* (Clarendon, Oxford 1961)
For nonlinear treatments see
A. Schlüter, D. Lortz, F. Busse: J. Fluid Mech. **23**, 129 (1965)
F. H. Busse: J. Fluid Mech. **30**, 625 (1967)
A. C. Newell, J. A. Whitehead: J. Fluid Mech. **38**, 279 (1969)
R. C. DiPrima, H. Eckhaus, L. A. Segel: J. Fluid Mech. **49**, 705 (1971)
F. H. Busse: J. Fluid Mech. **52**, 1 (1972)
F. H. Busse: Rep. Prog. Phys. **41**, 1929 (1978)
Nonlinearity *and* fluctuations are treated in
H. Haken: Phys. Lett. **46A**, 193 (1973); and, in particular, Rev. Mod. Phys. **47**, 67 (1975); and
H. Haken: *Synergetics*, Springer Ser. Synergetics, Vol. 1, 3rd. ed. (Springer, Berlin, Heidelberg, New York 1983)
R. Graham: Phys. Rev. Lett. **31**, 1479 (1973); Phys. Rev. **10**, 1762 (1974)
For recent experiments see also
G. Ahlers, R. Behringer: Phys. Rev. Lett. **40**, 712 (1978)
G. Ahlers, R. Walden: Phys. Rev. Lett. **44**, 445 (1981)
P. Bergé: In *Dynamical Critical Phenomena and Related Topics*, ed. by C. P. Enz, Lecture Notes Phys., Vol. 104 (Springer, Berlin, Heidelberg, New York 1979) p. 288
F. H. Busse, R. M. Clever: J. Fluid Mech. **102**, 75 (1981)
M. Giglio, S. Musazzi, U. Perini: Phys. Rev. Lett. **47**, 243 (1981)
E. L. Koschmieder, S. G. Pallas: Int. J. Heat Mass Transfer **17**, 991 (1974)
J. P. Gollub, S. W. Benson: J. Fluid Mech. **100**, 449 (1980)
J. Maurer, A. Libchaber: J. Phys. Paris Lett. **39**, 369 (1978); **40**, 419 (1979); **41**, 515 (1980)
G. Pfister, I. Rehberg: Phys. Lett. **83A**, 19 (1981)
H. Haken (ed.): *Chaos and Order in Nature*, Springer Ser. Synergetics, Vol. 11 (Springer, Berlin, Heidelberg, New York 1981), see in particular contributions by A. Libchaber and S. Fauve, E. O. Schulz-DuBois et al., P. Bergé, F. H. Busse
H. Haken (ed.): *Evolution of Order and Chaos*, Springer Ser. Synergetics, Vol. 17 (Springer, Berlin, Heidelberg, New York 1982)
H. L. Swinney, J. P. Gollub (eds.): *Hydrodynamic Instabilities and the Transition to Turbulence*, Topics Appl. Phys., Vol. 45 (Springer, Berlin, Heidelberg, New York 1981)

Texts and Monographs on Hydrodynamics

L. D. Landau, E. M. Lifshitz: *Course of Theoretical Physics*, Vol. 6 (Pergamon, London, New York 1959)
Chia-Shun-Yih: *Fluid Mechanics* (University Press, Cambridge 1970)
C. C. Lin: *Hydrodynamic Stability* (University Press, Cambridge 1967)
D. D. Joseph: *Stability of Fluid Motions*, Springer Tracts Nat. Phil., Vols. 27, 28 (Springer, Berlin, Heidelberg, New York 1976)

Metereology

R. Scorer: *Clouds of the World* (Lothian, Melbourne 1972)

1.2.2 Lasers: Coherent Oscillations

Early papers on laser theory including quantum fluctuations are
H. Haken: Z. Phys. **181**, 96 (1964); **190**, 327 (1966)
H. Risken: Z. Phys. **186**, 85 (1965)
R. D. Hempstead, M. Lax: Phys. Rev. **161**, 350 (1967)
W. Weidlich, H. Risken, H. Haken: Z. Phys. **201**, 396 (1967)
M. Scully, W. E. Lamb: Phys. Rev. **159**, 208 (1967); **166**, 246 (1968)
H. Haken: Rev. Mod. Phys. **47**, 67 (1975)
Laser-phase transition analogy

R. Graham, H. Haken: Z. Phys. **213**, 420 (1968)
R. Graham, H. Haken: Z. Phys. **237**, 31 (1970)
V. De Giorgio, M. O. Scully: Phys. Rev. **A2**, 117a (1970)

Ultra Short Pulses

R. Graham, H. Haken: Z. Phys. **213**, 420 (1968)
H. Risken, K. Nummedal: Phys. Lett. **26A**, 275 (1968); J. Appl. Phys. **39**, 4662 (1968)
H. Haken, H. Ohno: Opt. Commun. **16**, 205 (1976); Phys. Lett. **59A**, 261 (1976)
H. Knapp, H. Risken, H. D. Vollmer: Appl. Phys. **15**, 265 (1978)
M. Büttiker, H. Thomas: In *Solutions and Condensed Matter Physics*, ed. by A. R. Bishop, T. Schneider, Springer Ser. Solid-State Phys., Vol. 8 (Springer, Berlin, Heidelberg, New York 1981) p. 321
J. Zorell: Opt. Commun. **38**, 127 (1981)

Optical Bistability (some early and recent treatments):

S. L. McCall: Phys. Rev. **A9**, 1515 (1974)
R. Bonifacio, L. A. Lugiato: Opt. Commun. **19**, 172 (1976)
R. Sulomaa, S. Stenholm: Phys. Rev. **A8**, 2695 (1973)
A. Kossakowski, T. Marzalek: Z. Phys. **B23**, 205 (1976)
L. A. Lugiato, V. Benza, L. M. Narducci, J. D. Farina: Opt. Commun. **39**, 405 (1981)
L. A. Lugiato, V. Benza, L. M. Narducci: In *Evolution of Order and Chaos*, ed. by H. Haken Springer Ser. Synergetics, Vol. 17 (Springer, Berlin, Heidelberg, New York 1982) p. 120
M. G. Velarde: ibid., p. 132
L. A. Lugiato: In *Progress in Optics* (North-Holland, Amsterdam 1983)
R. Bonifacio (ed.): *Dissipative Systems in Quantum Optics*, Topics Curr. Phys., Vol. 27 (Springer, Berlin, Heidelberg, New York 1982)

Books: Laser Theory

H. Haken: Laser Theory, in *Encyclopedia of Physics,* Vol. XXV/2c, Light and Matter Ic, (Springer, Berlin, Heidelberg, New York 1970) and reprint edition *Laser Theory* (Springer, Berlin, Heidelberg, New York 1983)
M. Sargent, M. O. Scully, W. E. Lamb: *Laser Physics* (Addison-Wesley, Reading, MA 1974)

1.2.3 Plasmas: A Wealth of Instabilities

(We can give only a small selection of titles)

F. Cap: *Handbook on Plasma Instabilities*, Vols. 1, 2 (Academic, New York 1976 and 1978)
A. B. Mikhailowskii: *Theory of Plasma Instabilities*, Vols. 1, 2 (Consultants Bureau, New York, London 1974)
H. Wilhelmson, J. Weiland: *Coherent Non-Linear Interaction of Waves in Plasmas* (Pergamon, Oxford 1977)
S. G. Thornhill, D. ter Haar: Phys. Rep. **C43**, 43 (1978)

1.2.4 Solid State Physics: Multistability, Pulses, Chaos

Gunn Oscillator

J. B. Gunn: Solid State Commun. **1**, 88 (1963)
J. B. Gunn: IBM Res. Develop. **8**, 141 (1964)
K. Nakamura: J. Phys. Soc. Jpn. **38**, 46 (1975)

Tunnel Diodes

C. Zener: Proc. R. Soc. London **145**, 523 (1934)
L. Esaki: Phys. Rev. **109**, 603 (1958)
R. Landauer: J. Appl. Phys. **33**, 2209 (1962)
R. Landauer, J. W. F. Woo: In *Synergetics*, ed. by H. Haken (Teubner, Stuttgart 1973) p. 97

Thermoelastic Instabilities

C. E. Bottani, G. Caglioti, P. M. Ossi: J. Phys. F. **11**, 541 (1981)
C. Caglioti, A. F. Milone (eds.): *Mechanical and Thermal Behaviour of Metallic Materials.* Proc. Int. School of Physics Enrico Fermi (North-Holland, Amsterdam 1982)

Crystal Growth

J. S. Langer: In *Fluctuations, Instabilities and Phase Transitions*, ed. by T. Riste (Plenum, New York 1975) p. 82
J. S. Langer: Rev. Mod. Phys. **52**, 1 (1980)

1.3 Engineering

1.3.1 Civil, Mechanical, and Aero-Space Engineering: Post-Buckling Patterns, Flutter etc.

J. M. T. Thompson, G. W. Hunt: *A General Theory of Elastic Stability* (Wiley, London 1973)
K. Huseyn: *Nonlinear Theory of Elastic Stability* (Nordhoff, Leyden 1975)
D. O. Brush, B. D. Almroth: *Buckling of Bars, Plates and Shells* (McGraw-Hill, New York 1975)

1.3.2 Electrical Engineering and Electronics: Nonlinear Oscillations

A. A. Andronov, A. A. Vitt, S. E. Kaikin: *Theory of Oscillators* (Pergamon, Oxford, London 1966)
N. Minorsky: *Nonlinear Oscillations* (van Nostrand, Princeton 1962)
C. Hayashi: *Nonlinear Oscillations in Physical Systems* (McGraw-Hill, New York 1964)
P. S. Lindsay: Phys. Rev. Lett. **47**, 1349 (1981)

1.4 Chemistry: Macroscopic Patterns

C. H. Bray: J. Am. Chem. Soc. **43**, 1262 (1921)
B. P. Belousov: Sb. Ref. Radats. Med. Moscow (1959)
V. A. Vavalin, A. M. Zhabotinsky, L. S. Yaguzhinsky: *Oscillatory Processes in Biological and Chemical Systems* (Science Publ., Moscow 1967) p. 181
A. N. Zaikin, A. M. Zhabotinsky: Nature **225**, 535 (1970)
A. M. Zhabotinsky, A. N. Zaikin: J.Theor. Biol. **40**, 45 (1973)
A. M. Turing: Philos. Trans. R. Soc. London **B237**, 37 (1952)
G. Nicolis, I. Prigogine: *Self-Organization in Non-Equilibrium Systems* (Wiley, New York 1977)
H. Haken: Z. Phys. **B20**, 413 (1975)
G. F. Oster, A. S. Perelson: Arch. Rat. Mech. Anal. **55**, 230 (1974)
A. S. Perelson: G. F. Oster: Arch. Rat. Mech. Anal. **57**, 31 (1974/75)
G. Nicolis: Adv. Chem. Phys. **19**, 209 (1971)
B. Change, E. K. Pye, A. M. Ghosh, B. Hess (eds.): *Biological and Biochemical Oscillators* (Academic, New York 1973)
G. Nicolis, J. Portnow: Chem. Rev. **73**, 365 (1973)
R. M. Noyes, R. J. Field: Annu. Rev. Phys. Chem. **25**, 95 (1975)
J. J. Tyson: *The Belousov-Zhabotinsky Reaction.* Lecture Notes Biomath., Vol. 10 (Springer, Berlin, Heidelberg, New York 1976)
P. C. Fife: *Mathematical Aspects of Reacting and Diffusing Systems.* Lecture Notes Biomath., Vol. 28 (Springer, Berlin, Heidelberg, New York 1979)
A. Pacault, C. Vidal (eds.): *Synergetics. Far from Equilibrium.* Springer Ser. Synergetics, Vol. 3 (Springer, Berlin, Heidelberg, New York 1979)
C. Vidal, A. Pacault (eds.): *Nonlinear Phenomena in Chemical Dynamics.* Springer Ser. Synergetics, Vol. 12 (Springer, Berlin, Heidelberg, New York 1981)

1.5 Biology

1.5.1 Some General Remarks

T. H. Bullock, R. Orkand, A. Grinnell: *Introduction to Nervous Systems* (Freeman, San Francisco 1977)

A. C. Scott: *Neurophysics* (Wiley, New York 1977)
E. Basar: *Biophysical and Physiological System Analysis* (Addison Wesely, Reading MA 1976)
M. Conrad, W. Güttinger, M. Dal Chin (eds.): *Physics and Mathematics of the Nervous System.* Lecture Notes Biomath., Vol. 4 (Springer, Berlin, Heidelberg, New York 1974)
A. V. Holden: *Models of Stochastic Activity of Neurons.* Lecture Notes Biomath., Vol. 12 (Springer, Berlin, Heidelberg, New York 1976)
H. Shimizu: Adv. Biophys. **13**, 195 (1979)

1.5.2 Morphogenesis

A. M. Turing: Philos. Trans. R. Soc. London **B237**, 37 (1952)
L. Wolpert: J. Theor. Biol. **25**, 1 (1969)
A. Gierer, H. Meinhardt: Kybernetik **12**, 30 (1972); J. Cell. Sci. **15**, 321 (1974)
H. Haken, H. Olbrich: J. Math. Biol. **6**, 317 (1978)
J. P. Murray: J. Theor. Biol. **88**, 161 (1981)
C. Berding, H. Haken: J. Math. Biol. **14**, 133 (1982)

1.5.3 Population Dynamics

A. Lotka: Proc. Nat. Acad. Sci. (USA) **6**, 410 (1920)
V. Volterra: *Leçons sur la Théorie Mathématiques de la Lutte pour la Vie*, Paris (1931)
N. S. Goel, S. C. Maitra, E. W. Montroll: Rev. Mod. Phys. **43**, 231 (1971)
T. N. E. Greville (ed.): *Population Dynamics* (Academic, London 1972)
D. Ludwig: In *Stochastic Population Theories*, ed. by S. Levin, Lecture Notes Biomath., Vol. 3 (Springer, Berlin, Heidelberg, New York 1974)
R. B. May: Nature **261**, 459 (1976)

1.5.4 Evolution

M. Eigen: Naturwissenschaften **58**, 465 (1971)
M. Eigen, P. Schuster: Naturwissenschaften **64**, 541 (1977); **65**, 7 (1978); **65**, 341 (1978)
W. Ebeling, R. Feistel: *Physik der Selbstorganisation und Evolution* (Akademie-Verlag, Berlin 1982)

1.5.5 Immune System

F. M. Burnet: *Immunology, Aging, and Cancer* (Freeman, San Francisco 1976)
C. DeLisi: *Antigen Antibody Interactions*, Lecture Notes Biomath., Vol. 8 (Springer, Berlin, Heidelberg, New York 1976)
N. Dubin: *A Stochastic Model for Immunological Feedback in Carcinogenesis.* Lecture Notes Biomath., Vol. 9 (Springer, Berlin, Heidelberg, New York 1976)
P. H. Richter: Pattern formation in the immune system. Lect. Math. Life Sci. **11**, 89 (1979)

1.6 Computer Sciences

1.6.1 Self-Organization of Computers, in Particular Parallel Computing

R. W. Hockney, C. R. Jesshope: *Parallel Computers* (Hilger, Bristol 1981)

1.6.2 Pattern Recognition by Machines

K. S. Fu: *Digital Pattern Recognition*, 2nd ed. (Springer, Berlin, Heidelberg, New York 1980)
K. S. Fu: *Syntactic Pattern Recognition Applications* (Springer, Berlin, Heidelberg, New York 1976)
K. S. Fu: In *Pattern Formation by Dynamic Systems and Pattern Recognition*, ed. by H. Haken, Springer Ser. Synergetics, Vol. 5 (Springer, Berlin, Heidelberg, New York 1979) p. 176
T. Kohonen: *Associative Memory – A System Theoretical Approach* (Springer, Berlin, Heidelberg, New York 1978)
T. Kohonen: *Self-Organization and Associative Memory*, Springer Ser. Inf. Sci., Vol. 8 (Springer, Berlin, Heidelberg, New York 1983)
H. Haken (ed.): *Pattern Formation by Dynamic Systems and Pattern Recognition*, Springer Ser. Synergetics, Vol. 5 (Springer, Berlin, Heidelberg, New York 1979)

1.6.3 Reliable Systems from Unreliable Elements

H. Haken: unpublished material

1.7 Economy

G. Mensch, K. Kaasch, A. Kleinknecht, R. Schnopp: IIM/dp 80-5 *Innovation Trends, and Switching between Full- and Under-Employment Equilibria.* 1950–1978 Discussion Paper Series, International Institute of Management, Wissenschaftszentrum Berlin
H. Haken: *Synergetics*, Springer Ser. Synergetics, Vol. 1, 3rd. ed. (Springer, Berlin, Heidelberg, New York 1983)
W. Weidlich, G. Haag: *Quantitative Sociology*, Springer Ser. Synergetics, Vol. 14 (Springer, Berlin, Heidelberg, New York 1983)
H. Haken: *Erfolgsgeheimnisse der Natur* (Deutsche Verlagsanstalt, Stuttgart 1981)

1.8 Ecology

Ch. J. Krebs: *Ecology. The Experimental Analysis of Distribution and Abundance* (Harper and Row, New York 1972)
R. E. Rickleps: *Ecology* (Nelson, London 1973)

1.9 Sociology

S. E. Ash: *Social Psychology* (Prentice Hall, New York 1952) p. 452
W. Weidlich: Collect. Phenom. **1**, 51 (1972)
E. Noelle-Neumann: *Die Schweigespirale* (Piper, München 1980) [English transl. (to appear 1983): *The Spiral of Silence: Public Opinion – The Skin of Time* (Chicago, University Press)]
A. Wunderlin, H. Haken: Lecture Notes, Projekt Mehrebenenanalyse im Rahmen des Forschungsschwerpunkts Mathematisierung (Universität Bielefeld 1980)
H. Haken: *Erfolgsgeheimnisse der Natur* (Deutsche Verlagsanstalt, Stuttgart 1981)
W. Weidlich, G. Haag: *Quantitative Sociology*, Springer Ser. Synergetics, Vol. 14 (Springer, Berlin, Heidelberg, New York 1983)

1.11 The Kind of Equations we Want to Study

For a general background, see also

H. Haken: *Synergetics*, Springer Ser. Synergetics, Vol. 1, 3rd. ed. (Springer, Berlin, Heidelberg, New York 1983)

Because the individual topics of Sects. 1.11–17 will be dealt with in detail in later chapters, we refer the reader to the corresponding references belonging to those chapters.

Here we quote only those references which will not be quoted later.

1.11.1 Differential Equations

R. Courant, D. Hilbert: *Methods of Mathematical Physics*, Vols. 1, 2 (Wiley, New York 1962)
P. M. Morse, H. Feshbach: *Methods of Theoretical Physics*, Vols. 1, 2 (McGraw-Hill, New York 1953)
L. W. F. Elen: *Differential Equations*, Vols. 1, 2 (MacMillan, London 1967)

1.11.2 First Order Differential Equations

E. A. Coddington, N. Levinson: *Theory of Ordinary Differential Equations* (McGraw-Hill, New York 1955)

1.11.3 Nonlinearity

V. V. Nemytskii, V. V. Stepanov: *Qualitative Theory of Differential Equations* (University Press, Princeton 1960)
M. W. Hirsch, S. Smale: *Differential Equations, Dynamical Systems, and Linear Algebra* (Academic, New York 1974)

Z. Nitecki: *Differentiable Dynamics* (MIT Press, Cambridge, MA 1971)
R. Abraham, J. E. Marsden: *Foundations of Mechanics* (Benjamin/Cummings, Reading, MA 1978)
S. Smale: *The Mathematics of Time* (Springer, Berlin, Heidelberg, New York 1980)

1.11.4 Control Parameters

H. Haken: *Synergetics*, Springer Ser. Synergetics, Vol. 1, 3rd. ed. (Springer, Berlin, Heidelberg, New York 1983)

1.11.5 Stochasticity

J. L. Doob: *Stochastic Processes* (Wiley, New York 1953)
M. Loève: *Probability Theory* (van Nostrand, Princeton 1963)
R. von Mises: *Mathematical Theory of Probability and Statistics* (Academic, New York 1964)
Yu. V. Prokhorov, Yu. A. Rozanov: *Probability Theory*, Grundlehren der mathematischen Wissenschaften in Einzeldarstellungen, Vol. 157 (Springer, Berlin, Heidelberg, New York 1968)
R. C. Dubes: *The Theory of Applied Probability* (Prentice Hall, Englewood Cliffs, NJ 1968)
W. Feller: *An Introduction to Probability Theory and Its Applications*, Vol. 1 (Wiley, New York 1971)
Kai Lai Chung: *Elementary Probability Theory with Stochastic Processes* (Springer, Berlin, Heidelberg, New York 1974)
T. Hida: *Brownian Motion*, Applications of Mathematics, Vol. 11 (Springer, Berlin, Heidelberg, New York 1980)

Statistical Mechanics

L. D. Landau, E. M. Lifshitz: In *Course of Theoretical Physics*, Vol. 5 (Pergamon, London 1952)
R. Kubo: *Thermodynamics* (North-Holland, Amsterdam 1968)
D. N. Zubarev: *Non-Equilibrium Statistical Thermodynamics* (Consultants Bureau, New York 1974)

Quantum Fluctuations

H. Haken: Laser Theory, in *Encylopedia of Physics*, Vol. XXV/2c, Light and Matter Ic (Springer, Berlin, Heidelberg, New York 1970) and reprint edition *Laser Theory* (Springer, Berlin, Heidelberg, New York 1983)
with many further references.

Chaos

H. Haken: *Synergetics*, Springer Ser. Synergetics, Vol. 1, 3rd. ed. (Springer, Berlin, Heidelberg, New York 1978)
H. Haken (ed.): *Chaos and Order in Nature*, Springer Ser. Synergetics, Vol. 11 (Springer, Berlin, Heidelberg, New York 1981)
H. Haken: *Order and Chaos*, Springer Ser. Synergetics, Vol. 17 (Springer, Berlin, Heidelberg, New York 1982)

1.11.6 Many Components and the Mezoscopic Approach

H. Haken: unpublished material

1.12 How to Visualize Solutions

Y. Choquet-Bruhat, C. DeWitt-Morette, M. Dillard-Bleick: *Analysis, Manifolds and Physics* (North-Holland, Amsterdam 1982)
R. D. Richtmyer: *Principles of Advanced Mathematical Physics II* (Springer, Berlin, Heidelberg, New York 1981)

1.13 Qualitative Changes: General Approach

H. Haken: *Synergetics*, Springer Ser. Synergetics, Vol. 1, 3rd. ed. (Springer, Berlin, Heidelberg, New York 1983)
D'Arcy W. Thompson: *On Growth and Form* (Cambridge University Press, London 1961)

1.14 Qualitative Changes: Typical Phenomena

See H. Haken: *Synergetics*, Springer Ser. Synergetics, Vol. 1, 3rd. ed. (Springer, Berlin, Heidelberg, New York 1983) and references cited in later chapters. Here References are presented for

1.14.1 Lyapunov Exponents

V. I. Oseledec: A multiplicative ergodic theorem. Lyapunov characteristic number for dynamical systems. Tr. Mosk. Mat. Osc. **19**, 179 (1968) [English transl.: Trans. Moscow Math. Soc. **19**, 197 (1968)]
Ya. B. Pesin: Characteristic Lyapunov Exponents and Smooth Ergodic Theory. Russ. Math. Surv. **32(4)**, 55 (1977)
D. Ruelle: "Sensitive Dependence on Initial Conditions and Turbulent Behavior of Dynamical Systems", in *Bifurcation Theory and Its Applications in Scientific Disciplines*, ed. by O. Gurel, O. E. Rössler, New York Acad. of Sci. **316**, (1979)
J. D. Farmer: Physica **4D**, 366 (1982)
K. Tomita: Phys. Rep. **86**, 113 (1982)

1.15 The Impact of Fluctuations

See the references in Sect. 1.11.5 as well as those of later chapters

1.16 Evolution of Spatial Patterns

H. Haken: *Synergetics*, Springer Ser. Synergetics, Vol. 1, 3rd. ed. (Springer, Berlin, Heidelberg, New York 1983)

1.17 Discrete Maps: The Poincaré Map

Discrete Maps are treated in Chap. 11. The Poincaré map is discussed, e.g., in
R. Abraham, J. E. Marsden: *Foundations of Mechanics* (Benjamin/Cummings, Reading, MA 1978)

1.18 Discrete Noisy Maps

Compare Chap. 11

2. Linear Ordinary Differential Equations

2.2 Groups and Invariance

E. C. G. Sudarshan, M. Mukunda: *Classical Dynamics: A Modern Perspective* (Wiley, New York 1974)
R. D. Richtmyer: *Principles of Advanced Mathematical Physics II* (Springer, Berlin, Heidelberg, New York 1981)

1.19 Pathways to Self-Organization

1.19.1 Self-Organization Through Change of Control Parameters

H. Haken: *Synergetics*, Springer Ser. Synergetics, Vol. 1, 3rd. ed. (Springer, Berlin, Heidelberg, New York 1983)

1.19.2 Self-Organization Through Change of Number of Components

H. Haken: Prog. Theor. Phys. Suppl. **69**, 30 (1980)

1.19.3 Self-Organization Through Transients

H. Haken: Unpublished material

2.3 Driven Systems

O. Duffing: *Erzwungene Schwingungen bei veränderlicher Eigenfrequenz und ihre technische Bedeutung* (Vieweg, Braunschweig 1918)
C. Hayashi: *Nonlinear Oscillations in Physical Systems* (McGraw-Hill, New York 1964)
compare Sect. 2.1.1 also

2.4 General Theorems on Algebraic and Differential Equations

2.4.2 Jordan's Normal Form

R. Bellman, K. L. Cooke: *Introduction to Matrix Analysis* (McGraw-Hill, New York 1960)

2.4.3 Some General Theorems on Linear Differential Equations

A comprehensive treatment is provided by
N. Dunford, J. T. Schwartz: *Linear Operators*, Pure and Applied Mathematics, Vol. VII, Parts I–III (Wiley, Interscience, New York 1957)

2.4.4 Generalized Characteristic Exponents

see Sect. 2.4.3. For the Lyapunov exponents see Sect. 1.14.6.
Theorem on vanishing Lyapunov exponents:
H. Haken: Phys. Lett. **94A**, 71 (1983)

2.6 Linear Differential Equations with Constant Coefficients

e.g.,
E. A. Coddington, N. Levinson: *Theory of Ordinary Differential Equations* (McGraw-Hill, New York 1955)

2.7 Linear Differential Equations with Periodic Coefficients

G. Floquet: *Sur les équations différentielles linéaires à coefficients périodiques.* Ann. École Norm. Ser. 2 **12**, 47 (1883)

2.8 Group Theoretical Interpretation

compare Sect. 2.2

2.9 A Perturbation Approach

H. Haken: Unpublished material

3. Linear Ordinary Differential Equations with Quasiperiodic Coefficients

The results of this chapter, with the exception of Sect. 3.9, were obtained by the present author. The operator T in Eq. (3.1.6) was introduced in
H. Haken: Z. Naturforsch. **8A**, 228 (1954)
where the case of a unitary representation of T was treated and the form (3.1.20) proven. For further results see
H. Haken: In *Dynamics of Synergetic Systems*, Springer Ser. Synergetics, Vol. 6, ed. by H. Haken (Springer, Berlin, Heidelberg, New York 1980) p. 16
For the proof of Theorem 3.8.2 I used auxiliary theorems represented in
N. Dunford, J. T. Schwartz: *Linear Operators*, Pure and Applied Mathematics, Vol. VII, Parts I–III (Wiley, Interscience, New York 1957)
For attempts of other authors at this problem cf.
N. N. Bogoliubov, I. A. Mitropolskii, A. M. Samoilento: *Methods of Accelerated Convergence in*

Nonlinear Mechanics (Springer, Berlin, Heidelberg, New York 1976),
where further references are given.

The results of Sect. 3.9 are taken from N. N. Bogoliubov, I. A. Mitropolskii, A. M. Samoilento l. c.

4. Stochastic Nonlinear Differential Equations

For the original papers on the Itô and Stratonovich calculus see

K. Itô: *Lectures on Stochastic Processes* (Tata Institute of Fundamental Research, Bombay 1961)
K. Itô: *Stochastic Processes* (Universitet Matematisk Institut, Aarhus 1969)
K. Itô, H.P. McKean: *Diffusion Processes and Their Sample Paths* (Springer, Berlin, Heidelberg, New York 1965)
K. Itô: Nagoya Math. J. **1**, 35 (1950)
K. Itô: Nagoya Math. J. **3**, 55 (1951)
K.Itô: *On Stochastic Differential Equations* (Am. Math. Soc. New York, 1951)
P. Langevin: Sur la théorie du mouvement brownien. C. R. Acad. Sci. Paris **146**, 530 (1908)
R. L. Stratonovich: SIAM J. Control **4**, 362 (1966)

Recent texts and monographs include

I. I. Gihmann, A. V. Skorohod: *Stochastic Differential Equations* (Springer, Berlin, Heidelberg, New York 1972)
L. Arnold: *Stochastic Differential Equations* (Oldenbourg, München 1973)
N. G. van Kampen: *Stochastic Processes in Physics and Chemistry* (North-Holland, Amsterdam 1981)
C. W. Gardiner: *Handbook of Stochastic Methods.* Springer Ser. Synergetics, Vol. 13 (Springer, Berlin, Heidelberg, New York 1983)

5. The World of Coupled Nonlinear Oscillators

5.1 Linear Oscillators Coupled Together

This section gives only a sketch, for detailed treatments of nonlinear oscillators see

N. N. Bogoliubov, Y. A. Mitropolsky: *Asymptotic Methods in the Theory of Nonlinear Oscillations* (Hindustan Publ. Corp., New Delhi 1961)
N. Minorski: *Nonlinear Oscillations* (Van Nostrand, Toronto 1962)
A. Andronov, A. Vitt, S. E. Khaikin: *Theory of Oscillators* (Pergamon, London 1966)

5.2 Perturbations of Quasiperiodic Motion for Time-Independent Amplitudes (Persistence of Quasiperiodic Motion)

Our presentation is based on results reported in

N. N. Bogoliubov, I. A. Mitropolskii, A. M. Samoilento: *Methods of Accelerated Convergence in Nonlinear Mechanics* (Springer, Berlin, Heidelberg, New York 1976)

5.3 Some Considerations on the Convergence of the Procedure

see Sect. 5.2

6. Nonlinear Coupling of Oscillators: The Case of Persistence of Quasiperiodic Motion

A. N. Kolmogorov: Dokl. Akad. Nauk. USSR **98**, 527 (1954)
V. I. Arnol'd: Russ. Math. Surv. **18**, 9 (1963)
J. Moser: Math. Ann. **169**, 136 (1967)

In this chapter we essentially follow Moser's paper, but we use a somewhat different representation

Further reading
J. Moser: "Nearly Integrable and Integrable Systems", in *Topics in Nonlinear Dynamics*, ed. by S. Jorna (AIP Conf. Proc. **46**, 1 1978)
M. V. Berry: "Regular and Irregular Motion", in *Topics in Nonlinear Dynamics*, ed. by S. Jorna (AIP Conf. Proc. **46**, 16 1978)

7. Nonlinear Equations. The Slaving Principle

This chapter is based on a slight generalization of
H. Haken, A. Wunderlin: Z. Phys. **B47**, 179 (1982)
An early version, applied to quasiperiodic motion (in the case of laser theory) was developed by
H. Haken: Talk at the International Conference on Optical Pumping, Heidelberg (1962); also
H. Haken, H. Sauermann: Z. Phys. **176**, 47 (1963)

Here, by an appropriate decomposition of the variables into rapidly oscillating parts and slowly varying amplitudes, the atomic variables were expressed by the field modes (order parameters)

Other procedures are given in
H. Haken: Z. Phys. **B20**, 413 (1975); **B21**, 105 (1975); **B22**, 69 (1975); **B23**, 388 (1975) and
H. Haken: Z. Phys. **B29**, 61 (1978); **B30**, 423 (1978)
The latter procedures are based on rapidly converging continued fractions, at the expense that the slaved variables depend on the order parameters (unstable modes) at previous times (in higher order approximation). These papers included fluctuations of the Langevin type.

In a number of special cases (in particular, if the fluctuations are absent), relations can be established to other theorems and procedures, developed in mathematics, theoretical physics, or other disciplines.

Relations between the slaving principle and the center manifold theorem (and related theorems) are studied by
A. Wunderlin, H. Haken: Z. Phys. **B44**, 135 (1981)
For the center manifold theorem, see
V. A. Pliss: Izv. Akad. Nauk SSSR., Mat. Ser. **28**, 1297 (1964)
A. Kelley: In *Transversal Mappings and Flows*, ed. by R. Abraham, J. Robbin (Benjamin, New York 1967)
In contrast to the center manifold theorem, the slaving principle contains fluctuations, includes the surrounding of the center manifold, and provides a construction of $s(u, \varphi, t)$.

8. Nonlinear Equations. Qualitative Macroscopic Changes

In this chapter I present an approach initiated in 1962 (H. Haken: Talk at the International Conference on Optical Pumping, Heidelberg 1962), and applied to laser theory including quasiperiodic motion, e.g. bifurcation to tori (see, e.g.,
H. Haken, H. Sauermann: Z. Phys. **176**, 47 (1963)
H. Haken: Laser Theory, in *Encylopedia of Physics*, Vol. XXV, 2c, Light and Matter Ic (Springer, Berlin, Heidelberg, New York 1970) and reprint edition *Laser Theory* (Springer, Berlin, Heidelberg, New York 1983)
This author's approach is based on the slaving principle and represents, in modern language, "dynamic bifurcation theory" (which allows one to cope with transients and fluctuations). "Static" bifurcation theory was initiated in the classical papers by
H. Poincaré: *Les méthodes nouvelles de la mécanique céleste T. 1* (Gauthier-Villars, Paris 1892)
H. Poincaré: Acta Math. **7**, 1 (1885)
A. M. Lyapunov: *Sur le masse liquide homogène donnée d'un mouvement de rotation.* Zap. Acad. Nauk, St. Petersburg **1**, 1 (1906)
E. Schmidt: Zur Theorie der linearen und nichtlinearen Integralgleichungen, 3. Teil, Math. Annalen **65**, 370 (1908)

While this field seems to have been more or less dormant for a while (with the exception of bifurcation theory in fluid dynamics), the past decade has seen a considerable increase of interest as reflected by recent texts. We mention in particular

D. H. Sattinger: *Topics in Stability and Bifurcation Theory*, Lecture Notes Math., Vol. 309 (Springer, Berlin, Heidelberg, New York 1972)

D. H. Sattinger: *Group Theoretic Methods in Bifurcation Theory*, Lecture Notes Math., Vol. 762 (Springer, Berlin, Heidelberg, New York 1980)

G. Iooss: *Bifurcation of Maps and Applications*, Lecture Notes, Mathematical Studies (North-Holland, Amsterdam 1979)

G. Iooss, D. D. Joseph: *Elementary Stability and Bifurcation Theory* (Springer, Berlin, Heidelberg, New York 1980)

These authors deal in an elegant fashion with "static" bifurcation theory.

8.2 A Simple Real Eigenvalue Becomes Positive

H. Haken: *Synergetics*, Springer Ser. Synergetics, Vol. 1, 3rd. ed. (Springer, Berlin, Heidelberg, New York 1983)

8.3 Multiple Real Eigenvalue Becomes Positive

Here we follow

H. Haken: Unpublished material

References on catastrophe theory are

R. Thom: *Structural Stability and Morphogenesis* (Benjamin, Reading, MA 1975)

Further references on this subject can be found in

H. Haken: *Synergetics*, Springer Ser. Synergetics, Vol. 1, 3rd. ed. (Springer, Berlin, Heidelberg, New York 1983)

8.4 A Simple Complex Eigenvalue Crosses the Imaginary Axis. Hopf Bifurcation

The branching of oscillatory solutions was first treated in the classical paper

E. Hopf: *Abzweigung einer periodischen Lösung eines Differentialsystems.* Berichte der Mathematisch-Physikalischen Klasse der Sächsischen Akademie der Wissenschaften zu Leipzig XCIV, 1 (1942)

For recent treatments see

J. Marsden, M. McCracken: *The Hopf Bifurcation and Its Applications.* Lecture Notes Appl. Math. Sci., Vol. 18 (Springer, Berlin, Heidelberg, New York 1976)

D. D. Joseph: *Stability of Fluids Motion.* Springer Tracts Natural Philos., Vols. 27, 28 (Springer, Berlin, Heidelberg, New York 1976)

A. S. Monin, A. M. Yaglom: *Statistical Fluid Mechanics*, Vol. I (MIT Press, Cambridge, MA 1971)

H. Haken: *Synergetics*, Springer Ser. Synergetics, Vol. 1, 3rd. ed. (Springer, Berlin, Heidelberg, New York 1983)

8.5 Hopf Bifurcation, Continued

Compare Sect. 8.4

8.6 Frequency Locking Between Two Oscillators

See e. g.

H. Haken: Laser Theory, in *Encylopedia of Physics*, Vol. XXV, 2c, Light and Matter Ic (Springer, Berlin, Heidelberg, New York 1970) and reprint edition *Laser Theory* (Springer, Berlin, Heidelberg, New York 1983)

R. L. Stratonovich: *Topics in the Theory of Random Noise*, Vols. 1, 2 (Gordon and Breach, New York 1963, 1967)

8.7 Bifurcation from a Limit Cycle

We follow

H. Haken: Z. Phys. **B29**, 61 (1978)

8.8 Bifurcation from a Limit Cycle: Special Cases

Compare Sect. 8.7 and
H. Haken: Unpublished material

8.9 Bifurcation from a Torus (Quasiperiodic Motion)

Compare Sect. 8.7 and
H. Haken: Z. Phys. **B30**, 423 (1978)
and unpublished material. For different approaches cf.
A. Chenciner, G, Iooss: Arch. Ration. Mech. Anal. **69**, 109 (1979)
G. R. Sell: Arch. Ration. Mech. Anal. **69**, 199 (1979)
G. R. Sell: In *Chaos and Order in Nature*, Springer Ser. Synergetics, Vol. 11, ed. by H. Haken (Springer, Berlin, Heidelberg, New York 1980) p. 84

8.10 Bifurcation from a Torus: Special Cases

See Sect. 8.9

8.11 Instability Hierarchies, Scenarios, and Routes to Turbulence

8.11.1 The Landau-Hopf Picture

L. D. Landau, E. M. Lifshitz: In *Course of Theoretical Physics*, Vol. 6, Fluid Mechanics (Pergamon, London, New York 1959)
E. Hopf: Commun. Pure Appl. Math. **1**, 303 (1948)

8.11.2 The Ruelle and Takens Picture

D. Ruelle, F. Takens: Commun. Math. Phys. **20**, 167 (1971)
S. Newhouse, D. Ruelle, F. Takens: Commun. Math. Phys. **64**, 35 (1978)

8.11.3 Bifurcations of Tori. Quasiperiodic Motions

See Sects. 8.9, 11.3

8.11.4 The Period-Doubling Route to Chaos. Feigenbaum Sequence

S. Grossmann, S. Thomae: Z. Naturforsch. **32A**, 1353 (1977)
M. J. Feigenbaum: J. Stat. Phys. **19**, 25 (1978); Phys. Lett. **74A**, 375 (1979)
P. Collet, J. P. Eckmann: *Iterated Maps on the Interval as Dynamical Systems* (Birkhäuser, Boston 1980)
T. Geisel, J. Nierwetberg: In *Evolution of Order and Chaos*, Springer Ser. Synergetics, Vol. 17, ed. by H. Haken (Springer, Berlin, Heidelberg, New York 1982) p. 187

8.11.5 The Route via Intermittency

Y. Pomeau, P. Manneville: Commun. Math. Phys. **77**, 189 (1980)
G. Mayer-Kress, H. Haken: Phys. Lett. **82A**, 151 (1981)

9. Spatial Patterns

9.1 The Basic Differential Equations

Examples are provided by the Navier-Stokes Equations, e.g.
H. Haken: *Synergetics*, Springer Ser. Synergetics, Vol. 1, 3rd. ed. (Springer, Berlin, Heidelberg, New York 1983) and
O. A. Ladyzhenskaya: *The Mathematical Theory of Viscous Incompressible Flow* (Gordon and Breach, New York 1963)

D. D. Joseph: *Stability of Fluid Motions I and II*, Springer Tracts Natural Philos., Vols. 27, 28 (Springer, Berlin, Heidelberg, New York 1976)
Reaction Diffusion equations are treated, e.g., in
P. C. Fife: In *Dynamics of Synergetic Systems*, Springer Ser. Synergetics, Vol. 6, ed. by H. Haken (Springer, Berlin, Heidelberg, New York 1980) p. 97, with further references
J. S. Turner: Adv. Chem. Phys. **29**, 63 (1975)
J. W. Turner: Trans. NY Acad. Sci. **36**, 800 (1974), Bull. Cl. Sci. Acad. Belg. **61**, 293 (1975)
Y. Schiffmann: Phys. Rep. **64**, 87 (1980)
Compare also Sect. 1.5.2

9.2 The General Method of Solution

H. Haken: *Synergetics*, Springer Ser. Synergetics, Vol. 1, 3rd. ed. (Springer, Berlin, Heidelberg, New York 1983)

9.3 Bifurcation Analysis for Finite Geometries

H. Haken: *Synergetics*, Springer Ser. Synergetics, Vol. 1, 3rd. ed. (Springer, Berlin, Heidelberg, New York 1983)
See also the references cited in Sect. 9.1

9.4 Generalized Ginzburg-Landau Equations

H. Haken: Z. Phys. **B21**, 105 (1975)
H. Haken: Z. Phys. **B22**, 69 (1975); **B23**, 388 (1975)
For a different approach (for a more restricted class of problems) based on scaling see
Y. Kuramoto, T. Tsusuki: Prog. Theor. Phys. **52**, 1399 (1974)
A. Wunderlin, H. Haken: Z. Phys. **B21**, 393 (1975)

9.5 A Simplification of Generalized Ginzburg-Landau Equations. Pattern Formation in Bénard Convection

H. Haken: Unpublished material
Equation (9.5.15) with $A \cong 0$ was derived differently by
J. Swift, P. C. Hohenberg: Phys. Rev. **A15**, 319 (1977)
H. Haken: *Synergetics*, Springer Ser. Synergetics, Vol. 1, 3rd. ed. (Springer, Berlin, Heidelberg, New York 1983)

10. The Inclusion of Noise

10.1 The General Approach

Compare the references cited in Chap. 4 and
W. Horsthemke, R. Lefever: *Noise-Induced Transitions*, Springer Ser. Synergetics, Vol. 15 (Springer, Berlin, Heidelberg, New York 1983)

10.2 A Simple Example

H. Haken: *Synergetics*, Springer Ser. Synergetics, Vol. 1, 3rd. ed. (Springer, Berlin, Heidelberg, New York 1983)

10.3 Computer Solution of a Fokker-Planck Equation for a Complex Order Parameter

H. Risken: Z. Phys. **186**, 85 (1965); **191**, 302 (1966)
H. Risken, H. D. Vollmer: Z. Phys. **201**, 323 (1967); **204**, 240 (1967)
H. Risken: In *Progress in Optics*, Vol. VIII, ed by E. Wolf (North-Holland, Amsterdam 1970) p. 239

10.4 Some Useful General Theorems on the Solution of Fokker-Planck Equations

10.4.1 Time-Dependent and Time-Independent Solutions of the Fokker-Planck Equation, it the Drift Coefficients are Linear in the Coordinates and the Diffusion Coefficients Constant

G. E. Uhlenbeck, L. S. Ornstein: Phys. Rev. **36**, 823 (1930)
N. Wax (ed.): *Selected Papers on Noise and Statistical Processes* (Dover, New York 1954)

10.4.2 Exact Stationary Solution of the Fokker-Planck Equation for Systems in Detailed Balance

We follow essentially
R. Graham, H. Haken: Z. Phys. **248**, 289 (1971)
H. Risken: Z. Phys. **251**, 231 (1972)
For related work see
H. Haken: Z. Phys. **219**, 246 (1969)
H. Haken: Rev. Mod. Phys. **47**, 67 (1975)
R. Graham: Z. Phys. **B40**, 149 (1980)

10.4.4 Useful Special Cases

H. Haken: Z. Phys. **219**, 246 (1969)

10.5 Nonlinear Stochastic Systems Close to Critical points: A Summary

Compare (H. Haken: *Synergetics*, Springer Ser. Synergetics, Vol. 1, 3rd. ed. (Springer, Berlin, Heidelberg, New York 1983)), where also another approach is outlined. That approach starts right away from the master equation or Fokker-Planck equation and eliminates the slaved variables from these equations.

11. Discrete Noisy Maps

Basic works on discrete maps have been cited in Sect. 8.11.2. Scaling properties of discrete noisy maps have been analized in
J. P. Crutchfield, B. A. Huberman: Phys. Lett. **77A**, 407 (1980)
J. P. Crutchfield, M. Nauenberg, J. Rudnick: Phys. Rev. Lett. **46**, 935 (1981)
B. Shraiman, C. E. Wayne, P. C. Martin: Phys. Rev. Lett. **46**, 933 (1981)
We shall follow essentially
G. Mayer-Kress, H. Haken: J. Stat. Phys. **26**, 149 (1981)
H. Haken, G. Mayer-Kress: Z. Phys. **B43**, 185 (1981)
H. Haken: In *Chaos and Order in Nature*. Springer Ser. Synergetics, Vol. 11, ed. by H. Haken (Springer, Berlin, Heidelberg, New York 1981) p. 2
H. Haken, A. Wunderlin: Z. Phys. **B46**, 181 (1982)

12. Example of an Unsolvable Problem in Dynamics

K. Gödel: Monathsh. Math. Phys. **38**, 173 (1931)

Appendix

J. Moser: Convergent Series Expansions for Quasi-Periodic Motions. Math. Ann. **169**, 136 (1967)

Subject Index

H. Haken

Laser Theory

1983. 72 figures. XV, 320 pages
ISBN 3-540-12188-9
(Originally published as "Handbuch der Physik/Encyclopedia of Physics, Volume 25/2c", 1970)

Contents: Introduction. – Optical resonators. – Quantum mechanical equations of the light field and the atoms without losses. – Dissipation and fluctuation of quantum systems. The realistic laser equations. – Properties of quantized electromagnetic fields. – Fully quantum mechanical solutions of the laser equations. – The semiclassical approach and its applications. – Rate equations and their applications. – Further methods for dealing with quantum systems far from thermal equilibrium. – Appendix. Useful operator techniques. – Sachverzeichnis (Deutsch-Englisch). – Subject Index (English-German).

Hydrodynamic Instabilities and the Transition to Turbulence

Editors: **H. L. Swinney, J. P. Gollub**
1981. 81 figures. XII, 292 pages
(Topics in Applied Physics, Volume 45)
ISBN 3-540-10390-2

Contents: *H. L. Swinney, J. P. Gollub:* Introduction. – *O. E. Lanford:* Strange Attractors and Turbulence. – *D. D. Joseph:* Hydrodynamic Stability and Bifurcation. – *J. A. Yorke, E. D. Yorke:* Chaotic Behavior and Fluid Dynamics. – *F. H. Busse:* Transition to Turbulence in Rayleigh-Bénard Convection. – *R. C. DiPrima, H. L. Swinney:* Instabilities and Transition in Flow Between Concentric Rotating Cylinders. – *S. A. Maslowe:* Shear Flow Instabilities and Transition. – *D. J. Tritton, P. A. Davies:* Instabilities in Geophysical Fluid Dynamics. – *J. M. Guckenheimer:* Instabilities and Chaos in Nonhydrodynamic Systems.

Real-Space Renormalization

Editors: **T. W. Burkhardt, J. M. J. van Leeuwen**
1982. 60 figures. XIII, 214 pages
(Topics in Current Physics, Volume 30)
ISBN 3-540-11459-9

Contents: *T. W. Burkhardt, J. M. J. van Leeuwen:* Progress and Problems in Real-Space Renormalization. – *T. W. Burkhardt:* Bond-Moving and Variational Methods in Real-Space Renormalization. – *R. H. Swendsen:* Monte Carlo Renormalization. – *G. F. Mazenko, O. T. Valls:* The Real Space Dynamic Renormalization Group. – *P. Pfeuty, R. Jullien, K. A. Penson:* Renormalization for Quantum Systems. – *M. Schick:* Application of the Real-Space Renormalization to Adsorbed Systems. – *H. E. Stanley, P. J. Reynolds, S. Redner, F. Family:* Position-Space Renormalization Group for Models of Linear Polymers, Branched Polymers, and Gels. – Subject Index.

Structural Phase Transitions I

Editors: **K. A. Müller, H. Thomas**
1981. 61 figures. IX, 190 pages
(Topics in Current Physics, Volume 23)
ISBN 3-540-10329-5

Contents: *K. A. Müller:* Introduction. – *P. A. Fleury, K. Lyons:* Optical Studies of Structural Phase Transitions. – *B. Dorner:* Investigation of Structural Phase Transformations by Inelastic Neutron Scattering. – *B. Lüthi, W. Rehwald:* Ultrasonic Studies Near Structural Phase Transitions.

Springer-Verlag Berlin Heidelberg New York Tokyo

M. Eigen, P. Schuster

The Hypercycle

A Principle of Natural Self-Organization

1979. 64 figures, 17 tables. VI, 92 pages
ISBN 3-540-09293-5
(This book is a reprint of papers which were published in "Die Naturwissenschaften" issues 11/1977, 1/1978, and 7/1978)

Contents: Emergence of the Hypercycle: The Paradigm of Unity and Diversity in Evolution. What is a Hypercycle? Darwinian System. Error Threshold and Evolution. – The Abstract Hypercycle: The Concrete Problem. General Classification of Dynamic Systems. Fixed-Point Analysis of Self-Organizing Reaction Networks. Dynamics of the Elementary Hypercycle. Hypercycles with Translation. Hypercyclic Networks. – The Realistic Hypercycle: How to Start Translation. The Logic of Primordial Coding. Physics of Primordial Coding. The GC-Frame Code. Hypercyclic Organization of the Early Translation Apparatus. Ten Questions. Realistic Boundary Conditions. Continuity of Evolution.

G. Eilenberger

Solitons

Mathematical Methods for Physicists

1981. 31 figures. VIII, 192 pages
(Springer Series in Solid-State Sciences, Volume 19). ISBN 3-540-10223-X

Contents: Introduction. – The Korteweg-de Vries Equation (KdV-Equation). – The Inverse Scattering Transformation (IST) as Illustrated with the KdV. – Inverse Scattering Theory for Other Evolution Equations. – The Classical Sine-Gordon Equation (SGE). – Statistical Mechanics of the Sine-Gordon System. – Difference Equations: The Toda Lattice. – Appendix: Mathematical Details. – References. – Subject Index.

M. Toda

Theory of Nonlinear Lattices

1981. 38 figures. X, 205 pages
(Springer Series in Solid-State Sciences, Volume 20). ISBN 3-540-10224-8

Contents: Introduction. – The Lattice with Exponential Interaction. – The Spectrum and Construction of Solutions. – Periodic Systems. – Application of the Hamilton-Jacobi Theory. – Appendices A–J. – Simplified Answers to Main Problems. – References. – Bibliography. – Subject Index. – List of Authors Cited in Text.

Solitons

Editors: **R.K. Bullough, P. J. Caudrey**
1980. 20 figures. XVIII, 389 pages
(Topics in Current Physics, Volume 17)
ISBN 3-540-09962-X

Contents: *R. K. Boullough, P. J. Caudrey:* The Soliton and Its History. – *G. L. Lamb, Jr., D. W. McLaughlin:* Aspects of Soliton Physics. – *R. K. Bullough, P. J. Caudrey, H. M. Gibbs:* The Double Sine-Gordon Equations: A Physically Applicable System of Equations. – *M. Toda:* On a Nonlinear Lattice (The Toda Lattice). – *R. Hirota:* Direct Methods in Soliton Theory. – *A. C. Newell:* The Inverse Scattering Transform. – *V. E. Zakharov:* The Inverse Scattering Method. – *M. Wadati:* Generalized Matrix Form of the Inverse Scattering Method. – *F. Calogero, A. Degasperis:* Nonlinear Evolution Equations Solvable by the Inverse Spectral Transform Associated with the Matrix Schrödinger Equation. – *S. P. Novikov:* A Method of Solving the Periodic Problem for the KdV Equation and Its Generalizations. – *L. D. Faddeev:* A Hamiltonian Interpretation of the Inverse Scattering Method. – *A. H. Luther:* Quantum Solitons in Statistical Physics. – Further Remarks on John Scott Russel and on the Early History of His Solitary Wave. – Note Added in Proof. – Additional References with Titles. – Subject Index.